T0349026

Modal Analysis

Modal Analysis

Jimin He and Zhi-Fang Fu

Oxford   Auckland   Boston   Johannesburg   Melbourne   New Delhi

Butterworth-Heinemann
Linacre House, Jordan Hill, Oxford OX2 8DP
225 Wildwood Avenue, Woburn, MA 01801-2041
A division of Reed Educational and Professional Publishing Ltd

 A member of the Reed Elsevier plc group

First published 2001

British Library Cataloguing in Publication Data
He, Jimin
 Modal analysis
 1. Modal analysis
 I. Title II. Fu, Zhi-Fang
 620.3

Library of Congress Cataloguing in Publication Data
He, Jimin.
 Modal analysis/Jimin He and Zhi-Fang Fu.
 p. cm.
 ISBN 0 7506 5079 6
 1. Modal analysis. I. Fu, Zhi-Fang. II. Title.
 TA654.15.H4 2001
 624.1–dc21   2001037430

ISBN 0 7506 5079 6

For more information on all Butterworth-Heinemann
publications visit our website at www.bh.com

Typeset at Replika Press Pvt Ltd, 100% EOU, Delhi 110 040, India
Transferred to Digital Printing 2005
Printed and bound by Antony Rowe Ltd, Eastbourne

FOR EVERY TITLE THAT WE PUBLISH, BUTTERWORTH-HEINEMANN
WILL PAY FOR BTCV TO PLANT AND CARE FOR A TREE.

Contents

Preface

This book provides a detailed treatment of the theory of analytical and experimental modal analysis with applications. It is intended for several uses. The materials can be selectively used for undergraduate and postgraduate courses. The book can be a reference text for practising engineers whose work involves modal analysis, and for researchers in other engineering and science disciplines who use modal analysis as a tool in their research.

Though appearing to be specialized, modal analysis has been used in the last two to three decades in many engineering disciplines and technology fields to solve increasingly demanding structural dynamic problems. The path of modal analysis being incorporated into university teaching is paralleled with that taken by finite element analysis. Currently, the published literature in modal analysis is sparse without taking on board conference and journal papers. This is not conducive to users of this technology.

Chapter 1 of this book begins with a brief introduction to the current status of modal analysis in engineering fields and the practical background which gave rise to modal analysis as an effective and an advancing technology. This is followed by a concise clarification of modal analysis and modal testing, and the theoretical framework for modal analysis. It is succeeded by an outline of applications of modal analysis in dealing with different types of engineering problems and practical challenges. This chapter closes on a summary note on the historical development of modal analysis.

Chapter 2 reviews the mathematics tools used in the theory of modal analysis. In matrix theory, these include matrix inversion, linear simultaneous equations, eigenvalue problems, different decomposition of a real symmetrical matrix, matrix perturbation and the derivative of matrices. In addition, this chapter introduces poles and zeros of a polynomial function, Fourier series and transform, and the variable separation method for partial differential equations. These mathematics tools will be used in the following chapters of the book.

Chapter 3 reviews the basic vibration theory which forms the basis of modal analysis theory. This chapter is an epitome of a textbook for mechanical vibrations or dynamics of machines. It intends to review relevant materials rather than to offer a comprehensive coverage of vibration theory. This is to facilitate a more thorough and smooth excursion in modal analysis theory.

Chapters 4, 5 and 6 constitute the theoretical basis of modal analysis. In Chapter 4, modal analysis theory for an SDoF system is presented. It begins with the frequency response function (FRF) of an SDoF system, its concept and its different forms and various formats on which the FRF can be studied. This is followed by an in-depth disclosure of inherent properties of the FRF which will define modal analysis methods introduced in later chapters.

The FRF of an SDoF system is mathematically simple and its physical interpretation unambiguous. However, when extended to an MDoF system, both the concept and definition of an FRF require more stringent deliberation. Chapter 5 deals with the modal analysis theory of an MDoF system without damping. It begins with the concept of normal modes of an MDoF system and its orthogonality properties. This is followed by the definition and description of the dynamic stiffness matrix and frequency response function matrix of an MDoF system. The physical interpretation of the FRFs is conferred and the different graphical presentation of FRFs is given. Next, the chapter establishes the relationship between the modal model of an MDoF system and its FRFs. This relationship forms the kernel of theoretical modal analysis and the foundation for developing experimental modal analysis methods. The asymptote properties of an SDoF FRF are extended in this chapter for the FRFs of an MDoF system. The chapter then continues to explore the orthogonality properties of an MDoF system from more contemporary viewpoints. The harmonic response analysis prepares the reader for a more advanced forced response of an MDoF system. The last section of the chapter provides a sound understanding of the anti-resonance and minimum properties of an FRF. This will lead to the latest progress of modal analysis on structural modification and optimization in later chapters.

Chapter 6 is devoted to the modal analysis theory of an MDoF damped system. This is a natural extension of Chapter 5. In this approach, it is appropriate to begin with a proportional damping model which offers a special advantage of easy analysis. This is succeeded by an expansion of the theory of undamped FRFs into those for a damped MDoF system. Due to the existence of damping, the FRFs can be studied in the complex domain, allowing the study of their real and imaginary forms, Nyquist form and more. The force response of an MDoF damped system is analysed. The chapter ends on a note on complex vibration modes.

Chapter 7, on the measurement of frequency response function data, provides a brief coverage of all aspects involved in conducting a successful frequency response function measurement. This includes preparation of the tested structure, general measurement set-up, selection of excitation forces and locations, selection of measurement locations, calibration of measurement set-up and acquisition of FRF data. This chapter closes at the assessment of FRF data during measurement and before the analysis of these data is to proceed. This is an important step to ensure accurate and sufficient FRF data, and a successful outcome of the analysis. The assessment includes the linearity, reciprocity and time-invariance check of the measured data, and the use of some characteristics of an FRF.

Chapter 8 presents different modal analysis methods using measured FRF data. It begins with the identification of resonances from measured data. This is followed by the description of various modal analysis methods based on the SDoF assumption of MDoF assumption. This includes some latest modal analysis methods found in the literature and adopted by commercial modal analysis software. This full chapter coverage will provide a solid foundation for readers to conduct modal analysis on measured FRF data. Some of the figures are the outputs of a modal analysis software called ICATS.

In addition to the methods based on FRF data, Chapter 9 provides a basic introduction to modal analysis using free vibration or impulse responses of a structure – the time domain modal analysis.

Chapter 10 deals with multi-input multi-output (MIMO) modal analysis. The estimation of accurate FRF data in the MIMO circumstance is discussed. A number of frequency and time domain modal analysis methods for the MIMO test are introduced.

The volume of this book precludes the coverage of all the main applications of modal analysis. However, it would be useful to the reader to include some recent applications. Chapters 11, 12 and 13 serve this purpose. Chapter 11 discusses structural modification and its application to optimize the dynamic characteristics of a structure. This composes structural modification by changing the mass and stiffness properties and by changing physical properties such as cross-sectional changes of beam elements of a structure. This chapter embodies some late developments in structural modification and optimization.

Chapter 12 introduces neural network, its applications in identifying dynamic characteristics of a system and in modal analysis. The use of the artificial neural network is taking a rapid pace in many engineering disciplines, including structural dynamics. This chapter can only serve as a brief introduction to this technique.

Chapter 13 presents a number of practical applications of modal analysis. Many more can be found in the literature such as from the proceedings of the annual International Modal Analysis Conference. Materials presented here are selective for teaching and reference.

Modal analysis draws strength from many sister disciplines in engineering. A book on modal analysis undoubtedly covers materials explored by numerous researchers and practical engineers. In this sense, the literature list at the end of each chapter can only be indicative. It is impossible for the authors to acknowledge all the individual contributions without which this book could not exist. We hope that this book will serve as a gateway for the reader to appreciate these contributions and to discover many more relevant readings in the modal analysis literature.

Jimin He and Zhi-Fang Fu

1

Overview of modal analysis

1.1 Introduction

In the past two decades, modal analysis has become a major technology in the quest for determining, improving and optimizing dynamic characteristics of engineering structures. Not only has it been recognized in mechanical and aeronautical engineering, but modal analysis has also discovered profound applications for civil and building structures, biomechanical problems, space structures, acoustical instruments, transportation and nuclear plants. To appreciate its significance in the modern engineering arena and its potential for future science and technology, it is appropriate to capture some of the background facts which will help to underline this unique technology.

Contemporary design of complex mechanical, aeronautical or civil structures requires them to become increasingly lighter, more flexible and yet strong. For instance, resources have been devoted by car manufacturers to achieve microscopic reductions of product body weight. Aerospace structures such as satellite antennas do deserve weight reduction of every possible gram in order to minimize their inertial property during operation in space. These stringent demands on contemporary structures often made them more susceptible to unwanted vibrations.

Another relevant fact in modern life is the increasing demands of safety and reliability upon contemporary structures either defined by government regulations or accrued by consumers. These demands have created new challenges to the scientific understanding of engineering structures. Where the vibration of a structure is of concern, the challenge lies on better understanding its dynamic properties using analytical, numerical or experimental means, or a combination of them.

As the significance of dynamic behaviour of engineering structures is better appreciated, it becomes important to design them with proper consideration of dynamics. Finite element analysis as a computer modelling approach has provided engineers with a versatile design tool, especially when dynamic properties need to be perused. This numerical analysis requires rigorous theoretical guidance to ascertain meaningful outcomes in relation to structural dynamics. An important part of dynamic finite element analysis is modal analysis.

Computer modelling alone cannot determine completely the dynamic behaviour of structures, because certain structural properties such as damping and nonlinearity do not conform with traditional modelling treatment. There are also boundary condition uncertainties which modelling needs additional help to work. Substantial advances in experimental techniques have complemented modelling with the experimental determination of structural properties. A milestone of this endeavour is the advent of digital Fourier transform analysers. The experimental techniques are nurtured by the theory of modal analysis and in turn provide new impetus to it.

1.2 What is modal analysis?

Modal analysis is the process of determining the inherent dynamic characteristics of a system in forms of natural frequencies, damping factors and mode shapes, and using them to formulate a mathematical model for its dynamic behaviour. The formulated mathematical model is referred to as the modal model of the system and the information for the characteristics are known as its modal data.

The dynamics of a structure are physically decomposed by frequency and position. This is clearly evidenced by the analytical solution of partial differential equations of continuous systems such as beams and strings. Modal analysis is based upon the fact that the vibration response of a linear time-invariant dynamic system can be expressed as the linear combination of a set of simple harmonic motions called the natural modes of vibration. This concept is akin to the use of a Fourier combination of sine and cosine waves to represent a complicated waveform. The natural modes of vibration are inherent to a dynamic system and are determined completely by its physical properties (mass, stiffness, damping) and their spatial distributions. Each mode is described in terms of its modal parameters: natural frequency, the modal damping factor and characteristic displacement pattern, namely mode shape. The mode shape may be real or complex. Each corresponds to a natural frequency. The degree of participation of each natural mode in the overall vibration is determined both by properties of the excitation source(s) and by the mode shapes of the system.

Modal analysis embraces both theoretical and experimental techniques. The theoretical modal analysis anchors on a physical model of a dynamic system comprising its mass, stiffness and damping properties. These properties may be given in forms of partial differential equations. An example is the wave equation of a uniform vibratory string established from its mass distribution and elasticity properties. The solution of the equation provides the natural frequencies and mode shapes of the string and its forced vibration responses. However, a more realistic physical model will usually comprise the mass, stiffness and damping properties in terms of their spatial distributions, namely the mass, stiffness and damping matrices. These matrices are incorporated into a set of normal differential equations of motion. The superposition principle of a linear dynamic system enables us to transform these equations into a typical eigenvalue problem. Its solution provides the modal data of the system. Modern finite element analysis empowers the discretization of almost any linear dynamic structure and hence has greatly enhanced the capacity and scope of theoretical modal analysis. On the other hand, the rapid development over the last two decades of data acquisition and processing capabilities has given rise to major advances in the experimental realm of the analysis, which has become known as modal testing.

1.3 What is modal testing?

Modal testing is an experimental technique used to derive the modal model of a linear time-invariant vibratory system. The theoretical basis of the technique is secured upon establishing the relationship between the vibration response at one location and excitation at the same or another location as a function of excitation frequency. This relationship, which is often a complex mathematical function, is known as frequency response function, or FRF in short. Combinations of excitation and response at different locations lead to a complete set of frequency response functions (FRFs) which can be collectively represented by an FRF matrix of the system. This matrix is usually symmetric, reflecting the structural reciprocity of the system.

The practice of modal testing involves measuring the FRFs or impulse responses of a structure. The FRF measurement can simply be done by asserting a measured excitation at one location of the structure in the absence of other excitations and measure vibration responses at one or more location(s). Modern excitation technique and recent developments of modal analysis theory permit more complicated excitation mechanisms. The excitation can be of a selected frequency band, stepped sinusoid, transient, random or white noise. It is usually measured by a force transducer at the driving point while the response is measured by accelerometers or other probes. Both the excitation and response signals are fed into an analyser which is an instrument responsible for computing the FRF data.

A practical consideration of modal testing is how much FRF data need to be acquired in order to adequately derive the modal model of the tested object. When doing a simple hammer test, a fixed response location is used while alternately moving force excitation points. The measured FRF data constitute a row of the FRF matrix (explained in Chapter 7). These data would theoretically suffice for deriving the modal model. For a simple shaker test, a fixed force input location is used while alternately moving response collection points or simultaneous acquiring responses from points. The measured FRF data constitute a column of the FRF matrix. Again, the data should suffice theoretically. With sufficient data, numerical analysis will derive modal parameters by ways of curve fitting. This process is known as experimental modal analysis. The derived parameters will form the modal model for the test structure. Parameters can be extracted either from individual FRF curves or from a set of FRF curves.

In summary, experimental modal analysis involves three constituent phases: test preparation, frequency response measurements and modal parameter identification. Test preparation involves selection of a structure's support, type of excitation force(s), location(s) of excitation, hardware to measure force(s) and responses; determination of a structural geometry model which consists of points of response to be measured; and identification of mechanisms which could lead to inaccurate measurement. During the test, a set of FRF data is measured and stored which is then analysed to identify modal parameters of the tested structure.

1.4 Applications of modal analysis

Both theoretical and experimental modal analysis ultimately arrive at the modal model of a dynamic system. Compared with the FRF or the vibration response, the modal model explicitly portrays the dynamic characteristics of a system. Therefore, applications of modal analysis are closely related to utilizing the derived modal model in design, problem solving and analysis. Before embarking on the discussion of applications, it is important to refresh the two different paths from which a modal model is derived. Theoretical modal analysis relies on the description of physical properties of a system to derive the modal model. Such a description usually contains the mass, stiffness and damping matrices of the system. Thus, it is a path from spatial data to modal model. Experimental modal analysis obtains the modal model from measured FRF data or measured free vibration response data. Thus, it is a path from response data to modal model. Once the modal model is derived, a number of applications can be instigated. Some applications of modal analysis involve direct use of modal data from measurement while others use these data for further analysis. In the following, some of the applications of modal analysis are reviewed. Detailed theoretical analysis of these applications and more will be found later in the book or in the literature.

1.4.1 Troubleshooting

Troubleshooting using experimental modal analysis is to gain an insight into a dynamic structure which is problematic. This is a most popular application of experimental modal analysis since its emergence. It also often heralds further applications of modal analysis. Troubleshooting relies on experimentally derived natural frequencies, damping factors and mode shapes of the structure. These data provide a fundamental understanding of the structural characteristics and often reveal the root of dynamic problems encountered in real life.

1.4.2 Correlation of finite element model and experimental results

Many applications of structural dynamics rely their success entirely upon having an accurate mathematical model for a dynamic structure. Such a model can be derived from the finite element modelling. The resultant FE model, which is in the form of mass and stiffness matrices, can be essential for further applications such as sensitivity analysis and prediction due to proposed structural changes. However, owing to the complexity and uncertainty of the structure, it is unrealistic to expect such an FE model to be faithfully representative. An essential approach is to take a measurement of the structure, derive its modal model and use it to correlate with the existing FE model in order to update it. The philosophy behind this model correlation is that the modal model derived from measurement, though incomplete due to lack of sufficient numbers of vibration modes and measured locations, truly represents the structure's dynamic behaviour. Thus, it can be used to 'correct' the FE model, should any discrepancies occur between them. This philosophy may be challenged by some FE analysts but experience has justified its merits in most cases.

1.4.3 Structural modification

Structural modification makes changes on mass, stiffness or damping of a dynamic system. These physical changes will certainly alter the dynamic behaviour of the system. Using the modal model of the system, simulation and prediction of 'what if' can be conducted. The effect of hypothetical physical changes on the dynamic behaviour can be derived without another complete analysis or the actual structural changes. For instance, if a lumped mass is to be added to a part of the system, then the existing modal model and the mass together should predict a new modal model 'after' the structural modification. This is particularly useful in an early design stage to optimize dynamic characteristics of a new design, or to improve a structure's dynamic behaviour after its modal model is derived from measurement. However, this approach of structural modification often forbids large-scale structural changes.

1.4.4 Sensitivity analysis

A modal model of a dynamic system can be used to forecast the sensitivity of its modal parameters due to system physical parameter changes. This sensitivity analysis is intrinsically related to structural modification. However, the emphasis here is to identify which physical change is most effective to a proposed modal parameter change such as shifting a natural frequency. In contrast, that of structural modification is to study variations of modal parameters due to a selected physical change. This analysis is specifically useful in the redesigning of a dynamic structure when a target on dynamic characteristics is set and a most efficient way to accomplish it is sought.

1.4.5 Reduction of mathematical models

In FE modelling of a dynamic structure, the number of coordinates used determines the size of the FE model. Although employment of more coordinates does not necessarily translate into greater accuracy of the modelling results, it is often conducive. However, when only the dynamic behaviour of the modelled structure at low frequency range is of interest, a much reduced mathematical model will be preferred. Such a reduced model can be derived either from a modal model of the structure, or from the original FE model using various modelling reduction algorithms. In both cases, modal model is essential in the process or in the appraisal of the final outcome.

1.4.6 Forced response prediction

Another application of modal analysis is the prediction of vibration responses to a given force. With an established modal model, vibration responses due to a clearly defined force input can be computed. For instance, upon knowing the modal model of a vehicle, the measurement of test track vibrations can be used to predict the vehicle's response on the track before it is driven. Using the superposition principle, response due to several forces can also be predicted. The difficulty associated with this application often lies in accurately estimating or measuring the forces.

1.4.7 Force identification

In practice, forces which induce vibration of a system are not always measurable. However, it is possible to identify them using the response measurement and a modal model of the system. Identification of forces which induce severe vibration has paramount significance in some applications. For instance, a loosened bearing inside a turbine engine may breed an excitation force which instigates excessive vibration. This excitation force is also a potential source of causing catastrophic structural failure.

1.4.8 Response prediction

Once the modal model of a structure is adequately determined, it becomes possible to predict structural response for any coalition of input forces. This will provide a scientific basis for studying structural integrity with a known dynamic environment. When the structural vibration responses are in the form of dynamic strain, response prediction could be incisive in forecasting the fatigue life of the structure. An experimentally derived modal model usually provides damping factors of a structure which are crucial in accurate response prediction.

1.4.9 Substructure coupling

It is often required to predict the dynamic behaviour of a whole structure from the knowledge of the behaviour of its components. This process is known as substructure coupling. There are several practical reasons why the behaviour of the whole structure are measured or modelled directly. One is to break a complex dynamic problem into manageable parts. Many algorithms of substructure coupling are based on the modal parameters from components. Substructure coupling can also be effectively used in finite element analysis when the model of a structure is too complex for the computer capacities.

1.4.10 Structural damage detection

Detection of invisible structural damage has always been given due priority in industry. This is particularly evident in the aviation and aerospace industries. Recently, detection of structural damage has been applied to civil structures such as bridges. The theoretical basis of using modal analysis for damage detection lies in the fact that dynamic responses of a structure vary because of its inherent damage. This gives rise to the possibility of identifying the damage from the variation of structural responses before and after damage occurs. In particular, the damage detection formulates the relationship between the damage and modal parameter changes of a structure. A common practice is to obtain the 'fingerprint' or 'baseline' of modal parameters when a structure is in perfect dynamics health. Later, when the changes of these parameters occur, it is possible to investigate the structural damage which brings about the changes.

1.4.11 Active vibration control

A traditional control problem depends on a reliable plant model to derive control laws from observers. For active vibration control of a structure as a plant, it is imperative that an accurate mathematical model exists which delineates its dynamic characteristics. Experimental modal analysis is an ideal tool to serve this purpose. A good example is the active vibration control of a tall building under wind loading. With an accurate modal model of the building and proper filtering techniques, it is possible to devise actuators and sensors and form a feedback control loop so that selected vibration modes will be brought under control.

1.5 Practical applications of modal analysis

In the last two decades, there have been numerous applications of modal analysis reported in literature covering wide areas of engineering, science and technology. The application scope of modal analysis is expected to undergo significant expansion in the coming years. Practical applications of modal analysis are largely related to the advances in experimental technology. It is impossible to introduce all these applications; however, a narration of these practical applications embracing a number of typical areas will help to illuminate the understanding of modal analysis and its potential.

The majority of practical application cases reported in the literature have been those from aeronautical engineering, automotive engineering and mechanical engineering in particular. This is not to discredit the fact that the application of modal analysis is becoming more strongly interdisciplinary.

In automotive engineering, the enormous commercial and safety aspects associated with redesigning a vehicle oblige the best possible understanding of dynamic properties of vehicular structures and the repercussion of any design changes. Keen interest has been placed on combining both experimental modal analysis and finite element analysis in studying automotive components. Modern vehicle structures must be light in weight and high in strength. A combination of both analytical and experimental modal analysis enables improved design of automotive components and enhancement of dynamic properties of a vehicle. Experimental modal analysis as a troubleshooting tool also plays a crucial role in the study of vehicle noise and vibration harshness (NVH). A simple modal analysis of a body-in-white or a subframe structure is a typical application. However, more sophisticated applications have been achieved such as those involved in modal sensitivities of vehicle floor panels, structural optimization for vehicle comfort, vehicle fatigue life estimation, vehicle suspension with active vibration control mechanism and condition monitoring and diagnostic system for the vehicle engine. Another prominent application area for modal analysis is in the study of vehicle noise. Modal analysis is used as a tool to understand structure-borne noise from vehicle components or airborne sound transmission into vehicle cabins through door-like structures and the path noise takes to transmit in vehicles. A more advanced application of modal analysis involves interior noise reduction through structural optimization or redesign. As a whole, modal analysis has been an effective technique for automotive engineering, in its quest to improve a vehicle's NVH.

The rapid development in the aeronautical and astronautical industries has challenged many disciplines of engineering with diverse technological difficulties. The structural dynamics of both aircraft and spacecraft structures have been a significant catalyst to the development of modal analysis. Aircraft and spacecraft structures impose stringent requirements on structural integrity and dynamic behaviour which are shadowed by rigorous endeavours to reduce weight. The large dimension of spacecraft structures also asserts new stimulus for structural analysts. Establishing an accurate mathematical model is always a task for aircraft and spacecraft structures. Experimental modal analysis has provided an indispensable means of verifying a mathematical model derived from computer modelling. Modal testing has been reportedly carried out on assorted structures ranging from a scaled down aircraft frame, a whole satellite to an unmanned air vehicle. This asserts special significance in terms of damping properties, nonlinearities and accurate force and response estimation – which computer modelling alone often finds powerless to deal with. In fact, the majority of early publications in the 1980s on updating finite element models using experimental data originated from aeronautical industries. Large amplitudes of force and response experienced by aircraft and spacecraft structures stretched the linearity assumption of modal theory to its very limit. The necessity to enter the twilight zone of nonlinear dynamic behaviour became real. The complexity and dimension of aircraft structures were also responsible for rapid theoretical development of substructural coupling and synthesis. In some multi-disciplinary topics such as fluid induced vibration and flutter analysis, modal analysis has also been found to be a useful tool. Light and large-scale on-orbit space structures impose a control problem for structural engineers. The research in this area has led to notable advances in recent years on blending modal analysis and active control of space structures.

Modal analysis has also found increasing acceptance in the civil engineers' community where structural analysis has always been a critical area. The concern of dynamic behaviour of civil structures under seismic and wind loading warrants the application of modal analysis. Civil structures are usually much larger than those mechanical and aeronautical structures for which modal testing was originally developed. A large number of applications are concerned with the prediction of responses of a civil construction due to ambient vibration or external loadings. This response prediction endeavour relies on an accurate mathematical model which can be derived by modal analysis. Examples of such applications range from tall buildings, soil-structure interaction to a dam-foundation system. The real life force inputs involved in civil structures are earthquake waves, wind, ambient vibration, traffic load, etc. Recent years have seen an upsurge of modal testing on bridges to complement traditional visual bridge inspection and static testing. Modal testing has been used as an effective non-destructive testing technique to locate promptly the presence of critical defects. This can provide invaluable information for bridge maintenance and budgetary decision making. Such modal testing can be done using either traffic loading, a shaker or impact input, depending on the span and size of the bridge. Bridge testing is strongly related to the research of structural damage detection using modal test data.

Modal analysis has been successfully used outside traditional engineering fields. A good example is acoustics and musical instruments. Acoustic modal analysis has been used to analyse the dynamic characteristics of speaker cabinets. This analysis provides crucial information in the design of new speakers with improved sound quality. Experimental modal analysis has been used to study a number of musical instruments such as violins and guitars. This experiment helps to decipher the mystery behind the generations' craftsmanship of instrument makers and provide manufacturers of musical instruments with useful scientific data for improving product quality. Moreover, it assists in establishing an effective procedure of scientifically evaluating a musical instrument.

1.6 Historical development of modal analysis

The essential idea of modal analysis is to describe complex phenomena in structural dynamics with simple constituents, i.e. natural vibration modes. This idea is very much like atomism which attempts to find the most basic elements for varieties of different substances, or the concept is of Fourier series which represents a complicated waveform by a combination of simple sine and cosine waves. In this sense, the origin of modal analysis may be traced way back in history. However, there are two landmarks of such atomistic comprehension in the recent history of science which paved the way for the birth of modal analysis. Newton, from his observation of the spectrum of sunlight, confirmed its composition of colour components. Fourier, based on earlier mathematical wisdom, claimed that an arbitrary periodic function with finite interval can always be represented by a summation of simple harmonic functions. The Fourier series and spectrum analysis laid a solid foundation for the development of modal analysis.

Theoretical modal analysis can be closely identified with the wave equation which describes the dynamics of a vibrating string. From the solution, we can determine its natural frequencies, mode shapes and forced responses – constituents so accustomed by today's modal analysts. This stage of modal analysis, developed during the nineteenth century, was largely dependent upon mathematics to solve partial differential equations which describe different continuous dynamic structures. The elegance of the solution is evident while the scope of solvable structures is limited.

The concept of discretization of an object and introduction of matrix analysis brought about a climax in theoretical modal analysis early in the last century. Theory was developed such that structural dynamic analysis of an arbitrary system can be carried out when knowing its mass and stiffness distribution in matrix forms. However, the theory could only be realized after computers became available. In that aspect, theoretical (or analytical) modal analysis is very much numerical modal analysis.

The foundation of experimental modal analysis, a name contrived long after the engineering practice it embodies, was laid early in the last century. The core of experimental modal analysis is system identification. As a result, it was nourished by the development in electrical engineering. The analogy of electric circuits and a mechanical system enabled the application of some circuit analysis theory into the study of mechanical systems. This gave forth system identification, mechanical impedance and sub-system analysis in structural dynamics.

The invention of the fast Fourier transform (FFT) algorithm by J.W. Cooly and J.W. Tukey in 1965 finally paved the way for rapid and prevalent application of experimental technique in structural dynamics. With FFT, frequency responses of a structure can be computed from the measurement of given inputs and resultant responses. The theory of modal analysis helps to establish the relationship between measured FRFs and the modal data of the tested specimen. Efforts were focused on deriving modal data from measured FRF data. The first, and perhaps the most significant, method of experimental modal analysis was proposed by C.C. Kennedy and C.D. Pancu in 1947 before FFT was conceived. Their method was largely forgotten until FFT gave life to experimental modal analysis. Since then, numerous methods have been proposed and many have been computerized, including those time domain methods which are based on free vibration of a structure rather than its frequency responses.

The experimental development also helped to advance the theory of modal analysis. Traditional analytical modal analysis based on the proportional damping model was expanded into the non-proportional damping model. The theory of complex vibration modes was developed. Modal analysis evolves more in parallel with control theory. Inverse structural dynamic problems such as force identification from measured responses were actively pursued. Nonlinear dynamic characteristics were studied experimentally.

Today, modal analysis has entered many fields of engineering and science. Applications range from automotive engineering, aeronautical and astronautical engineering to bioengineering, medicine and science. Numerical (finite element) and experimental modal analysis have become two pillars in structural dynamics.

Literature

1. Allemang, R.J. 1984: Experimental Modal Analysis Bibliography. *Proceedings of the 2nd International Modal Analysis Conference*, Orlando, Florida, 1085–1097.
2. Allemang, R.J. 1993: Modal Analysis – Where Do We Go from Here? Keynote address, *Proceedings of the 11th International Modal Analysis Conference*.
3. Brown, D.L. 2000: Structural dynamics/modal analysis – from here to. *Sound and Vibration*, **34**(1), 6–9.
4. Cooley, J.W. and Tukey, J.W. 1965: An algorithm for the machine calculation of complex Fourier series. *Mathematics and Computations*, **19**(90), 297–301.
5. Dossing, O. 1995: Going Beyond Modal Analysis or IMAC in a New Key. Keynote address, *Proceedings of the 13th International Modal Analysis Conference*, Nashville, USA.
6. Dovel, G. 1989: A dynamic tool for design and troubleshooting. *Mechanical Engineering*, 82–86.
7. Kennedy, C.C. and Pancu, C.D.P. 1947: Use of Vectors in Vibration Measurement and Analysis. *Journal of the Aeronautical Sciences*, **14**(11), 603–625.
8. Leuridan, J. 1992: Modal Analysis: A Perspective on Integration. Keynote address, *Proceedings of the 10th International Modal Analysis Conference*, San Diego, California.
9. Mitchell, L.D. 1984: Modal Analysis Bibliography – An Update – 1980–1983. *Proceedings of the 2nd International Modal Analysis Conference*, Orlando, Florida, 1098–1114.

10. Mitchell, L.D. 1988: A Perspective View of Modal Analysis. Keynote Address, *Proceedings of the 6th International Modal Analysis Conference*, Kissimmee, Florida, xvii–xxi.
11. Rogers, P. 1988: Genuine modal testing of rotating machinery. *Sound and Vibration*, January, 36–42.
12. Snoeys, R. and Sas, P. 1987: Reflections of Modal Analysis and its Applications. Keynote address, *Proceedings of the 5th International Modal Analysis Conference*, London, UK, xviii–xxii.
13. Stevens, K.K. 1985: Modal analysis – an old procedure with a new look. *Mechanical Engineering*, 52–56.

2

Mathematics for modal analysis

Modal analysis relies on mathematics to establish theoretical models for a dynamic system and to analyse data in various forms. The mathematics involved is wide ranging, partly because modal analysis involves both time domain and frequency domain analysis. It deals with discretized dynamic systems and continuous structures. The analytical and numerical study carried out extends to curve fitting, matrix manipulation, statistical analysis, parameter identification and more. As a result, necessary mathematical preparation is essential. Instead of leaving the reader to obtain the knowledge every time when the needs arise, this chapter provides the mathematics to be used in this book.

Matrix theory plays a pivotal role in modal analysis. This is because modal analysis theory is largely based on the analysis of a multi-degree-of-freedom (MDoF) dynamic system which relies essentially on matrix theory. The main matrix analyses involved are the solution of linear equations, matrix inverse and eigenvalue problems. More specific topics such as matrix derivatives are also used in modal analysis and its applications.

The Fourier transform is fundamental to the signal processing in modal analysis. It is customary to believe that the Fast Fourier Transform is the landmark for the development of modal analysis technology. Without it, modal analysis would still remain as an academic and analytical venture.

During its course of evolution, modal analysis has been nourished by modern control theory. This is not surprising, as the subjects of control theory are systems while modal analysis deals with dynamic systems. The common theme of system identification enables modal analysis to adopt a number of useful techniques from modern control theory. Among them are state–space theory, Laplace transform, time series analysis and Hilbert transform.

Because the content in this chapter is meant to serve for the following chapters, many conclusions are presented without rigorous mathematical proof. Interested readers can find that from books or other publications.

2.1 Basic matrix concepts

Matrix theory is fundamental to the theory of modal analysis. In the following, some basic concepts of matrix theory are refreshed.

2.1.1 Trace of a matrix

The trace of a square matrix is a scalar function. For a matrix $[A]$ of order $n \times n$, its trace is denoted as $tr[A]$ and is given as:

$$tr[A] = \sum_{i=1}^{n} a_{ii} \tag{2.1}$$

The following relations for matrix trace can be verified easily.

$$tr([A] + [B]) = tr[A] + tr[B] \tag{2.2}$$

$$tr([A][B]) = tr([B][A]) \tag{2.3}$$

$$tr([A]^T) = tr[A] \tag{2.4}$$

$$tr(c[A]) = c(tr[A]) \text{ (c is a constant)} \tag{2.5}$$

2.1.2 Determinant of a matrix

The determinant of a square matrix $[A]$ of dimension $n \times n$ is denoted as $\|[A]\|$. It is easy to show that

$$\|[A][B]\| = \|[A]\| \, \|[B]\| \tag{2.6}$$

$$\|[A]\| = \|[A]\|^T \tag{2.7}$$

A minor $[M]_{ij}$ of the element a_{ij} in $[A]$ is a determinant of the matrix formed by deleting the ith row and jth column of $[A]$.

The cofactor $[C]_{ij}$ of the element a_{ij} in $[A]$ is defined by:

$$[C]_{ij} = (-1)^{i+j} [M]_{ij} \tag{2.8}$$

For example
$$[A] = \begin{bmatrix} 2 & 1 & 5 \\ 4 & 2 & 1 \\ 2 & 0 & 3 \end{bmatrix} \quad \|[A]\| = \begin{bmatrix} 2 & 1 & 5 \\ 4 & 2 & 1 \\ 2 & 0 & 3 \end{bmatrix}$$

$$[M]_{21} = \begin{bmatrix} 1 & 5 \\ 0 & 3 \end{bmatrix} \quad \|[M]\|_{21} = \begin{vmatrix} 1 & 5 \\ 0 & 3 \end{vmatrix} = 3$$

$$\|[C]\|_{21} = (-1)^{2+1} \begin{vmatrix} 1 & 5 \\ 0 & 3 \end{vmatrix} = -3$$

The order of determinant can be reduced by one by expanding any row or column in terms of its cofactors. For $[A]$ shown above, select the second column, then:

$$\begin{vmatrix} 2 & 1 & 5 \\ 4 & 2 & 1 \\ 2 & 0 & 3 \end{vmatrix} = 1(-1)^{1+2} \begin{vmatrix} 4 & 1 \\ 2 & 3 \end{vmatrix} + 2(-1)^{2+2} \begin{vmatrix} 2 & 5 \\ 2 & 3 \end{vmatrix} + 0(-1)^{3+2} \begin{vmatrix} 2 & 5 \\ 4 & 1 \end{vmatrix} = -18$$

Generally, for $[A]$ of dimension $n \times n$, select the rth column, then

$$\|[A]\| = \sum_{i=1}^{n} a_{ir} \|[C]\|_{ir} \tag{2.9}$$

A square matrix is said to have rank r if at least one of its r-square minors is different from zero while every $(r + 1)$ square minor, if any, is zero.

2.1.3 Norm of a matrix

When we deal with a vector or a matrix, it is sometimes desirable to have a measure which quantifies their 'magnitude' in the similar way to the magnitudes of a real or complex scalar value. It is obvious that the 'magnitude' of a vector or a matrix is going to be very different from the magnitudes of any scalar quantities.

The technical name of the magnitude of a matrix $[A]$ is called the *norm* and is denoted as $\|[A]\|$. The estimation of the norm of a matrix can be resembled from the norm of a vector which is defined as a function that assigns to the vector a non-negative real number. Typically, the p-norm of a vector $\{x\}$ of dimension $n \times 1$ is defined as:

$$\|\{x\}\|_p = \left(\sum_{i=1}^{n} |x_i|^p \right)^{\frac{1}{p}} \tag{2.10}$$

It can be seen readily from this definition that the following norms can be derived:

$$\|\{x\}\|_1 = \sum_{i=1}^{n} |x|_i \quad \text{Sum of magnitude} \tag{2.11}$$

$$\|\{x\}\|_2 = \sqrt{\sum_{i=1}^{n} x_i^2} \quad \text{Euclidean norm} \tag{2.12}$$

$$\|\{x\}\|_\infty = \max |x|_i \quad \text{Maximum magnitude norm} \tag{2.13}$$

The norm of $[A]$ can now be developed with the reference to vector norms.

$$\|[A]\|_{1c} = \max_{1<j<n} \sum_{i=1}^{n} |a_{ij}| \quad \text{Maximum column sum} \tag{2.14}$$

$$\|[A]\|_{1r} = \max_{1<i<n} \sum_{j=1}^{n} |a_{ij}| \quad \text{Maximum row sum} \tag{2.15}$$

$$\|[A]\|_e = \sqrt{\sum_{i=1}^{n} \sum_{j=1}^{n} a_{ij}^2} \quad \text{Euclidean (also called Frobenius) norm} \tag{2.16}$$

As an example, consider the matrix:

$$[A] = \begin{bmatrix} 5 & -5 & -7 \\ -4 & 2 & -4 \\ -7 & -4 & 5 \end{bmatrix}$$

then we have

$$\|[A]\|_{1c} = 16, \|[A]\|_{1r} = 17 \quad \text{and} \quad \|[A]\|_e = 15$$

One application of matrix trace is to determine the Euclidean norm of a matrix. It is easy to verify the following:

$$\|[A]\|_e = \sqrt{tr([A][A]^T)} \tag{2.17}$$

2.1.4 The rank of a matrix

A non-singular matrix is a square one whose determinant is not zero. The rank of a matrix $[A]$ is equal to the order of the largest non-singular submatrix of $[A]$. It follows that a non-singular square matrix of $n \times n$ has a rank of n. Thus, a non-singular matrix is also known as a full rank matrix. For a non-square $[A]$ of $m \times n$, where $m > n$, full rank means only n columns are independent.

There are many other ways to describe the rank of a matrix. In linear algebra, it is possible to show that all these are effectively the same. For example, the rank of a matrix can be said as the number of independent rows or columns the matrix has (whichever is smaller). Alternatively, the rank is the number of non-zero rows (columns) of the matrix after Gaussian elimination.

For an MDoF structural system, usually the mass matrix is a full rank matrix. The stiffness matrix is the same if the system is not allowed to have rigid body motion. A damping matrix that represents damping at a few isolated degrees of freedom (DoFs) will have a sparse matrix that is not of full rank (known as rank deficient). If the system does not have repeated eigenvalues, the mode shape matrix is a full rank matrix.

The rank of a matrix product does not exceed the rank of any individual matrix.

2.1.5 Similarity of matrices

Two real matrices $[A]$ and $[B]$ are similar if there exists a non-singular matrix $[P]$ such that

$$[P]^{-1}[A][P] = [B] \tag{2.18}$$

It can be verified that similar matrices have the same eigenvalues.

2.2 Linear simultaneous equations

The solution of a set of linear simultaneous equations (hereafter called linear equations) is the core of linear algebra. Many essential topics in linear algebra stem from this solution. Examples include matrix inversion and some eigenvalue solution methods. The solution is also one of the most important numerical endeavours for modal analysis where many problems and applications can be formulated ultimately using linear equations.

The problem we face is a set of linear equations:

$$\begin{cases} a_{11}x_1 + a_{12}x_2 + \ldots a_{1n}x_n = b_1 \\ a_{21}x_1 + a_{22}x_2 + \ldots a_{2n}x_n = b_2 \\ \qquad \ldots \\ a_{m1}x_1 + a_{m2}x_2 + \ldots a_{mn}x_n = b_m \end{cases} \qquad (2.19)$$

Here, there are n unknown $(x_1, x_2, \ldots, x_n)$; $m \times n$ coefficients a_{ij} $(1, 2, \ldots, n, 1, 2, \ldots, m)$ and m constants $(b_1, b_2, \ldots, b_m)$. The equations can be succinctly written in matrix form as:

$$[A]_{m \times n}\{x\}_{n \times 1} = \{b\}_{m \times 1} \qquad (2.20)$$

where

$$[A] = \begin{bmatrix} a_{11} & a_{12} & \ldots & a_{1n} \\ a_{21} & a_{22} & \ldots & a_{2n} \\ \ldots & \ldots & \ldots & \ldots \\ a_{m1} & a_{m2} & \ldots & a_{mn} \end{bmatrix} \qquad (2.21)$$

$$\{x\} = \begin{Bmatrix} x_1 \\ x_2 \\ \ldots \\ x_n \end{Bmatrix} \qquad (2.22)$$

and

$$\{b\} = \begin{Bmatrix} b_1 \\ b_2 \\ \ldots \\ b_m \end{Bmatrix} \qquad (2.23)$$

Depending on the numbers of rows and columns, and the condition of $[A]$, there are three possible solutions of the linear algebraic equations:

$$\begin{cases} m = n & \text{unique solution possible} \\ m > n & \text{optimized solution possible} \\ m < n & \text{non-unique solutions possible} \end{cases}$$

The primary interest for modal analysis lies on the first solution, i.e. when $m = n$ and when the equations are linearly independent.

The Gaussian elimination method is the basic pattern of a large number of direct methods for the solution. The basic idea of the Gaussian method is to eliminate the number of unknowns to one so that the last unknown can be solved. Later, back substitute the solved unknown to the remaining equations to solve another unknown. The Gaussian elimination method is summarized below.

$$\begin{bmatrix} a_{11}^{(1)} & a_{12}^{(1)} & \ldots & a_{1n}^{(1)} \\ a_{21}^{(1)} & a_{22}^{(1)} & \ldots & a_{2n}^{(1)} \\ \ldots & \ldots & \ldots & \ldots \\ a_{n1}^{(1)} & a_{n2}^{(1)} & \ldots & a_{nn}^{(1)} \end{bmatrix} \begin{bmatrix} x_1 \\ x_2 \\ \ldots \\ x_n \end{bmatrix} \begin{Bmatrix} b_1^{(1)} \\ b_2^{(1)} \\ \ldots \\ b_n^{(1)} \end{Bmatrix} \quad \text{or} \quad [A]^{(1)}\{x\} = \{b\}^{(1)} \qquad (2.24)$$

Step 1:

Multiplying the first row of the augmented matrix by $-\dfrac{a_{i1}^{(1)}}{a_{11}^{(1)}}$ $(i = 2, 3, \ldots, n)$ and adding the resultant row to the ith row will convert all the elements (except element $a_{11}^{(1)}$) to zero. Thus,

$$
\begin{bmatrix}
a_{11}^{(1)} & a_{12}^{(1)} & \cdots & a_{1n}^{(1)} \\
0 & a_{22}^{(2)} & \cdots & a_{2n}^{(2)} \\
\cdots & \cdots & \cdots & \cdots \\
0 & a_{n2}^{(2)} & \cdots & a_{nn}^{(2)}
\end{bmatrix}
\begin{Bmatrix}
x_1 \\ x_2 \\ \cdots \\ x_n
\end{Bmatrix}
=
\begin{Bmatrix}
b_1^{(1)} \\ b_2^{(2)} \\ \cdots \\ b_n^{(2)}
\end{Bmatrix}
\quad \text{or} \quad [A]^{(2)}\{x\} = \{b\}^{(2)}
$$

(2.25)

Repeat the process until $[A]$ becomes an upper triangular matrix:

$$
\begin{bmatrix}
a_{11}^{(1)} & a_{12}^{(1)} & a_{13}^{(1)} & \cdots & a_{1n}^{(1)} \\
 & a_{22}^{(2)} & a_{23}^{(2)} & \cdots & a_{2n}^{(2)} \\
 & & a_{33}^{(3)} & \cdots & a_{3n}^{(3)} \\
\cdots & \cdots & \cdots & \cdots & \cdots \\
 & & & \cdots & a_{nn}^{(n)}
\end{bmatrix}
\begin{Bmatrix}
x_1 \\ x_2 \\ x_3 \\ \cdots \\ x_n
\end{Bmatrix}
=
\begin{Bmatrix}
b_1^{(1)} \\ b_2^{(2)} \\ b_3^{(3)} \\ \cdots \\ b_n^{(n)}
\end{Bmatrix}
\quad \text{or} \quad [A]^{(n)}\{x\} = \{b\}^{(n)}
$$

(2.26)

Effectively, the elimination process is a series of elementary row operations. Step 1 is equivalent to the following matrix process:

$$[A]^{(2)} = [M]_1[A]^{(1)} \quad \text{and} \quad \{b\}^{(2)} = [M]_1\{b\}^{(1)} \tag{2.27, 28}$$

$$
\text{where} \, [M]_1 =
\begin{bmatrix}
1 & 0 & \cdots & 0 \\
-\dfrac{a_{21}^{(1)}}{a_{11}^{(1)}} & 1 & \cdots & 0 \\
\cdots & \cdots & \cdots & \cdots \\
-\dfrac{a_{n1}^{(1)}}{a_{11}^{(1)}} & 0 & \cdots & 1
\end{bmatrix}
=
\begin{bmatrix}
1 & 0 & \cdots & 0 \\
-m_{21} & 1 & \cdots & 0 \\
\cdots & \cdots & \cdots & \cdots \\
-m_{n1} & 0 & \cdots & 1
\end{bmatrix}
$$

(2.29)

Likewise, Step 2 is equated to the following matrix process:

$$[A]^{(3)} = [M]_2[A]^{(2)} \tag{2.30}$$

and

$$\{b\}^{(3)} = [M]_2\{b\}^{(2)} \tag{2.31}$$

where

$$[M]_2 = \begin{bmatrix} 1 & 0 & 0 & \cdots & 0 \\ 0 & 1 & 0 & \cdots & 0 \\ 0 & -\dfrac{a_{32}^{(2)}}{a_{22}^{(2)}} & 1 & \cdots & 0 \\ \cdots & \cdots & \cdots & \cdots & \cdots \\ 0 & -\dfrac{a_{n2}^{(1)}}{a_{22}^{(2)}} & 0 & \cdots & 1 \end{bmatrix} = \begin{bmatrix} 1 & 0 & 0 & \cdots & 0 \\ 0 & 1 & 0 & \cdots & 0 \\ 0 & -m_{32} & 1 & \cdots & 0 \\ \cdots & \cdots & \cdots & \cdots & \cdots \\ 0 & -m_{n2} & 0 & \cdots & 1 \end{bmatrix}$$

(2.32)

Therefore, the Gaussian elimination process to triangularize $[A]$ (or $[A]^{(1)}$) effectively means to conduct the following matrix operations:

$$\begin{cases} [A]^{(n)} = [M]_{n-1}[M]_{n-2} \cdots [M]_1 [A]^{(1)} \\ \{b\}^{(n)} = [M]_{n-1}[M]_{n-2} \cdots [M]_1 \{b\}^{(1)} \end{cases}$$

(2.33)

2.3 Matrix inversion

Matrix inversion is a common problem in engineering. For instance, in static finite element analysis, the inverse of the stiffness matrix is needed in order to obtain structural analysis results. Matrix inversion is used in modal analysis. The matrices involved can be a non-singular real matrix, a complex square matrix or a singular matrix.

2.3.1 Inverse of a non-singular real matrix

If the product of two non-singular matrices is the identity matrix, the two matrices are said to be inverse to each other and both are invertible. If $[A][B] = [I]$, then $[A]^{-1} = [B]$ and $[B]^{-1} = [A]$. Inverse commutes on multiplication, which is not true for matrices in general.

The inverse of $[A]$ can be estimated by:

$$[A]^{-1} = \frac{[a]}{|A|}$$

(2.34)

where $[a]$ is the adjoint matrix of $[A]$. Each element in $[a]$ is the cofactor of the same element in $[A]$. However, unless $[A]$ is sparse, this method is numerically most inefficient and therefore is only used for the inverse of a small matrix.

The problem of inverting a matrix is fundamentally the same as solving a set of linear simultaneous equations. If the inverse of $[A]$ (denoted as $[X]$) is sought, then

$$[A][X] = [I]$$

(2.35)

This is effectively to solve n sets of linear equations:

$$[A]\{X\}_i = \{e\}_i \quad (i = 1, 2, \ldots, n) \tag{2.36}$$

where $\{e\}_i = \{0 \quad 0 \quad \ldots \quad 1 \quad \ldots \quad 0\}^T$ consists of zero elements except the ith element is one.

The Gaussian elimination algorithm used for solving linear equations can be employed to invert a square matrix. Basically, the process is similar to a series of elementary row operations. However, there is a difference this time. For each column, after eliminating all the elements under the diagonal a_{ii}, the elements above a_{ii} are eliminated, too. This is effectively to pre-multiply $[A]$ with matrix:

$$[\overline{M}]_k = \begin{bmatrix} 1 & \ldots & -\dfrac{a_{1k}^{(k)}}{a_{kk}^{(k)}} & \ldots & 0 \\ \ldots & \ldots & \ldots & \ldots & \ldots \\ 0 & \ldots & \dfrac{1}{a_{kk}^{(k)}} & \ldots & 0 \\ \ldots & \ldots & \ldots & \ldots & \ldots \\ 0 & \ldots & -\dfrac{a_{nk}^{(k)}}{a_{kk}^{(k)}} & \ldots & 1 \end{bmatrix} \tag{2.37}$$

Therefore, the elimination method for matrix inversion means to perform a series of row operations such that $[A]$ is converted to a unity matrix:

$$[\overline{M}]_n [\overline{M}]_{n-1} \ldots [\overline{M}]_2 [\overline{M}]_1 [A] = [I] \tag{2.38}$$

The inverse of $[A]$ can be easily derived as:

$$[A]^{-1} = [\overline{M}]_n [\overline{M}]_{n-1} \ldots [\overline{M}]_2 [\overline{M}]_1 \tag{2.39}$$

If a non-singular real matrix $[A]$ is symmetric, it can be decomposed into a product of two triangular matrices:

$$[A] = [U]^T [U] \tag{2.40}$$

where $[U]$ is an upper triangular matrix. Therefore, it can be shown that

$$[A]^{-1} = ([U]^T [U])^{-1} = [U]^{-1} ([U]^{-1})^T \tag{2.41}$$

This shows that by decomposing the symmetric matrix $[A]$ into $[U]^T [U]$, we can determine the inverse of $[A]$ by merely inverting an upper-triangular matrix $[U]$ that can be estimated using equation (2.34).

2.3.2 Inverse of a square complex matrix

A square complex matrix $[C]$ can be inverted using the inverse of a real matrix. The idea is to transform the inverse into that of a real matrix with double-sized order. If:

$$[C] = [A] + j[B] \tag{2.42}$$

and

$$[C]^{-1} = [X] + j[Y] \tag{2.43}$$

then the multiplication of these two inverse pairs will yield:

$$[A][X] - [B][Y] = [I] \tag{2.44}$$

and
$$[A][Y] + [B][X] = [0] \tag{2.45}$$

Combining these two equations will lead to the solution of the real and imaginary parts of the complex inverse in a real matrix inverse:

$$\frac{[X]}{[Y]} = \frac{[A] \quad -[B]}{[B] \quad \ [A]}^{-1} \quad \frac{[I]}{[0]} \tag{2.46}$$

Alternatively, matrices $[X]$ and $[Y]$ can be determined directly from the definition of matrix inverse. If $[A]$ is non-singular, then it can be found that:

$$\begin{cases} [X] = ([A] + [B][A]^{-1}[B])^{-1} \\ [Y] = -[A]^{-1}[B]([A] + [B][A]^{-1}[B])^{-1} \end{cases} \tag{2.47}$$

If $[B]$ is non-singular, then it can be found that:

$$\begin{cases} [X] = [B]^{-1}[A]([B] + [A][B]^{-1}[A])^{-1} \\ [Y] = -([B] + [A][B]^{-1}[A])^{-1} \end{cases} \tag{2.48}$$

If both $[A]$ and $[B]$ are non-singular, then equations (2.47) and (2.48) should produce the same results.

2.3.3 Pseudo inverse of a matrix

When a matrix is either non-square, or rank-deficient or both, normal inverse does not apply anymore. It requires a pseudo inverse. The simplest case of pseudo inverse occurs when solving a set of linear equations with a rectangular coefficient matrix $(m > n)$:

$$[A]_{m \times n} \{x\}_{n \times 1} = \{b\}_{m \times 1} \tag{2.49}$$

If the columns of $[A]$ are independent, then the solution is easy to obtain. Pre-multiplying the transpose of $[A]$ on both sides of the equation leads to:

$$[A]^T[A]\{x\} = [A]^T\{b\} \tag{2.50}$$

Since $[A]^T[A]$ is invertible, the solution of the linear equations becomes:

$$\{x\} = ([A]^T[A])^{-1}[A]^T\{b\} \tag{2.51}$$

Hence, the pseudo inverse of $[A]$, denoted as $[A]^+$, is determined by:

$$[A]^+ = ([A]^T[A])^{-1}[A]^T \tag{2.52}$$

Likewise, the pseudo inverse of $[A]$ of dimension $m \times n$, where $m < n$, and rank m is given as:

$$[A]^+ = [A]^T([A][A]^T)^{-1} \tag{2.53}$$

2.4 Decomposition of a matrix

The theory of modal analysis sometimes requires a matrix to be decomposed. There are several ways of decomposition which are used in the analysis.

2.4.1 *LU* decomposition

The *LU* decomposition is to decompose a square matrix into a product of lower triangular matrix and an upper triangular one. This decomposition can be obtained from Gaussian elimination for the solution of linear equations. It can be verified that the inverse of $[M]_1$ in equation (2.29) takes a very simple form:

$$[M]_1^{-1} = \begin{bmatrix} 1 & 0 & 0 & \dots & 0 \\ m_{21} & 1 & 0 & \dots & 0 \\ m_{31} & 0 & 1 & \dots & 0 \\ \dots & \dots & \dots & \dots & \dots \\ m_{n1} & 0 & 0 & \dots & 1 \end{bmatrix} \qquad (2.54)$$

Thus, from equation (2.33) we have:

$$[A]^{(1)} = [M]_1^{-1}[M]_2^{-1}\dots[M]_{n-1}^{-1}[A]^{(n)} \qquad (2.55)$$

Since the final outcome of Gaussian elimination is an upper triangular matrix $[A]^{(n)}$ and the product of all $[M]_i^{-1}$ matrices will yield a lower triangular matrix, the *LU* decomposition is realized:

$$[A] = [L][U] = ([M]_1^{-1}[M]_2^{-1}\dots[M]_{n-1}^{-1})[A]^{(n)} \qquad (2.56)$$

Here, $[L] = [M]_1^{-1}[M]_2^{-1}\dots[M]_{n-1}^{-1} = \begin{bmatrix} 1 & 0 & 0 & \dots & 0 \\ m_{21} & 1 & 0 & \dots & 0 \\ m_{31} & m_{32} & 1 & \dots & 0 \\ \dots & \dots & \dots & \dots & \dots \\ m_{n1} & m_{n2} & m_{n3} & \dots & 1 \end{bmatrix}$

$$(2.57)$$

$$[U] = [A]^{(n)} = \begin{bmatrix} a_{11}^{(1)} & a_{12}^{(1)} & a_{13}^{(1)} & \dots & a_{1n}^{(1)} \\ 0 & a_{22}^{(2)} & a_{23}^{(2)} & \dots & a_{2n}^{(2)} \\ 0 & 0 & a_{33}^{(3)} & \dots & a_{3n}^{(3)} \\ \dots & \dots & \dots & \dots & \dots \\ 0 & 0 & 0 & \dots & a_{nn}^{(n)} \end{bmatrix} \qquad (2.58)$$

The following example shows the process of using Gaussian elimination to solve the linear equations to obtain the LU decomposition of $[A]$.

$$[A] = \begin{bmatrix} 3 & -1 & 2 \\ 1 & 2 & 3 \\ 2 & -2 & -1 \end{bmatrix} \quad \{b\} = \begin{Bmatrix} 12 \\ 11 \\ 2 \end{Bmatrix}$$

$$[A|b] = \begin{bmatrix} 3 & -1 & 2 & 12 \\ 1 & 2 & 3 & 11 \\ 2 & -2 & -1 & 2 \end{bmatrix} \rightarrow \begin{bmatrix} 3 & -1 & 2 & 12 \\ 0 & 7/3 & 7/3 & 7 \\ 0 & -4/3 & -7/3 & -6 \end{bmatrix} \rightarrow \begin{bmatrix} 3 & -1 & 2 & 12 \\ 0 & 7/3 & 7/3 & 7 \\ 0 & 0 & -1 & -2 \end{bmatrix}$$

where $\quad [M]_1 = \begin{bmatrix} 1 & 0 & 0 \\ -1/3 & 1 & 0 \\ -2/3 & 0 & 1 \end{bmatrix} \quad [M]_2 = \begin{bmatrix} 1 & 0 & 0 \\ 0 & 1 & 0 \\ 0 & 4/7 & 1 \end{bmatrix}$

Back transformation yields the solution for the linear equations:

$$\begin{Bmatrix} x_1 \\ x_2 \\ x_3 \end{Bmatrix} = \begin{Bmatrix} 3 \\ 1 \\ 2 \end{Bmatrix}$$

Meanwhile, the following LU decomposition has been realized:

$$[L] = [M]_1^{-1}[M]_2^{-1} = \begin{bmatrix} 1 & 0 & 0 \\ 1/3 & 1 & 0 \\ 2/3 & -4/7 & 1 \end{bmatrix} \quad \text{and} \quad [U] = \begin{bmatrix} 3 & -1 & 2 \\ 0 & 7/3 & 7/3 \\ 0 & 0 & -1 \end{bmatrix}$$

2.4.2 *QR* decomposition

QR decomposition of a square matrix $[A]$ of dimension $p \times p$ and rank p yields:

$$[A] = [Q][R] \tag{2.59}$$

Here, $[Q]$ is a $p \times p$ orthogonal matrix with full rank and $[R]$ is a $p \times p$ upper triangular matrix with full rank.

If $[A]$ is non-square and rank-deficient, QR decomposition still exists. Assume an $[A]$ of dimension $p \times q$ and rank k. The QR decomposition appears to be the same except this time $[Q]$ is a $p \times k$ orthonormal column matrix and $[R]$ is a $k \times q$ upper triangular matrix with rank k.

2.4.3 Schur decomposition

Without imposing symmetry, full rank, or realness, a square matrix $[A]$ of dimension $p \times p$ can be decomposed using Schur decomposition. Mathematically, it can be shown that $[A]$ is similar to an upper-triangular matrix $[T]$ whose diagonal elements are the eigenvalues of $[A]$. Therefore,

$$[T] = [P]^H[A][P] \qquad (2.60)$$

Here, $[P]$ is a square unitary matrix. $[T]$ is known as the Schur form of $[A]$ and the following decomposition is Schur decomposition:

$$[A] = [P][T][P]^H \qquad (2.61)$$

The matrices we deal with in modal analysis are often either real symmetric or Hermitian. They are known as normal matrices such that:

$$[A]^H[A] = [A][A]^H \qquad (2.62)$$

Schur decomposition of a normal matrix renders a diagonal matrix $[T]$. Since the eigenvalues of a normal matrix are real, $[T]$ will be a real matrix.

2.4.4 Spectrum decomposition

Spectrum decomposition of a matrix relies on its eigenvalue solution. We can start by a real symmetric $n \times n$ matrix $[A]$ with distinct eigenvalues λ_r $(r = 1, 2, \ldots, n)$. The eigenvectors $\{\varphi\}_r$ $(r = 1, 2, \ldots, n)$ are all real vectors. It is possible to normalize the eigenvectors such that they will be orthogonal to each other. Thus, an orthogonal matrix $[Q]$ that consists of eigenvectors will be found and the following decomposition of $[A]$ is available:

$$[A] = [\{\varphi\}_1 \{\varphi\}_2 \cdots \{\varphi\}_n] \, \text{diag} \, [\lambda_r][\{\varphi\}_1\{\varphi\}_2 \cdots \{\varphi\}_n]^T = [\Phi][\Lambda][\Phi]^T$$

and $[A] = \sum_{r=1}^{n} \lambda_r \{\varphi\}_r \{\varphi\}_r^T \qquad (2.63)$

If $[A]$ is an $n \times n$ Hermitian matrix, its eigenvalues λ_r $(r = 1, 2, \ldots, n)$ are still real quantities. The eigenvectors $\{u\}_r$ $(r = 1, 2, \ldots, n)$ will form a unitary matrix $[U]$. Thus, the spectrum decomposition of a Hermitian matrix is as follows:

$$[A] = [U][\Lambda][U]^H = \sum_{r=1}^{n} \lambda_r \{u\}_r \{u\}_r^H \qquad (2.64)$$

For a general complex $n \times n$ matrix $[A]$, assume its unique eigenvalues are λ_r $(r = 1, 2, \ldots, n)$. The independent eigenvectors of matrices $[A]$ and $[A]^H$ are $\{u\}_r$ and $\{v\}_r$ $(r = 1, 2, \ldots, n)$ respectively. The spectrum decomposition of $[A]$ can be derived as:

$$[A] = \sum_{r=1}^{n} \lambda_r \{u\}_r \{v\}_r^H \qquad (2.65)$$

2.4.5 Submatrix decomposition

Submatrix decomposition of a matrix is useful in modal analysis to discover or utilize structural connectivity of a dynamic system. The concept of submatrix approach can be outlined using a 3 DoF mass–spring system shown in Figure 2.1.

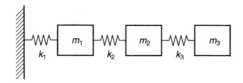

Figure 2.1 A 3 degree-of-freedom system

The stiffness matrix of the system can be decomposed using the stiffness values of each of the three springs:

$$[K] = \begin{bmatrix} k_1 & -k_1 & 0 \\ -k_1 & k_1 + k_2 & -k_2 \\ 0 & -k_2 & k_3 \end{bmatrix} = [K]_1 + [K]_2 + [K]_3 \qquad (2.66)$$

where three submatrices of the springs are:

$$[K]_1 = k_1 \begin{bmatrix} 1 & -1 & 0 \\ -1 & 1 & 0 \\ 0 & 0 & 0 \end{bmatrix} \qquad (2.67)$$

$$[K]_2 = k_2 \begin{bmatrix} 0 & 0 & 0 \\ 0 & 1 & -1 \\ 0 & -1 & 1 \end{bmatrix} \qquad (2.68)$$

$$[K]_3 = k_3 \begin{bmatrix} 0 & 0 & 0 \\ 0 & 0 & 0 \\ 0 & 0 & 1 \end{bmatrix} \qquad (2.69)$$

The numerical advantage of submatrix decomposition is obvious. Since the physical connectivity has been factored in, $[K]$ is now determined by three stiffness values (k_1, k_2 and k_3) while the $[K]$ without the connectivity information would be determined by six non-zero elements.

2.4.6 Singular value decomposition (SVD)

Singular value decomposition (SVD) has found many applications in modal analysis. For a real matrix $[A]$ of dimension $p \times q$, there exists a $p \times p$ orthogonal matrix $[U]$, a $q \times q$ orthogonal matrix $[V]$ and a $p \times q$ diagonal matrix $[\Sigma]$ (possibly a diagonal square matrix augmented with zero rows or columns), so that the following decomposition holds:

$$[A] = [U][\Sigma][V]^T \qquad (2.70a)$$

For example:

$$\begin{bmatrix} 1 & 2 & 3 & 4 \\ 2 & 3 & 4 & 5 \end{bmatrix} = \begin{bmatrix} 0.5969 & 0.8023 \\ 0.8023 & -0.5969 \end{bmatrix} \begin{bmatrix} 9.1521 & 0 & 0 & 0 \\ 0 & 0.4886 & 0 & 0 \end{bmatrix}$$

$$\times \begin{bmatrix} 0.2405 & -0.8013 & 0.5477 & 0 \\ 0.3934 & -0.3811 & -0.7303 & 0.4082 \\ 0.5463 & 0.0392 & -0.1826 & -0.8165 \\ 0.6992 & 0.4585 & 0.3651 & 0.4082 \end{bmatrix}^T$$

Matrices $[U]$ and $[V]$ consist of the left and right singular vectors of $[A]$ and the diagonal elements of $[\Sigma]$ are its singular values. The SVD can be computed by using some existing mathematical software. Analytically, it is useful to know (without proof) the composition of the matrices that form the SVD. The eigenvectors of $[A][A]^T$ constitute $[U]$ and the eigenvalues of it constitute $[\Sigma]^T [\Sigma]$. Likewise, the eigenvectors of $[A]^T[A])$ constitute $[V]$ and the eigenvalues of it (the same as the eigenvalues of $[A][A]^T$) constitute $[\Sigma][\Sigma]^T$. Therefore, SVD is closely linked to the eigenvalue solution.

If $[A]$ of dimension $p \times q$ is a complex matrix, then equation (2.70a) becomes:

$$[A] = [U][\Sigma][V]^H \qquad (2.70b)$$

Here, both the left and right singular vectors $[U]$ and $[V]$ are unitary matrices. The singular values in the diagonal matrix $[\Sigma]$ are still real. The eigenvectors of $[A][A]^H$ constitute $[U]$ and the eigenvalues of it constitute $[\Sigma]^H[\Sigma]$. The eigenvectors of $[A]^H[A]$ constitute $[V]$ and the eigenvalues of it (the same as the eigenvalues of $[A][A]^H$) constitute $[\Sigma][\Sigma]^H$. This means that although $[U]$ and $[V]$ are complex, the singular values of a complex matrix still remain as real quantities.

SVD reveals useful information about $[A]$. For instance, the number of non-zero singular values (therefore the rank of $[\Sigma]$) coincides with the rank of $[A]$. Once the rank is known, the first k columns of $[U]$ form an orthogonal basis for the column space of $[A]$.

In numerical analysis, zero singular values can become small quantities due to numerical errors, measurement errors, noise and ill-conditioning of the matrix.

2.4.7 Eigenvalue decomposition

For a square matrix $[A]$ of dimension $n \times n$, assume its eigenvalues are λ_r and corresponding eigenvector $\{\phi\}_r$, $(r = 1, 2, \ldots, n)$. Also, assume the eigenvector family consists of independent vectors. The eigenvalue matrix and eigenvector matrix can be formed as:

$$[\Lambda] = \text{diag } [\lambda_1 \ \lambda_2 \ldots \lambda_n] \tag{2.71}$$

and
$$[\Psi] = [\{\phi\}_1 \ \{\phi\}_2 \ldots \{\phi\}_n] \tag{2.72}$$

The eigenvalue decomposition pronounces that:

$$[A] = [\Psi][\Lambda][\Psi]^{-1} \tag{2.73}$$

regardless how each eigenvector is scaled. Equation (2.73) also reveals that the square matrix $[A]$ is similar to a diagonal matrix. Thus, $[A]$ can be converted into a diagonal matrix:

$$[\Lambda] = [\Psi]^{-1}[A][\Psi] \tag{2.74}$$

2.4.8 Cholesky decomposition

This decomposition only applies to a symmetric and positive definite matrix. The mass and stiffness matrices of a dynamic system are usually of this type. If $[A]$ of dimension $n \times n$ is such a matrix, then from its LU decomposition we have:

$$[L][U] = [U]^T[L]^T \tag{2.75}$$

and
$$[U][L]^{-T} = [L]^{-1}[U]^T \tag{2.76}$$

Since matrices $[U][L]^{-T}$ and $[L]^{-1}[U]^T$ are upper triangular and lower triangular matrices, equation (2.76) shows that both should actually be a diagonal matrix (denoted as $[D]$ for convenience). This leads to:

$$[A] = [L][D][L]^T = ([L][D]^{1/2})([L][D]^{1/2})^T = [\bar{L}][\bar{L}]^T \tag{2.77}$$

2.5 The matrix eigenvalue problem

The eigenvalue problem is a commonly encountered problem in engineering but it is particularly important in modal analysis. The solution of an eigenvalue problem bears important physical meanings to a dynamic system. Let us consider the standard (and the simplest) eigenvalue problem first. An $n \times n$ real matrix $[A]$ with full rank is said to have eigenvalues λ_r and corresponding non-zero eigenvectors $\{\varphi\}_r$ $(r = 1, 2, \ldots, n)$ if:

$$([A] - \lambda_r[I])\{\varphi\}_r = \{0\} \tag{2.78}$$

or
$$[A]\{\varphi\}_r = \lambda_r\{\varphi\}_r \tag{2.79}$$

Clearly, any multiples of $\{\varphi\}_r$ will also satisfy these equations and thus are a valid replacement of $\{\varphi\}_r$.

From linear algebra, it is known that equation (2.78) holds only if:

$$|[A] - \lambda_r[I]| = \begin{vmatrix} a_{11} - \lambda_1 & a_{12} & \cdots & a_{1n} \\ a_{21} & a_{22} - \lambda_2 & \cdots & a_{2n} \\ \cdots & \cdots & \cdots & \cdots \\ a_{n1} & a_{n2} & \cdots & a_{nn} - \lambda_n \end{vmatrix} = 0 \tag{2.80}$$

This determinant can be expanded, forming an nth order polynomial for λ. The roots of this polynomial are the eigenvalues of $[A]$. Therefore, if equation (2.80) holds, then $[A]$ should always have n (not necessarily distinct or non-zero) eigenvalues. For each eigenvalue, a corresponding eigenvector $\{\varphi\}_r$ can be derived from equation (2.78). Therefore, $[A]$ has n eigenvectors.

If $[A]$ is the system matrix of an MDoF undamped structural system ($[M]^{-1}[K]$), then the square roots of the eigenvalues are the natural frequencies and the eigenvectors are known to represent the mode shapes of the system. Zero eigenvalues indicate rigid body vibration modes for a system not physically grounded. Repeated eigenvalues signify identical mode shapes – a phenomenon usually occurs for a physically symmetrical structure.

If we add a term $\tau\{\varphi\}$ (where τ is a real constant) to both sides of equation (2.79), we will have:

$$([A] + \tau[I])\{\varphi\} = (\lambda + \tau)\{\varphi\} \tag{2.81}$$

This means that the eigenvalues of $[A]$ are changed or shifted by a constant τ, if adding to $[A]$ the $\tau[I]$. The eigenvectors are unchanged. The shifting process is an important feature of some algorithms for computing eigensolutions.

Eigenvalue solution is rarely done by solving the roots of the nth order polynomial in equation (2.80) since it is a horrific numerical task that usually does not result in accurate answers. Using Gaussian elimination can change $[A]$ into a triangular matrix but such row operations do not result in similar matrices and therefore eigenvalues would have changed. The LR and QR methods preserve matrix similarity and are perhaps the best methods currently used for eigensolutions.

The LR algorithm requires that $[A]$ be decomposed first as a product of lower- and upper-triangular matrices $[L]$ and $[R]$ (the same as the LU decomposition). Then, a new matrix is created by exchanging the two triangular matrices to form $[R][L]$. The decomposition is again performed on $[R][L]$. The repetition of this process will eventually lead to a product of a unit matrix and an upper-triangular matrix whose diagonal elements are the eigenvalues of the original $[A]$. It is known that every operation brings about a similar matrix. Therefore, the product has the same eigenvalues of $[A]$. The algorithm is:

$$[A] = [A]_1 = [L]_1[R]_1$$
$$[A]_2 = [R]_1[L]_1 = [L]_2[R]_2 \tag{2.82}$$
$$\ldots$$
$$[A]_{k+1} = [R]_k[L]_k = [L]_{k+1}[R]_{k+1}$$
$$\ldots$$

After sufficient iterations, $[L]_m$ approaches a unit matrix and $[R]_m$ an upper-triangular matrix. Therefore, $[R]_m$ contains in its diagonal the eigenvalues of $[A]$. This is because $[A]_k$ in the LR algorithm is similar to $[A]_{k-1}$. Thus their eigenvalues are the same. For example,

$$[A]_2 = [R]_1[L]_1 = [L]_1^{-1}[L]_1[R]_1[L]_1 = [L]_1^{-1}[A]_1[L]_1 \tag{2.83}$$

The LR method is noticeably simple to implement. When eigenvalues are not well separated, the triangular decomposition of $[A]$ becomes numerically difficult. This leads to a modification of the *LR* method by placing the lower triangular matrix $[L]$ with an orthogonal matrix $[Q]$. It is known in linear algebra that a non-singular matrix can always be decomposed into an orthogonal matrix and an upper-triangular matrix. With similar operation to the *LR* algorithm, the *QR* method will result in an upper-triangular matrix $[R]$ that has the same eigenvalues as $[A]$.

The problem of finding the eigenvectors is simpler than finding eigenvalues. This is solved by the substitution of each eigenvalue found in equation (2.78) to determine the non-trivial solutions of vector $\{\varphi\}_r$.

For an MDoF undamped structural system with mass matrix $[M]$ and stiffness matrix $[K]$, the eigenvalue problem derived from the differential equations is as follows:

$$[K]\{\varphi\} = \lambda[M]\{\varphi\} \tag{2.84}$$

This is known as a generalized eigenvalue problem. Since $[M]$ is usually positive definite, it can be decomposed using square root decomposition as $[M] = [L][L]^T$. Equation (2.84) will then be recast into:

$$([L]^{-1}[K][L]^{-T})([L]^T\{\varphi\}) = \lambda([L]^T\{\varphi\}) \tag{2.85}$$

Now, the generalized eigenvalue problem for $[M]$ and $[K]$ becomes a standard eigenvalue problem for $([L]^{-1}[K][L]^{-T})$.

A more complicated eigenvalue problem than that presented in equation (2.78) is called a higher order eigenvalue problem. In this case, the eigenvalues Δ and eigenvectors $\{\theta\}$ are associated with a series matrices such that:

$$\Delta^p[A]_p\{\theta\} + \Delta^{p-1}[A]_{p-1}\{\theta\} + \ldots + \Delta[A]_1\{\theta\} + [I]\{\theta\} = \{0\} \tag{2.86}$$

Here, p is an integer. The solution of this eigenvalue problem is aided by the concept of state space transformation. Assume that:

$$\{\theta\} = \{\Theta\}e^{\Delta\tau} \tag{2.87}$$

where $\{\Theta\}$ is a constant vector and Δ is the eigenvalue. The rth derivative of the eigenvector $\{\theta\}$ takes a simple form:

$$\frac{\partial^r}{\partial\tau^r}\{\theta\} = \{\theta\}^{(r)} = \Delta^r\{\theta\} \tag{2.88}$$

If we define a state space vector:

$$\{V\} = \{\{\theta\}^{(p-1)} \{\theta\}^{(p-2)} \ldots \{\theta\}^{(0)}\}^T \tag{2.89}$$

then equation (2.86) will become:

$$([A]_p[A]_{p-1} \ldots [A]_1)\{V\}^{(1)} + ([0] [0] \ldots [I])\{V\} = \{0\} \tag{2.90}$$

Using the state space concept, this equation can be appended by a number of equalities to form the following eigenvalue equation:

$$\Delta \begin{bmatrix} [A]_p & [A]_{p-1} & [A]_{p-2} & \cdots & [A]_1 \\ [0] & [I] & [0] & \cdots & [0] \\ [0] & [0] & [I] & \cdots & [0] \\ \cdots & \cdots & \cdots & \cdots & \cdots \\ [0] & [0] & [0] & \cdots & [0] \end{bmatrix} \{V\}$$

$$+ \begin{bmatrix} [0] & [0] & [0] & \cdots & [I] \\ -[I] & [0] & [0] & \cdots & [0] \\ [0] & -[I] & [0] & \cdots & [0] \\ \cdots & \cdots & \cdots & \cdots & \cdots \\ [0] & [0] & \cdots & -[I] & [0] \end{bmatrix} \{V\} = \{0\} \qquad (2.91)$$

This is a standard eigenvalue problem. The eigenvalues Δ solved from this equation will be that for the series $([A]_p\,[A]_{p-1} \cdots [A]_1)$. The eigenvectors $\{\theta\}$ can be derived from the solution of vector $\{V\}$.

2.6 Derivatives of matrices

Let $[A]$ be an $m \times n$ matrix and each element $a_{ij}(i = 1, 2, \ldots m$ and $j = 1, 2, \ldots n)$ be a function of time t. Then the derivative of $[A]$ with respect to time is defined as:

$$\frac{d[A]}{dt} = \begin{bmatrix} \dfrac{da_{11}}{dt} & \dfrac{da_{12}}{dt} & \cdots & \dfrac{da_{1n}}{dt} \\ \dfrac{da_{21}}{dt} & \dfrac{da_{22}}{dt} & \cdots & \dfrac{da_{2n}}{dt} \\ \cdots & \cdots & \cdots & \cdots \\ \dfrac{da_{m1}}{dt} & \dfrac{da_{m2}}{dt} & \cdots & \dfrac{da_{mn}}{dt} \end{bmatrix} \qquad (2.92)$$

Thus, the following elementary operations can be easily deducted:

$$\frac{d([A] \pm [B])}{dt} = \frac{d[A]}{dt} \pm \frac{d[B]}{dt} \qquad (2.93)$$

$$\frac{d([A][B])}{dt} = \frac{d[A]}{dt} [B] + [A] \frac{d[B]}{dt} \qquad (2.94)$$

In particular,

$$\frac{d([A][A]^{-1})}{dt} = \frac{d[A]}{dt} [A]^{-1} + [A] \frac{d[A]^{-1}}{dt} = \frac{[I]}{dt} = [0] \qquad (2.95)$$

Therefore, the derivative of the inverse of a matrix can be expressed as:

$$\frac{d[A]^{-1}}{dt} = -[A]^{-1}\frac{d[A]}{dt}[A]^{-1} \tag{2.96}$$

The derivative of a matrix product in equation (2.94) can be generalized for a product of more than two matrices. It is important to keep the original order of the matrices in the derivative. For instance,

$$\frac{d([A][B]\dots[Z])}{dt} = \frac{d[A]}{dt}[B]\dots[Z] + [A]\frac{d[B]}{dt}\dots[Z] + \dots + [A][B]\dots\frac{d[Z]}{dt} \tag{2.97}$$

The derivative of the power of a matrix is a special case for equation (2.97). For example, the derivative of a matrix power is:

$$\frac{d[A]^3}{dt} = \frac{d[A]}{dt}[A]^2 + [A]\frac{d[A]}{dt}[A] + [A]^2\frac{d[A]}{dt} \tag{2.98}$$

Sometimes the derivative of a matrix with respect to a variable other than time is required. For instance, the matrix derivative with respect to an element within a matrix. Nevertheless, this element can be a function of time. Several such derivatives useful in vibration and modal analysis are described below.

2.6.1 Derivatives of a bilinear form

The derivative of a bilinear form is used in modal analysis. In linear algebra, a bilinear form is given as:

$$p = \{y\}^T[A]\{x\} \tag{2.99}$$

where $\{y\}$ and $\{x\}$ are an $n \times 1$ and $m \times 1$ vector respectively and $[A]$ is an $n \times m$ constant matrix. Assume the elements in vectors $\{x\}$ and $\{y\}$ are functions of time, then,

$$\frac{dp}{dy_1} = \{1 \quad 0 \quad \dots \quad 0\}^T[A]\{x\} \tag{2.100}$$

It is evident from (2.100) that

$$\begin{Bmatrix} \dfrac{\partial p}{\partial y_1} \\[2mm] \dfrac{\partial p}{\partial y_2} \\[1mm] \dots \\[1mm] \dfrac{\partial p}{\partial y_n} \end{Bmatrix} = [A]\{x\} \quad \text{and} \quad \left\{\frac{\partial p}{\partial y_1} \quad \frac{\partial p}{\partial y_2} \quad \dots \quad \frac{\partial p}{\partial y_n}\right\}^T = \{x\}^T[A]^T \tag{2.101}$$

Likewise,

$$\begin{Bmatrix} \dfrac{\partial p}{\partial x_1} \\[4pt] \dfrac{\partial p}{\partial x_2} \\[4pt] \cdots \\[4pt] \dfrac{\partial p}{\partial x_m} \end{Bmatrix} = [A]\{y\} \quad \text{and} \quad \left\{ \dfrac{\partial p}{\partial x_1} \ \dfrac{\partial p}{\partial x_2} \ \cdots \ \dfrac{\partial p}{\partial x_m} \right\}^T = \{y\}^T [A]^T \qquad (2.102)$$

If the bilinear form in (2.99) becomes a quadratic form, i.e. $\{y\}=\{x\}$, and $[A]$ becomes symmetric, then the derivative given in (2.101) and (2.102) takes a simpler form as:

$$\begin{Bmatrix} \dfrac{\partial p}{\partial x_1} \\[4pt] \dfrac{\partial p}{\partial x_2} \\[4pt] \cdots \\[4pt] \dfrac{\partial p}{\partial x_m} \end{Bmatrix} = 2[A]\{x\} \quad \text{and} \quad \left\{ \dfrac{\partial p}{\partial x_1} \ \dfrac{\partial p}{\partial x_2} \ \cdots \ \dfrac{\partial p}{\partial x_m} \right\}^T = 2\{x\}^T [A] \qquad (2.103)$$

The derivative of this bilinear form with respect to time becomes:

$$\frac{d(\{x\}^T [A]\{x\})}{dt} = 2\{x\}^T [A] \frac{d\{x\}}{dt} \qquad (2.104)$$

2.6.2 Derivatives of matrix traces

For an $n \times m$ $[Y]$ and an $m \times m$ $[X]$, the following derivative exists:

$$\frac{\partial(tr([Y][X][Y]^T))}{\partial[Y]} = 2[Y][X]_{n \times m} \qquad (2.105)$$

For an $n \times m$ $[Y]$ and an $n \times m$ $[X]$, the following derivative exists:

$$\frac{\partial(tr([Y][X]^T))}{\partial[Y]} = [X]_{n \times m} \qquad (2.106)$$

For the derivative of the trace of $[A]$ with respect to a variable z, we have:

$$\frac{\partial tr[A]}{\partial z} = tr \frac{\partial[A]}{\partial z} \qquad (2.107)$$

2.7 Perturbation

For a variable x, its perturbation can be denoted as εx. Here, ε is a constant scalar quantity much less than unity and is known as the perturbation factor. Perturbation theory aims to study the behaviour of a system subjected to small perturbations without having to resolve the problem with new variable $x + \varepsilon x$. In fact, the theory deals with the deviation of the solution when a variable (or more) perturbs.

As an example, if the system is represented by a set of linear equations in the form of $[A]\{x\} = \{b\}$, then the theory is to determine the solution of the equations when $[A]$ and $\{b\}$ become $[A] + \varepsilon[A]_1$ and $\{b\} + \varepsilon\{b\}_1$ respectively. The exact solution of the new system is given in:

$$([A] + \varepsilon[A]_1)\{x\}_{new} = \{b\} + \varepsilon\{b\}_1 \qquad (2.108)$$

Perturbation theory assumes that the new solution can be expressed as a convergent series of factor ε. Thus, we have:

$$\{x\}_{new} = \{x\} + \varepsilon\{x\}_1 + \varepsilon^2\{x\}_2 + \ldots \qquad (2.109)$$

By substituting equation (2.109) into (2.108) and clustering like terms, we have the following solutions:

$$\begin{cases} \{x\}_1 = -[A]^{-1}([A]_1\{x\} - \{b\}_1) \\ \{x\}_2 = -[A]^{-1}[A]_1\{x\}_1 \\ \qquad \vdots \\ \{x\}_{n+1} = -[A]^{-1}[A]_1\{x\}_n \\ \qquad \vdots \end{cases} \qquad (2.110)$$

Since we have already had the solution for $\{x\}$ and, therefore, the inverse $[A]^{-1}$, the solution provided in equation (2.110) is much simpler and economical than solving equation (2.108) to find $\{x\}_{new}$.

The same solution can be applied to an eigenvalue problem. The eigenvalue problem of $[A]$ is defined as:

$$[A]\{X\} = \lambda\{X\} \qquad (2.111a)$$

and
$$\{Y\}^T[A]^T = \lambda\{Y\}^T \qquad (2.111b)$$

Here, $\{X\}$ is the right eigenvector and $\{Y\}$ is the left eigenvector. With matrix perturbation, this problem becomes:

$$([A] + \varepsilon[A]_1)\{\overline{X}\} = \overline{\lambda}\{\overline{X}\} \qquad (2.112)$$

The eigenvalue solution can be expressed as:

$$\overline{\lambda} = \lambda + \varepsilon\lambda^{(1)} + \varepsilon^2\lambda^{(2)} + \ldots \qquad (2.113)$$

$$\{\overline{X}\} = \{X\} + \varepsilon\{X\}^{(1)} + \varepsilon^2\{X\}^{(2)} + \ldots \qquad (2.114)$$

Substituting these two equations into equation (2.112) and clustering the like terms, we find the solutions below:

$$\lambda_r^{(1)} = \{Y\}_r^T[A]_1\{X\}_r \qquad (2.115)$$

and
$$\{X\}_r^{(1)} = \sum_{\substack{k=1 \\ k \neq r}}^{n} \frac{\{Y\}_r^T[A]_1\{X\}_r}{\lambda_r - \lambda_k}\{X\}_k \qquad (2.116)$$

Thus, if $[A]$ is perturbed by $\varepsilon[A]_1$, we can expect that the eigenvalues and eigenvectors be perturbed by the amount defined in equations (2.115) and (2.116).

2.8 The least-squares method

The least-squares method is behind many numerical and analytical investigation of modal analysis. Assume a variable y forms a linear relationship with n independent variables x_i ($i = 1, 2, \ldots, n$) so that:

$$y = \{a\}^T\{x\} \tag{2.117}$$

where $\quad \{a\} = \{a_1, a_2, \ldots, a_n\}^T \quad$ and $\quad \{x\} = \{x_1, x_2, \ldots, x_n\}^T$

$\{a\}$ is a constant vector while $\{x\}$ is a time-dependent variable vector. For m different observations, equation (2.117) can be used m times to form a matrix equation:

$$\{y\} = [X]\{a\} \quad \text{where} \quad [X] = [\{x(1)\}, \{x(2)\}, \ldots, \{x(m)\}]^T \tag{2.118}$$

If the number of samples m exceeds the number of variables in $\{x\}$, then there is a need to determine a set of constants in $\{a\}$ such that the total squares of errors between the predicted values and observations will be the least. The error vector between the predicted and the observations can be defined as:

$$\{\varepsilon\} = \{y\} - [X]\{a\} \tag{2.119}$$

It includes measurement or modelling errors. The total error is:

$$E = \{\varepsilon\}^T\{\varepsilon\} = \{y\}^T\{y\} - \{a\}^T[X]^T\{y\} - \{y\}^T[X]^T\{a\} + \{a\}^T[X]^T[X]\{a\} \tag{2.120}$$

To minimize E, take its derivative with respect to vector $\{a\}$ and assign it to zero.

$$\frac{\partial E}{\partial\{a\}} = -2[X]^T\{y\} + 2[X]^T[X]\{a\} = 0 \tag{2.121}$$

This leads to the estimation of the vector

$$\{\hat{a}\} = ([X]^T[X])^{-1}[X]^T\{y\} \tag{2.122}$$

This estimate is the least-squares estimation for $\{a\}$.

This least-squares method can be expanded. When we know that errors from different measurement locations are of different extents, it is possible to weight errors in the least-squares analysis in order to optimize the outcome. This will lead to a 'weighted least-squares' method. Assume the known weighting matrix is $[W]$. The total error becomes:

$$E = \{\varepsilon\}^T[W]\{\varepsilon\} = (\{y\} - [X]\{a\})^T[W](\{y\} - [X]\{a\}) \tag{2.123}$$

By taking the derivative of E with respective to $\{a\}$ and equating it to zero, the solution of the weighted least-squares method yields:

$$\{a\} = ([X]^T[W][X])^{-1}[X]^T[W]\{y\} \tag{2.124}$$

2.9 Partial fraction expansion

When a ratio of two polynomials $f(s)$ and $g(s)$, where s is the Laplace operator (see next section) and the order of $g(s)$ is *greater* than that of $f(s)$, is expressed as a sum of two or more simpler ratios, the ratio is said to be resolved into *partial fractions*. To illustrate the principle of the expansion, assume we are dealing with polynomials with real coefficients only. For example, let:

$$\frac{f(s)}{g(s)} = \frac{b_0 s^m + b_1 s^{m-1} + \ldots + b_{m-1} s + b_m}{s^n + a_1 s^{n-1} + \ldots + a_{n-1} s + a_n}$$

$$= \frac{b_0 s^m + b_1 s^{m-1} + \ldots + b_{m-1} s + b_m}{(s + s_1)(s + s_2)^p (s^2 + as + b)(s^2 + cs + d)^q} \ldots (n > m) \qquad (2.125)$$

Then the partial fraction expansion will be of the form:

$$\frac{f(s)}{g(s)} = \frac{A_1}{(s + s_1)} + \frac{B_1}{(s + s_2)} + \frac{B_2}{(s + s_2)^2} + \ldots + \frac{B_p}{(s + s_2)^p} + \frac{C_1 s + D_1}{s^2 + as + b}$$

$$+ \frac{E_1 s + F_1}{s^2 + cs + d} + \frac{E_2 s + F_2}{(s^2 + cs + d)^2} + \ldots + \frac{E_q s + F_q}{(s^2 + cs + d)^q} \qquad (2.126)$$

where $A_1; B_1, \ldots B_p; C_1, D_1; E_1, F_1, \ldots E_q, F_q$ are constants whose values are to be determined.

Two methods can be used to determine these constants. Method 1 is based on the fact that equations (2.125) and (2.126) are identical. Hence, by giving suitable values to s or by equating coefficients of like terms, the constants may be determined. Method 2 is concerned with calculating the *residue* of each factor and in many cases is less complicated than method 1. A combination of the two approaches can be employed.

We can use an example to show how both methods work in expanding partial fractions. For method 1, we have:

$$\frac{(s^2 + 2s - 5)}{s(s + 1)(s + 5)^2} = \frac{A}{s} + \frac{B}{s + 1} + \frac{C}{s + 5} + \frac{D}{(s + 5)^2}$$

$$= \frac{A(s+1)(s+5)^5 + Bs(s+5)^2 + Cs(s+1)(s+5)^2 + Ds(s+1)}{s(s + 1)(s + 5)^2}$$

This leads to the following identity equality:

$$s^2 + 2s - 5 = A(s + 1)(s+ 5)^2 + Bs(s + 5)^2 + Cs(s + 1)(s + 5)^2 + Ds(s + 1)$$

Equating the like terms, all unknown coefficients can be found.

For method 2, examine the following example:

$$\frac{f(s)}{g(s)} = \frac{f(s)}{(s + s_1)(s + s_2)(s + s_3)^k}$$

then the partial fraction expansion will be:

$$\frac{f(s)}{(s + s_1)(s + s_2)(s + s_3)^k} = \frac{A}{s + s_1} + \frac{B}{s + s_2} + \frac{C_1}{s + s_3} + \cdots \frac{C_k}{(s + s_3)^k}$$

where constants $A, B, C_1, \ldots C_k$ are called residues. They can be conveniently found by the following operations:

$$A = \frac{f(s)}{(s + s_1)(s + s_2)(s + s_3)^k}(s + s_1)\Big|_{s=-s_1}$$

$$B = \frac{f(s)}{(s + s_1)(s + s_2)(s + s_3)^k}(s + s_2)\Big|_{s=-s_2}$$

$$C_k = \frac{f(s)}{(s + s_1)(s + s_2)(s + s_3)^k}(s + s_3)^k\Big|_{s=-s_3}$$

$$C_{k-1} = \frac{d}{ds}\frac{f(s)}{(s + s_1)(s + s_2)(s + s_3)^k}(s + s_3)^k\Big|_{s=-s_3}$$

If complex numbers are to be used, then polynomials such as $s^2 + cs + d$ can be factorized further to be in $(s + x + jy)(s + x - jy)$ form. Thus, polynomial $g(s)$ will become a product of first order polynomials only.

2.10 Laplace transform and transfer function

In modal analysis, we are concerned with differential equations that inherit physical meanings. There is a systematic technique for finding the solutions of differential equations. This technique is called the *Laplace transform*. For a time domain function $f(t)$, the transform is defined and denoted by:

$$F(s) = \mathcal{L}(f(t)) = \lim_{n \to \infty} \int_0^n f(t)e^{-st}\, dt \qquad (2.127)$$

In the context of vibration and modal analysis, this limit *exists*. The Laplace transform is a function of s that is called the Laplace variable. In fact, the integration constitutes a transformation from the *time domain* signal $f(t)$ to the s domain.

For example,

$$\mathcal{L}(e^{-at}) = \lim_{n \to \infty} \int_0^n e^{-at}e^{-st}\, dt = \frac{1}{s + a} \qquad (2.128)$$

The actual Laplace transform is often done using the Laplace transform table. Therefore, there is no need to be stringently fluent with finding the Laplace transform of a time domain function. However, some properties of the Laplace transform are useful in modal analysis. If $\mathcal{L}\{f(t)\} = F(s)$, then the following properties exist:

(1) Linearity $\quad \mathcal{L}(c_1 f_1(t) + c_2 f_2(t)) = c_1 F_1(s) + c_2 F_2(s)$

(2) Shifting $\quad \mathcal{L}(e^{at}f(t)) = F(s - a)$

(3) Scale $\qquad \mathcal{L}(f(at)) = \dfrac{1}{a} F\left(\dfrac{s}{a}\right)$

(4) Derivatives

$$\mathcal{L}\left(\frac{d(f(t))}{dt}\right) = sF(s) - F(0), \ \mathcal{L}\left(\frac{d^2(f(t))}{dt^2}\right) = s^2 F(s) - sF(0) - \frac{dF(0)}{ds}$$

(5) Integrals $\qquad \mathcal{L}\left(\displaystyle\int_0^t f(h)\,dh\right) = \dfrac{F(s)}{s}$

For example, if $\mathcal{L}(\cos 2t) = \dfrac{s}{s^2 + 4}$, then $\mathcal{L}(e^{-t}\cos 2t) = \dfrac{s + 1}{(s + 1)^2 + 4}$.

The inverse Laplace transform is defined as:

$$f(t) = \mathcal{L}^{-1}(F(s)) = \lim_{n \to \infty} \int_0^n F(s)e^{st}\,ds$$

(1) Shifting $L^{-1}(F(s - a)) = e^{at} f(t)$

(2) Scale $L^{-1}(F(as)) = \dfrac{1}{a} f\left(\dfrac{t}{a}\right)$

(3) Derivatives $L^{-1}\left(\dfrac{d^n f(s)}{ds^n}\right) = (-1)^n t^n f(t)$

(4) Integrals $L^{-1}\left(\displaystyle\int_0^s f(h)\,dh\right) = \dfrac{f(t)}{t}$

For example, $L^{-1}\dfrac{s + 1}{s^2 + 2s + 5} = L^{-1}\dfrac{s + 1}{(s + 1)^2 + 4} = e^{-t}\cos 2t$.

An important use of Laplace transform in modal analysis is to convert a differential equation into an algebraic equation. For a SDoF system as shown in Figure 2.2:

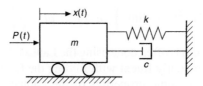

Figure 2.2 A SDoF system

The equation of motion can be written as:

$$m\ddot{x}(t) + c\dot{x}(t) + kx(t) = P(t) \tag{2.129}$$

Assuming all initial conditions are zero, we can apply the Laplace transform on the equation, yielding an algebraic equation:

$$ms^2X(s) + csX(s) + kX(s) = P(s) \tag{2.130}$$

The response of the system becomes:

$$X(s) = \frac{1}{ms^2 + cs + k}P(s) \tag{2.131}$$

The transfer function of the system, which defines the relationship between the force input and response in displacement, can be found as:

$$G(s) = \frac{X(s)}{P(s)} = \frac{1}{ms^2 + cs + k} \tag{2.132}$$

This transfer function becomes the frequency response function when only the imaginary part of the s operator is considered:

$$G(j\omega) = \frac{X(j\omega)}{P(j\omega)} = \frac{1}{-m\omega^2 + cj\omega + k} \tag{2.133}$$

This is the same result one would get by solving the equation of motion as a differential equation. The Laplace transform can also be applied to differential equations for an MDoF system.

2.11 Fourier series and Fourier transform

Fourier series is an ingenious representation of a periodic function. For a periodic time domain function $x(t)$ with period T, we have:

$$x(t) = x(t + nT) \tag{2.134}$$

Mathematically, it can be shown that $x(t)$ consists of a number of sinusoids with frequencies multiple to a fundamental frequency. This fundamental frequency f is dictated by the period such that $f = \frac{1}{T}$. The contribution to $x(t)$ by a sinusoid with frequency f_k is $X(f_k)e^{\frac{j2\pi kt}{T}}$. The amplitude of the kth sinusoid can be determined by:

$$X(f_k) = \frac{1}{T}\int_{-\frac{T}{2}}^{\frac{T}{2}} x(t)e^{\frac{-j2\pi kt}{T}}\,dt \tag{2.135}$$

This component is usually a complex quantity with its amplitude and phase. The term $e^{\frac{-j2\pi kt}{T}}$ represents a unit vector rotating at a frequency of $-\frac{k}{T}$. This integral shows that the component in signal $x(t)$ that has frequency as $kf = \frac{k}{T}$ will be 'frozen' at the rotating frequency of the unit vector, thus posing a non-zero value after the integration. Those other components will become zero after the integration over the whole period.

A periodic signal consists of the summation of the components at all frequencies:

$$x(t) = \sum_{k=-\infty}^{\infty} X(f_k)e^{\frac{j2\pi kt}{T}} = \sum_{k=-\infty}^{\infty} \left(\frac{1}{T} \int_{-\frac{T}{2}}^{\frac{T}{2}} x(t)e^{\frac{-j2\pi kt}{T}} dt \right) e^{\frac{j2\pi kt}{T}} \qquad (2.136)$$

Each frequency component $X(f_k)$ is a complex quantity. However, $x(t)$ is a real quantity. This is because $X(f_k)$ and $X(-f_k)$ are a pair of complex conjugates. The product of these two is the power the signal has at frequency f_k:

$$P_X(f_k) = X(f_k)X(-f_k) = X(f_k)X^*(f_k) \qquad (2.137)$$

Summation of the power at different frequencies will produce the total power of the signal. Each power component will have amplitude information only. The phase information vanishes.

When the period approaches infinity, $x(t)$ becomes a non-periodic signal. The Fourier series defined in equation (2.135) becomes the Fourier transform:

$$X(f) = \int_{-\infty}^{\infty} x(t)e^{-j2\pi ft} dt \qquad (2.138)$$

If we separate the real and imaginary parts of $X(f)$, we have:

$$X_{\text{Re}}(f) = \int_{-\infty}^{\infty} x(t) \cos 2\pi ft \, dt \quad X_{\text{Im}}(f) = \int_{-\infty}^{\infty} x(t) \sin 2\pi ft \, dt \qquad (2.139, 140)$$

The inverse Fourier transform returns $X(f)$ to time domain:

$$x(t) = \int_{-\infty}^{\infty} X(f)e^{j2\pi ft} \, df \qquad (2.141)$$

2.12 Variable separation method for partial differential equations

The variable separation method is a very useful method for dealing with differential equations of continuous vibratory structures such as beams, plates and strings. The method can be described conveniently by using an example of a vibrating beam. Consider the dynamic diagram of a small section of a beam under load $f(x, t)$, as shown in Figure 2.3. The element endures shear force Q and bending moment M.

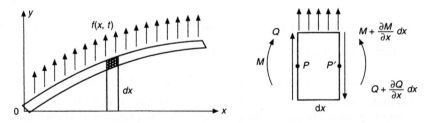

Figure 2.3 A small element of the beam and its dynamic diagram

The equation of motion at the 'y' direction can be derived using Newton's second law as:

$$Q - \left(Q + \frac{\partial Q}{\partial x}dx\right) + f(x, t)\, dx = dx\rho A(x)\frac{\partial^2 y}{\partial t^2} \qquad (2.142)$$

or

$$\rho A(x)\frac{\partial^2 y}{\partial t^2} + \frac{\partial Q}{\partial x} = f(x, t) \qquad (2.143)$$

Here, ρ is the density of the material and $A(x)$ the cross-section. Likewise, the equation of motion for moments about the 'y' axis passing through point P' can be derived as:

$$\frac{\partial M}{\partial x} = Q \qquad (2.144)$$

Therefore, we have the following equation for the beam section:

$$\rho A(x)\frac{\partial^2 y}{\partial t^2} + \frac{\partial^2 M}{\partial x^2} = f(x, t) \qquad (2.145)$$

From the elementary bending theory of a beam, we know that the relationship between bending moment and deflection can be expressed as:

$$M = EI(x)\frac{\partial^2 y}{\partial x^2} \qquad (2.146)$$

Here, E is Young's modulus and $I(x)$ the second moment of area of the beam cross-section about the y axis. This leads to:

$$\rho A(x)\frac{\partial^2 y}{\partial t^2} + \frac{\partial^2}{\partial x^2}\left(EI(x)\frac{\partial^2 y}{\partial x^2}\right) = f(x, t) \qquad (2.147)$$

For a uniform beam with no external forces, this equation reduces to:

$$c^2\frac{\partial^2}{\partial x^2}\left(\frac{\partial^2 y}{\partial x^2}\right) + \frac{\partial^2 y}{\partial t^2} = 0 \qquad c^2 = \frac{EI}{\rho A} \qquad (2.148)$$

The free vibration solution of the beam can be found from equation (2.148) by using the variable separation method. The essence of the method is to assume that the solution of the differential equation, which is a function of both spatial variable x and time variable t, can be separated into a product of a spatial function and time function only:

$$y(x, t) = Y(x)T(t) \qquad (2.149)$$

This separation, when introduced into equation (2.148), leads to:

$$\frac{c^2}{Y(x)}\frac{d^4 Y(x)}{dx^4} = \omega^2 \qquad (2.150)$$

$$\frac{-1}{T(t)}\frac{d^2 T(t)}{dt^2} = \omega^2 \qquad (2.151)$$

These equations can be written as:

$$\frac{d^4 Y(x)}{dx^4} - \beta^4 Y(x) = 0 \quad \left(\beta^4 = \frac{\omega^2}{c^2} = \frac{\rho A \omega^2}{EI} \right) \tag{2.152}$$

$$\frac{d^2 T(t)}{dt^2} + \omega^2 T(t) = 0 \tag{2.153}$$

Thus, equation (2.148) as a partial differential equation is now converted into two normal differential equations with spatial and time functions separated into one equation each. The solution of equation (2.152) will reveal the mode shape of the continuous structure and the solution of equation (2.153) provides the physical animation of the mode shape in the time domain.

2.13 Poles and zeros of a polynomial function

Poles and zeros of a polynomial function dictate its behaviour. For a dynamic system, the function is most often the frequency response function (with frequency ω as the variable) or the transfer function (with Laplace operator s as the variable). Poles of a frequency response function are the frequency roots of the denominator – the natural frequencies. Zeros are those frequency roots of the numerator that signify the presence of a nodal point for the system.

If a frequency response function can be written as the ratio of two polynomials:

$$H(\omega) = \frac{N(\omega)}{D(\omega)} \tag{2.154}$$

then the roots of $N(\omega)$ are the zeros of $H(\omega)$ and the roots of $D(\omega)$ are the poles of $H(\omega)$. Analytically it can be shown that the poles of the FRFs of a dynamic system are identical while the zeros from different FRFs of the same dynamic system are different.

If $H(\omega)$ is expressed as:

$$H(\omega) = \frac{N_1(\omega)}{D(\omega)} + \frac{N_2(\omega)}{D(\omega)} \tag{2.155}$$

then the roots of neither $N_1(\omega)$ nor $N_2(\omega)$ are the zeros for $H(\omega)$. Only $N_1(\omega) + N_2(\omega)$ gives the zeros.

2.14 State–space concept

A state–space model is a different representation of the input–output relationship compared to the transfer or frequency response function approach. This model was developed in the 1960s to satisfy increasingly stringent requirements to study large-scale dynamic systems with computers. The model was based on the concept of *state* which had been in existence in the field of classical dynamics but had not been used in the same way for the state–space model.

The state of a dynamic system is the smallest set of variables (state variables) which, together with the future inputs to the system, can determine the dynamic behaviour of the system. In analytical terms, this means that the state at time t is uniquely determined by the state at time t_0 and the inputs for $t \geq t_0$. It is independent of the state and inputs before t_0. If we choose the reference time t_0 to be zero, this means that a linear time-variable dynamic system is causal.

There are advantages of using a state–space model for studying a dynamic system. This model is able to represent the internal characteristics (this will be made clear in the following). It formulates equations of simplicity in a form which is not suitable for computation. In addition, it is often easier to optimize a dynamic system using this model.

It is convenient to use a simple SDoF system to illustrate the establishment of the state–space model for a dynamic system. The SDoF system is governed by:

$$m\ddot{y} + c\dot{y} + ky = f(t) \tag{2.156}$$

or

$$m\ddot{y} = f(t) - c\dot{y} - ky \tag{2.157}$$

Using the Laplace transform and ignoring initial conditions, the transfer function of the system can be derived as:

$$G(s) = \frac{Y(s)}{F(s)} = \frac{1}{ms^2 + cs + k} \tag{2.158}$$

This is the classical representation of the dynamic system. However, by selecting two new variables as: $x_1(t) = y(t)$ and $x_2(t) = \dot{x}_1(t)$, equation (2.157) becomes:

$$\dot{x}_2(t) = \frac{1}{m}f(t) - \frac{c}{m}x_2(t) - \frac{k}{m}x_1(t) \tag{2.159}$$

Then, the following equations, which use only the first derivatives, can be used to represent the dynamics of the system. They are called the *state equations*:

$$\begin{bmatrix} \dot{x}_1(t) \\ \dot{x}_2(t) \end{bmatrix} = \begin{bmatrix} 0 & 1 \\ -\dfrac{k}{m} & -\dfrac{c}{m} \end{bmatrix} \begin{bmatrix} x_1(t) \\ x_2(t) \end{bmatrix} + \begin{bmatrix} 0 \\ \dfrac{1}{m} \end{bmatrix} f(t) \tag{2.160}$$

$$y(t) = \begin{bmatrix} 1 & 0 \end{bmatrix} \begin{bmatrix} x_1(t) \\ x_2(t) \end{bmatrix} \tag{2.161}$$

This example has shown the steps of establishing the state–space model for a dynamic system. The model is described using the state–space equations. These equations comprise state variables. For a multi-input and multi-output system we can define n state variables $x_1(t), x_2(t), \dots, x_n(t)$ such that each variable is the integrator of the one after it in the time domain. In addition, assume the system has r inputs as $u_1(t), u_2(t), \dots, u_r(t)$ and p outputs as $y_1(t), y_2(t), \dots, y_p(t)$. We can derive the following state–space equations:

$$\{\dot{x}(t)\} = [A]\{x(t)\} + [B]\{u(t)\} \tag{2.162}$$

$$\{y(t)\} = [C]\{x(t)\} + [D]\{u(t)\} \tag{2.163}$$

Here, $\{x(t)\}$ is the $n \times 1$ state vector; $[A]$ is the $n \times n$ system matrix; $[B]$ is the $n \times r$ input matrix; $\{u(t)\}$ is the $r \times 1$ input vector; $\{y(t)\}$ is the $p \times 1$ output vector; $[C]$ is the $p \times n$ output matrix; $[D]$ is the $p \times r$ coupling matrix between inputs and outputs.

Equation (2.162) is known as the system (or state) equation while equation (2.163) as the output equation. The system equation only contains the first order derivative and the output equation has no derivatives. A real life dynamic system such as a vibratory structure usually has insignificant or no coupling between the inputs and outputs. Thus, $[D]$ is null. For a single input case, $[B]$ is reduced to a column vector. For a single output case (this can be for a multi-output system but only one particular output is of interest), $[C]$ is reduced to a row vector.

The state–space model can also be derived from the transfer function of a system rather than from its differential equations, as done before. This needs a tool called a simulation diagram (Figure 2.4). The basic element of this diagram is the integrator which represents the time and frequency domains respectively as:

$$y(t) = \int x(t)\, dt \qquad (2.164)$$

$$Y(s) = \frac{1}{s} X(s) \qquad (2.165)$$

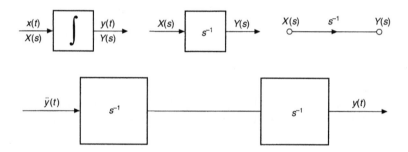

Figure 2.4 Simulation diagram

For the SDoF system, the differential equation can be recast as:

$$\ddot{y}(t) = -\frac{B}{M}\dot{y}(t) - \frac{K}{M} y(t) + \frac{1}{M} f(t) \qquad (2.166)$$

The transfer function can be written as:

$$G(s) = \frac{\dfrac{1}{M} s^{-2}}{1 + \dfrac{B}{M} s^{-1} + \dfrac{K}{M} s^{-2}} \qquad (2.167)$$

The simulation diagram in Figure 2.5 represents this transfer function.

The transfer function can be derived from the state–space equations. For a system with the following state–space equations:

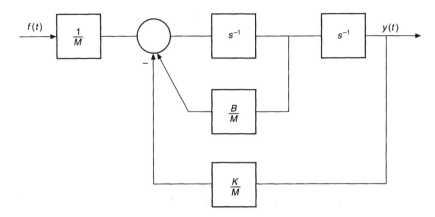

Figure 2.5 Simulation diagram representing the transfer function in equation (2.167)

$$\begin{cases} \{\dot{x}\} = [A]\{x\} + [B]\{u\} \\ \{y\} = [C]\{x\} \end{cases} \tag{2.168}$$

the Laplace transform (ignoring initial conditions) reveals:

$$s\{X(s)\} = [A]\{X(s)\} + [B]\{U(s)\} \tag{2.169}$$

or
$$\{X(s)\} = (s[I] - [A])^{-1}[B]\{U(s)\} \tag{2.170}$$

Since $\{Y(s)\} = [C]\{X(s)\}$, the equation becomes:

$$\{Y(s)\} = [C](s[I] - [A])^{-1}[B]\{U(s)\} \tag{2.171}$$

This leads to the following expression for the transfer function:

$$\{G(s)\} = [C](s[I] - [A])^{-1}[B] \tag{2.172}$$

For a single input–single output

$$G(s) = [C](s[I] - [A])^{-1}[B] \tag{2.173}$$

The eigenvalues of $[A]$ are the poles of $G(s)$.

2.15 Time series analysis

Time series analysis is an alternative to spectral analysis. For some vibration studies, spectral analysis may be impaired by the short time record of the data or by the energy leakage caused by windowing. Time series analysis establishes a mathematical model from the measured time domain data based on the finite difference concept. It is not restricted by a short data record. A number of mathematical models can be established from time series data. We shall start the introduction from the linear regression model of measured data.

A linear regression model of two groups of measured data reveals the correlation between them. Assume for a dynamic system, the observation of the input are a set of independent measurements x_i and the output another set of independent and random measurements y_i ($i = 1, 2, \ldots, N$). The simplest linear regression model between the input and output of the system can be described mathematically as a linear relation:

$$y_i = \beta_0 + \beta_1 x_i + \varepsilon_i \tag{2.174}$$

Here, the residue $\varepsilon_i (i = 1, 2, \ldots, N)$ deserves special attention. First, they are independent of each other. Second, it is independent of variable x_i. Third, it is assumed to be a random variable normally distributed. Equation (2.174) shows that the random output of the system y_i consists of two parts: the deterministic part as $\beta_1 x_i$ and the random part as ε_i. When ε_i becomes zero, we see the familiar linear relation between the two sets of data. The least square solution for parameters is:

$$\hat{\beta}_1 = \frac{\sum\limits_{i=1}^{N} (y_i - \bar{y})(x_i - \bar{x})}{\sum\limits_{i=1}^{N} (x_i - \bar{x})} \tag{2.175}$$

$$\hat{\beta}_0 = \bar{y} - \hat{\beta}_1 \bar{x} \tag{2.176}$$

The model described in equation (2.174) can be extended if the single variable x_i becomes multi-variables ($x_{1i}, x_{2i}, \ldots, x_{ri}$):

$$y_i = \beta_0 + \beta_1 x_{1i} + \beta_2 x_{2i} + \ldots + \beta_r x_{ri} + \varepsilon_i \tag{2.177}$$

The output y_i also consists of two parts: the deterministic part as $\beta_1 x_{1i} + \beta_2 x_{2i} + \ldots + \beta_r x_{ri}$ and the random part as ε_i. When ε_i becomes zero, the least squares solution for the parameters is:

$$\{\hat{\beta}\} = ([X]^T [X])^{-1} [X]^T \{Y\} \tag{2.178}$$

where

$$\{Y\} = \begin{Bmatrix} y_1 \\ y_2 \\ \ldots \\ y_N \end{Bmatrix}, \tag{2.179}$$

$$[X] = \begin{bmatrix} 1 & x_{11} & x_{21} & \ldots & x_{r1} \\ 1 & x_{12} & x_{22} & \ldots & x_{r2} \\ \ldots & \ldots & \ldots & & \ldots \\ 1 & x_{1N} & x_{2N} & \ldots & x_{rN} \end{bmatrix}, \tag{2.180}$$

$$\{\hat{\beta}\} = \begin{Bmatrix} \hat{\beta}_1 \\ \hat{\beta}_2 \\ \ldots \\ \hat{\beta}_N \end{Bmatrix} \tag{2.181}$$

2.15.1 AR model

The linear regression model presented in equation (2.174) was for the correlation between two sets of time series data. Let us now consider a random process quantified by a smooth time series x_i ($i = 1, 2, \ldots, N$) only and use the equation for the correlation of the data in the series, then equation (2.174) can be rewritten as:

$$x_i = \phi_1 x_{i-1} + a_i \tag{2.182}$$

This equation, although similar in appearance to equation (2.174), represents a crucial change. It describes the internal dynamic correlation of a random process. It is a typical finite difference equation. This model is known as an auto-regression (AR) model. Since x_i is correlated only with one x_{i-1}, the model is also called an AR(1) model. The auto-regression coefficient ϕ_1 is a constant. The residue a_i ($i = 1, 2, \ldots, N$) is independent within its series $\{a_i\}$ and of the time series $\{x_i\}$. Using the regression operator B such that $x_{i-1} = Bx_i$, equation (2.182) can be recast as:

$$x_i = \frac{1}{1 - \phi_1 B} a_i \tag{2.183}$$

This is equivalent to a first order system with white noise input a_i and random output x_i. The transfer function of the system is $\frac{1}{1 - \phi_1 B}$.

The AR(1) model can be extended if x_i is correlated with more than one previous record. For example, an AR(2) model means that x_i is correlated with both x_{i-1} and x_{i-2}:

$$x_i = \phi_1 x_{i-1} + \phi_2 x_{i-2} + a_i \tag{2.184}$$

This AR model harbours a fundamental concern, i.e. how can x_i be correlated with x_{i-1} yet does not rely on a_{i-1}, as x_{i-1} does. To address this concern, we need to expand the AR model to its new dimension, as described below.

2.15.2 ARMA model

The AR(1) model was designed for a smooth and normally distributed time series with zero average $\{x_i\}$. It assumes that at a given moment the value in the time series x_i only correlates with the value a moment before, i.e. x_{i-1} (or more). For a dynamic system, this is a bold assumption, since the behaviour of the system such as the impulse response will tend to carry on for a while. If x_i is not correlated with x_{i-1} only and the residue series $\{a_i\}$ is not white noise, then the AR(1) model becomes invalid.

If x_i is correlated with not only x_{i-1}, but also x_{i-2}, then it will also be correlated with a_{i-1}. Therefore, this time we can express x_i as:

$$x_i = \phi_1 x_{i-1} + \phi_2 x_{i-2} - \theta_1 a_{i-1} + a_i \tag{2.185}$$

or

$$x_i - \phi_1 x_{i-1} - \phi_2 x_{i-2} = -\theta_1 a_{i-1} + a_i \tag{2.186}$$

Equation (2.186) shows that the value x_i in the time series $\{x_i\}$ depends on two immediately previous values x_{i-1} and x_{i-2}, and two residues a_i and a_{i-1}. Here the residues still follow the characteristics outlined in the AR(1) model. The left-hand side of equation (2.186) is an AR model with ϕ_1 and ϕ_2 being the regression coefficients. The right-hand side is called the moving average (MA) model with θ_1 being the moving average coefficient. Since x_i correlates with two previous values and relies on one previous residue, the model in equation (2.186) is called an ARMA(2,1) model. Obviously, this model can be extended if more correlated quantities are involved.

The inclusion of the term $-\theta_1 a_{i-1}$ signifies a fundamental change from the AR model. It means a dynamic system moving from moment '$i-1$' to moment 'i' is able to memorize the input happened at moment '$i-1$', i.e. a_{i-1}. As a result, the output at moment 'i', x_i, is determined not only by the current input a_i but also by the previous input a_{i-1}. Therefore, if the AR model is seen as a static model, then the ARMA model is a dynamic model with memory to the past.

The essence of an ARMA model is that the random value x_i is decomposed into two parts: the deterministic and dependent part characterized by ($\phi_1 x_{i-1} + \phi_2 x_{i-2} - \theta_1 a_{i-1}$), and the random and independent part denoted as a_i. By using the regression operator B, we can transform the ARMA model, as we did for the AR model, into:

$$x_i = \frac{1 - \theta_1 B}{1 - \phi_1 B - \phi_2 B^2} a_i \tag{2.187}$$

This means that physically the ARMA model transforms a self-dependent smooth time series $\{x_i\}$ into a mutually independent time series $\{a_i\}$. If time series $\{a_i\}$ is the white noise type of random input to a linear system with transfer function $\dfrac{1 - \theta_1 B}{1 - \phi_1 B - \phi_2 B^2}$, then the output of the system will be time series $\{x_i\}$.

The ARMA(2,1) model can be expanded into an ARMA(n, m) model as characterized by the following equation:

$$x_i = \sum_{r=1}^{n} \phi_r x_{i-r} - \sum_{s=1}^{m} \theta_s a_{i-s} + a_i \tag{2.188}$$

The transfer function counterpart of this ARMA(n, m) model using the regression operator B will be:

$$x_i = \frac{1 - \sum\limits_{s=1}^{m} \theta_s B^s}{1 - \sum\limits_{r=1}^{n} \phi_r B^r} \tag{2.189}$$

The physical interpretation of the ARMA(n, m) model is also important to understand the model. As before, such a model decomposes the response x_i into two parts: the deterministic and dependent part characterized by $\sum\limits_{r=1}^{n} \phi_r x_{i-r} - \sum\limits_{s=1}^{m} \theta_s a_{i-s}$ and the random and independent part denoted as a_i. For a dynamic system, at time 'i', $\sum\limits_{r=1}^{n} \phi_r x_{i-r} - \sum\limits_{s=1}^{m} \theta_s a_{i-s}$ consists of the quantities already happened in history. Therefore, they are deterministic. At the same time, a_i represents a current random input or disturbance. The nature of normal distribution of the output x_i is due to the same nature exhibited by a_i.

2.16 The z-transform

The z-transform takes discrete time domain signals into a complex-variable frequency domain. This resembles the Laplace transform which takes continuous time domain signals into a complex frequency domain. For signal processing, discrete time domain signals are often the data to be processed.

Unlike many other transforms, the z-transform is named after a letter of the alphabet rather than a mathematician. It was initially proposed to solve linear and constant difference equations from sampled signal or sequence. With the advent of digital computers, this transform found its niche in engineering.

Assume we deal with a time domain real function $f(t)$ through its sample values at $t = 0, 1, 2, \ldots, n$:

$$f(0)\, f(1)\, f(2) \ldots f(n)$$

Then the following polynomial is referred to as the z-transform of $f(t)$:

$$F(z) = f(0) + f(1)z^{-1} + f(2)\, z^{-2} + \ldots + f(n)z^{-n} \tag{2.190}$$

Here, z is a complex variable.

The sampling of function $f(t)$ can be seen as the work of a series of impulse strings. Thus, the discretized function $f(t)$ can be represented by:

$$x(t) = f(0)\, \delta(t) + f(1)\, \delta(t-1)\, f(2)\, \delta(t-2) + \ldots + f(n)\, \delta(t-n) \tag{2.191}$$

Taking the Laplace transform of this function, we have:

$$X(s) = \int_{-\infty}^{\infty} x(t)e^{-st}\, dt = \sum_{n=0}^{\infty} f(n)e^{-nst} \tag{2.192}$$

Let $z = e^{-st}$, we have:

$$X(s) = \int_{-\infty}^{\infty} x(t)e^{-st}\, dt = \sum_{n=0}^{\infty} f(n)z^{-n} \tag{2.193}$$

This happens to be the z-transform of function $f(t)$. Therefore, we define the z-transform of a time series $x(n)$ as:

$$Z\{x(n)\} = X(z) = \sum_{n=-\infty}^{\infty} x(n)z^{-n} \tag{2.194}$$

where z is a complex variable. If the time function is causal (in most cases the signals from a dynamic system are causal), then the z-transform becomes one-sided:

$$Z\{x(n)\} = X(z) = \sum_{n=0}^{\infty} x(n)z^{-n} \tag{2.195}$$

For example, a unit step function is characterized as $x(n) = 1$ for $n \geq 0$. So its z-transform becomes:

$$X(z) = \sum_{n=0}^{\infty} z^{-i} = \frac{1}{1 - z^{-1}}$$

Literature

1. Bracewell, R.N. 1965: *The Fourier Transform and its Applications*. McGraw-Hill.
2. Brandon, J.A. and Cremona, C.F. 1990: Singular value decomposition: sufficient but not necessary. *Proceedings of the 8th International Modal Analysis Conference*, Orlando, FL, 1376–1380.
3. Brandon, J.A. 1989: Derivation and application of the Choleski decomposition for the positive semi-definite matrices used in structural dynamics re-analysis. *Proceedings of the Modern Practice of Stress and Vibration Analysis*, 225–233.
4. Deif, S.S. 1982: *Advanced Matrix Theory for Scientists and Engineers*. Abacus Press, ISBN 0-85626-327-3.
5. Frazer, R.A. and Collar, A.R. 1965: *Elementary Matrices*. The Syndics of the Cambridge University Press.
6. Golub, G.H. and Van Loan, C.F. 1983: *Matrix Computations*. North Oxford Academic, Oxford.
7. Golub, G.H. and Kahan, W. 1965: Calculating the Singular Values and Pseudo-Inverse of a Matrix. *SIAM Journal Numerical Analysis*, Series B, **2**(2).
8. Greville, T.N.E. 1959: The pseudoinverse of a rectangular or singular matrix and its application to the solution of systems of linear equations. *SIAM Review*, **1**, 38–43.
9. Leung, A.Y.T. and Liu, Y.F. 1992: A Generalised Complex Symmetric Engensolver. *Computers & Structures*, **43**(6), 1183–1186.
10. Maia, N.N.M. 1991: Fundamentals of Singular Value Decomposition. *Proceedings of the 9th International Modal Analysis Conference*, Firenze, Italy, 1515–1521.
11. Nashed, M.Z. 1976: *Generalised Inverses and Applications*. Academic Press, Inc., New York, USA.
12. Ogata, K. 1990: *Modern Control Engineering*. Prentice-Hall.
13. Paige, C.C. and Saunders, M.A. 1981: Towards a Generalized Singular Value Decomposition. *SIAM Journal of Numerical Analysis*, **18**(3), 398–405.
14. Penrose, R. 1955: A Generalised Inverse for Matrices. *Proceedings of the Cambridge Philosophical Society*, **51**, 406–413.
15. Randall, R. 1977: *Frequency Analysis*. Bruel & Kjaer.
16. Stengel, R.F. 1986: *Stochastic Optimal Control – Theory and Application*. John Wiley & Sons.
17. Wilkinson, J.H. 1965: *The Algebraic Eigenvalue Problem*. Oxford University Press.
18. Zhang, Q. and Lallement, G. 1985: New Method of Determining the Eigensolutions of the Associated Conservative Structure from the Identified Eigensolutions. *Proceedings of the 3rd International Modal Analysis Conference*, Orlando, Florida, 322–328.

3

Basic vibration theory

Vibration problems can be simple or complicated. Vibration theory rarely applies directly to a real structure with the exception of those structures whose dynamics can be described accurately by a finite number of partial differential equations. This is because theory is usually developed for an idealized version of a real problem with various assumptions. Not all of these assumptions can be spelled out clearly. The process of idealizing a real structure before analysis can proceed is mathematical modelling for the structure.

The model from the process of idealization for a real structure is not unique. The need to derive a model that is simple, effective and easy for theoretical analysis is a challenge to be met by experiences as well as by the depth of comprehension of dynamics. For modal analysis, the mathematical model is often a discretized one with a finite number of coordinates. The continuous behaviour, such as mode shapes, is represented using a selected spatial resolution. Thus, the model is often presented by a number of ordinary differential equations. Upon Laplace or Fourier transform, these equations can be converted to algebraic equations. This highlights the need for matrix algebra in modal analysis. The finiteness of a mathematical model also explains the reliance of modal analysis on the theory of an MDoF system.

3.2 Basic concepts of vibration

Vibration is a motion that repeats itself. This repetition may or may not perpetuate. The repetition also does not have to be a literal duplication. Some vibration can repeat itself in a statistical sense.

Contrary to a general perception, the vital elements for vibration are not the presence of inertial and elastic components such as mass and spring. Since vibration can be regarded as the transfer between the kinetic energy and potential energy, a vibratory system has to include a means of storing (and releasing) both energies. The former is often done by a mass and the latter by a spring. A mass connected to a horizontal spring shown in Figure 3.1 is a typical vibratory system. The mass is the component responsible for kinetic energy while the spring is that for potential energy. A pendulum is also a typical example of a vibratory system. This system does

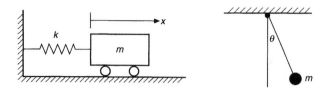

Figure 3.1 Two simple vibratory systems

not have a spring-like component for potential energy. In fact, the mass plays a dual role for both kinetic energy and potential energy. This is an example to contend the perception that a vibratory system has to have both mass and stiffness components.

To be able to study the vibration of a dynamic system, it is essential to know how many degrees of freedom the system has before analysis proceeds. The number of the degrees of freedom of a vibratory system is defined as the minimum number of independent coordinates required to determine completely the motion of all parts of the system at any instant of time.

For example, the mass–spring system in Figure 3.2 has two mass blocks, each being able to move independently. A simple test of 'fixing' one mass will see clearly that the other can still move physically. Therefore, there are 2 degrees of freedom for this system. The degrees of freedom can be represented by different coordinates. For the mass–spring system, the two coordinates used for analysis can be either x_1 or x_2, or they can be x_1 and $(x_1 - x_2)$, or even other choices. A different selection of coordinates will lead to different equations of motion. However, since we are dealing with the same system, we should expect the natural frequencies derived from these different equations to be the same, regardless the choice of coordinates.

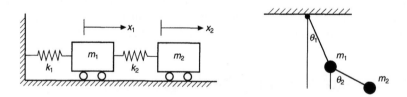

Figure 3.2 Vibratory systems having two degrees of freedom

There are different ways of classifying the types of vibration. These classifications may overlap. Table 3.1 summarizes these vibrations and their brief descriptions.

3.3 Free vibration of an SDoF system

The study of the free vibration of an SDoF system can commence from its equation of motion. The simple mass–spring system shown in Figure 3.3 is an example.

Table 3.1 Different types of vibration

Reference terms	Vibration type	Description
External excitation	Free vibration	Vibration induced by initial input(s) only
	Forced vibration	Vibration subjected to one or more continuous external inputs
Presence of damping	Undamped vibration	Vibration with no energy loss or dissipation
	Damped vibration	Vibration with energy loss
Linearity of vibration	Linear vibration	Vibration for which superposition principle holds
	Nonlinear vibration	Vibration that violates superposition principle
Predictability	Deterministic vibration	The value of vibration is known at any given time
	Random vibration	The value of vibration is not known at any given time but the statistical properties of vibration are known

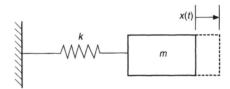

Figure 3.3 A simple mass and spring system

With the displacement x, Newton's second law of motion derives the following equation:

$$m\ddot{x}(t) = -kx(t)$$

or

$$m\ddot{x} + kx = 0 \tag{3.1}$$

where dots on the top of the variable represent derivatives with respect to time. The square root of the ratio of stiffness and mass is defined as the natural frequency of the system:

$$\omega_0 = \sqrt{\frac{k}{m}} \quad \text{rad/sec} \tag{3.2}$$

Contrary to the perception, it is not accurate to say that the natural frequency of such a system depends on its mass and stiffness. The frequency actually depends only on the ratio of them. Therefore, we can have systems sharing the same natural frequency when mass and stiffness quantities differ. Equation (3.1) can also be derived using the energy approach. Since the system does not have energy dissipation or input, the total amount of energy at any given moment is a constant. By estimating both the potential energy $U = \frac{1}{2}kx^2$ and kinetic energy $T = \frac{1}{2}m\dot{x}^2$ at a moment with displacement x, we can have the following equation:

$$\frac{1}{2}m\dot{x}^2 + \frac{1}{2}kx = \text{const} \tag{3.3}$$

The derivative of the equation with respect to time yields equation (3.1). The same outcome will arrive using the Lagrange equation (introduced later in this chapter).

The free vibration response of the SDoF system is dictated by the initial displacement $x(0)$ and initial velocity $\dot{x}(0)$:

$$x(t) = \frac{\dot{x}(0)}{\omega_0} \sin \omega_0 t + x(0) \cos \omega_0 t \tag{3.4}$$

This solution shows that the SDoF system always chooses its own natural frequency for its free vibration. This conclusion is also correct for a linear MDoF system except the system tends to choose all its natural frequencies for its free vibration.

3.4 Harmonic vibration of an SDoF system

Harmonic vibration of an SDoF system is the most fundamental type of vibration and is the building blocks for more sophisticated types of vibration. From Fourier series we know that a periodic vibration consists of a finite number of simple harmonic vibrations whose frequencies are multiples of a fundamental frequency. From Fourier transform we appreciate that a non-periodic vibration, which can be viewed as a periodic vibration whose period is infinity, consists of an infinite number of harmonic vibrations covering every frequency within a range.

Before discussing the harmonic vibration of an SDoF system, it is useful to refresh the features of a simple harmonic signal. A time domain harmonic signal $x(t)$ with period T and circular frequency $\omega \left(\omega = \dfrac{2\pi}{T} \right)$ can be seen as the projection along the time axis of a vector of length X rotating counterclockwise, as shown in Figure 3.4.

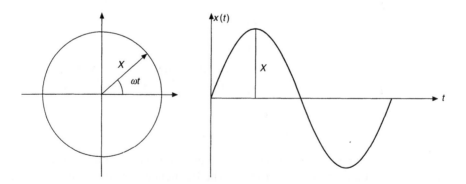

Figure 3.4 A harmonic signal generated by a rotating vector

This time domain signal can be conveniently expressed as:

$$x(t) = X \sin \omega t = X \sin (\omega t + nT) \quad n = 1, 2, \ldots \tag{3.5}$$

If $x(t)$ represents a displacement vibration signal, then the velocity and acceleration can be determined from its first and second order derivatives with respect to time.

Owing to the simplicity of the sinusoidal function, both the amplitudes and the relative phases of the velocity and acceleration can be determined from those of the displacement. The amplitudes and relative phases between the displacement, velocity and acceleration can be written as:

$$\dot{x} = \omega x \qquad (3.6a)$$

$$\ddot{x} = \omega \dot{x} \qquad (3.6b)$$

$$\theta_{\dot{x}} = \theta_x + \frac{\pi}{2} \qquad (3.7a)$$

and $$\theta_{\ddot{x}} = \theta_{\dot{x}} + \frac{\pi}{2} = \theta_x + \pi \qquad (3.7b)$$

Using complex variables, equations (3.6) and (3.7) can be combined. As a result, the displacement, velocity and acceleration can be related conveniently as follows:

$$\dot{x}(t) = j\omega x(t) \qquad (3.8a)$$

and $$\ddot{x}(t) = j\omega \dot{x}(t) = - X\omega^2 x(t) \qquad (3.8b)$$

Here, 'j' is the imaginary unit.

Pure harmonic vibration of an SDoF system occurs in two possible occasions. The first is the free vibration of an SDoF system without damping (therefore no energy dissipation). This has been solved. The second is the vibration when the system is subjected to an external harmonic force. Assume now the system also has a viscous damper with damping value c. This value is equal to the damping force quantity for a unit velocity. The system is subjected to an external harmonic force characterized as $F_0 \sin \omega t$, as shown in Figure 3.5.

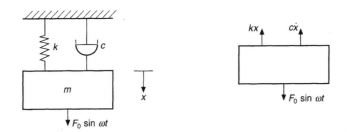

Figure 3.5 An SDoF system subjected to a harmonic force input

The equation of motion of the system can be derived as:

$$m\ddot{x}(t) + c\dot{x}(t) + kx(t) = F_0 \sin \omega t \qquad (3.9)$$

We know from mathematics that the solution of this equation for the displacement vibration of the system is a harmonic function that is in the form of:

$$x(t) = X \sin (\omega t + \varphi) \qquad (3.10)$$

This solution can be found conveniently using the features for harmonic vibration defined earlier in this section and vector summation. Since equation (3.9) is a vector summation, a vector diagram can be constructed in Figure 3.6.

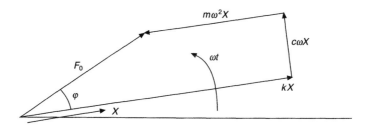

Figure 3.6 Vector diagram of the harmonic vibration of an SDoF system

The displacement amplitude X and relative phase between the input force and vibration displacement φ can be determined from the vector diagram. The displacement response of the SDoF system for the harmonic force input becomes:

$$x(t) = \frac{F_0}{\sqrt{(k - m\omega^2)^2 + (c\omega^2)}} \sin\left(\omega t + \tan^{-1}\frac{c\omega}{k - m\omega^2}\right) \qquad (3.11)$$

Structural damping is an alternative damping model in vibration analysis. This damping is originated from the hysteresis property of stress–strain curves for metal materials. The damping force is proportional to the response amplitude with a 90 degrees phase difference. Equation (3.9) with structural damping will become:

$$m\ddot{x}(t) + kx(t) + jhx(t) = F_0 e^{j\omega t} \qquad (3.12)$$

where 'j' is the imaginary unit. The response is also harmonic. We can define $(k + jh)$ as the complex stiffness of the system and denote it k_c. Then equation (3.12) takes a very simple form:

$$m\ddot{x}(t) + k_c x(t) = F_0 e^{j\omega t} \qquad (3.13)$$

3.5 Vibration of an SDoF system due to an arbitrary force

The SDoF system may also be subjected to a non-harmonic excitation force. There are several mathematical methods available to derive the solution of the resultant vibration. The Laplace transform is a convenient method for deriving it. For an arbitrary force $f(t)$, the equation of motion can be derived as:

$$m\ddot{x}(t) + c\dot{x}(t) + kx(t) = f(t) \qquad (3.14)$$

Taking the Laplace transform and considering initial conditions will lead to the following solution for the displacement vibration:

$$X(s) = \frac{F(s)}{ms^2 + cs + k} + \frac{(ms + c)x(0) + m\dot{x}(0)}{ms^2 + cs + k} \tag{3.15}$$

Here, $X(s)$ and $F(s)$ are the Laplace transforms of the time function $x(t)$ and $f(t)$, respectively. For all zero conditions, the equation is reduced to:

$$X(s) = \frac{F(s)}{ms^2 + cs + k} = G(s)F(s) \tag{3.16}$$

This is a typical equation for a linear dynamic system where $G(s)$ is called the *transfer function* of the system. The time domain vibration $x(t)$ can be determined from the inverse Laplace transform of $X(s)$.

3.6 Free and harmonically forced vibration of an MDoF system

Whenever possible, it is always helpful to compare the study of an MDoF system with that of an SDoF system. The study of free vibration of an MDoF system commences at its equations of motion, as it did for an SDoF system. For the sake of simplicity but without losing generality, we can use the simple 2DoF system shown in Figure 3.7 as the system for studying the free vibration of an MDoF system.

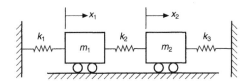

Figure 3.7 A 2DoF system

When selecting x_1 and x_2 as the coordinates to describe the displacement vibration of the system, the equations of motion of the system can be derived as:

$$\begin{cases} m_1\ddot{x}_1 + (k_1 + k_2)x_1 - k_2x_2 = 0 \\ m_2\ddot{x}_2 - k_2x_1 + (k_2 + k_3)x_2 = 0 \end{cases} \tag{3.17}$$

When combined together, this set of differential equations becomes a single matrix equation:

$$\begin{bmatrix} m_1 & 0 \\ 0 & m_2 \end{bmatrix}\begin{Bmatrix} \ddot{x}_1 \\ \ddot{x}_2 \end{Bmatrix} + \begin{bmatrix} (k_1 + k_2) & -k_2 \\ -k_2 & (k_2 + k_3) \end{bmatrix}\begin{Bmatrix} x_2 \\ x_2 \end{Bmatrix} = \begin{Bmatrix} 0 \\ 0 \end{Bmatrix} \tag{3.18}$$

This matrix equation heralds a general form of matrix representation for an MDoF system:

$$[M]\{\ddot{x}\} + [K]\{x\} = \{0\} \tag{3.19}$$

The free vibration solution is mathematically the non-trivial solution of equation (3.19). It should take the form as:

$$\{x\} = \{X\} \sin \omega t \tag{3.20}$$

This form of solution, when substituted into equation (3.19), will change it into a simple algebraic matrix equation:

$$([K] - \omega^2[M])\{X\} = \{0\} \tag{3.21}$$

For this equation to have non-zero solution $\{X\}$, matrix $([K] - \omega^2[M])$ has to be singular so that:

$$\|[K] - \omega^2[M]\| = 0 \tag{3.22}$$

This is the characteristic equation of the system. The solutions of this equation are its natural frequencies. In fact, equation (3.21) represents an eigenvalue problem where ω^2 is the eigenvalue and $\{X\}$ the eigenvector. The eigenvalue is actually the square of the natural frequency of the system and the eigenvector the mode shape. From equation (3.21) it is evident that mode shape $\{X\}$ is not unique since any multiples of it satisfy the equation.

Using the 2DoF system as an example, equation (3.21) becomes:

$$\begin{bmatrix} k_1 + k_2 - \omega^2 m_1 & -k_2 \\ -k_2 & k_2 + k_3 - \omega^2 m_2 \end{bmatrix} \begin{Bmatrix} X_1 \\ X_2 \end{Bmatrix} = \begin{Bmatrix} 0 \\ 0 \end{Bmatrix} \tag{3.23}$$

The characteristic equation of the system is a quadratic equation for ω^2. The two roots of the equation ω_1^2 and ω_2^2 are the squares of the two natural frequencies ω_1 and ω_2. The mode shapes of the system can be determined as:

$$\begin{Bmatrix} X_1^{(1)} \\ X_2^{(1)} \end{Bmatrix} = \begin{Bmatrix} \dfrac{k_1 + k_2 - \omega_1^2 m_1}{k_2} \\ 1 \end{Bmatrix} \tag{3.24}$$

$$\begin{Bmatrix} X_1^{(2)} \\ X_2^{(2)} \end{Bmatrix} = \begin{Bmatrix} \dfrac{k_1 + k_2 - \omega_2^2 m_1}{k_2} \\ 1 \end{Bmatrix} \tag{3.25}$$

The free vibration of the system becomes:

$$\begin{cases} x_1(t) = X_1^{(1)} \sin \omega_1 t + X_1^{(2)} \sin \omega_2 t \\ x_2(t) = X_2^{(1)} \sin \omega_1 t + X_2^{(2)} \sin \omega_2 t \end{cases} \tag{3.26}$$

This free vibration is a linear combination of two harmonic vibrations with frequencies ω_1 and ω_2. Since two natural frequencies are usually not identical, the free vibration of a 2DoF system is periodic with two frequencies. The 'amplitudes' in equation (3.26) $X_i^{(j)}$ ($i, j = 1, 2$) are not fully determined because mode shapes are not uniquely determined. These 'amplitudes' will be quantified fully with the help of initial conditions. Since the free vibration $x_i(t)$ ($i = 1, 2$) consists of the contributions of mode 1 and mode 2, each being harmonic and SDoF like, the composition of SDoF behaviour in an MDoF system response is evident. This concept is important throughout the modal analysis development.

It is possible to design initial conditions such that only one of the two modes is excited. In this case, we see harmonic free vibration for the 2DoF system. However, this bears no generality. These observations for the 2DoF system can be extended to an MDoF system.

Having discussed the free vibration, it is now useful to look into the forced vibration of an MDoF system. Compared with an SDoF system, there are now varieties in forced vibration. For example, it is possible that the forces applied have different frequencies, phase angles, amplitudes, or a combination of them. As more than one coordinate is used, the relative phase among the coordinates becomes possible. To understand the essence of the forced vibration, we consider a simple case where the forcing functions have the same frequency and the phases among them are zero. This does not preclude the case that some of the forces may even have zero amplitude. If the forcing functions do have different frequencies, then we can always use the superposition principle to separate the case into several different cases of excitation frequency and sum up the response solutions.

The equation of motion for an MDoF system with a set of forcing functions having the same frequency ω and zero phases can be written as:

$$[M]\{\ddot{x}\} + [K]\{x\} = \{F\} \sin \omega t \qquad (3.27)$$

Here, vector $\{F\}$ contains the amplitudes of the forcing functions. The assumption of harmonic response in equation (3.20) can still be used so that equation (3.27) can be converted into an algebraic equation:

$$([K] - \omega^2[M])\{X\} = \{F\} \qquad (3.28)$$

The forced vibration response of the MDoF system is then given by:

$$\{x(t)\} = ([K] - \omega^2[M])^{-1}\{F\} \sin \omega t \qquad (3.29)$$

Unlike free vibration, this time the vibration frequency is identical to the frequency of the forcing functions.

3.7 Energy approach

The energy approach is an alternative to the force approach in studying the vibration of a dynamic system. While the force approach is based on Newton's second law of motion, the energy approach depends on the energy conservation principle. To understand the energy approach, we begin with the theory of equilibrium for a dynamic system. The principle of virtual work is essentially a statement for the equilibrium definition of a dynamic system. It is also the foundation of the energy approach used for the dynamics of a system. We begin with the virtual work principle for the statics of a dynamic system and later extend it to its dynamics.

Like the earlier discussion, we deal with an MDoF discretized conservative system so that the main findings are not diluted by the presence of damping. In this analysis, it is convenient to use bold print to denote a vector quantity in contrast with a scalar quantity.

The virtual displacements of a dynamic system are defined as *infinitesimal* and *arbitrary* changes in the coordinates of the system. They are infinitesimal. Thus, system's physical constraints are preserved. They are also arbitrary. Thus, they are *not* true displacements, and there are no time changes associated with them. If the system's displacement r (for convenience bold print is used for vectors instead of brackets) is given in terms of 'n' generalized coordinates q_i ($i = 1, 2, \ldots, n$):

$$r_k = r_k(q_1, q_2, \ldots, q_n) \tag{3.30}$$

then the virtual displacements are denoted as δq_i ($i = 1, 2, \ldots, n$). The symbol 'δ' is used here to distinguish from the symbol 'd' or '∂' for differentiation. These virtual displacements are small variations from the true position of the system and must be compatible with the constraints of the system. For example, a single pendulum stays in equilibrium. For the virtual displacements to be compatible with the constraints, the pendulum can only be swung by a small angle $\delta\theta$. It cannot be bent or moved up and down.

3.7.1 Virtual work principle

For a system with 'n' DoFs to be in static equilibrium, we can divide all the forces acting at DoF 'k' of the system (R_k) into two groups. One is the *applied forces* (F_k) and the other the *constraint forces* (f_k):

$$R_k = F_k + f_k \tag{3.31}$$

The true (not the virtual) displacement of coordinate 'k' in terms of generalized coordinates is:

$$r_k = r_k(q_1, q_2, \ldots, q_n, t) \tag{3.32}$$

and the work done by all the forces on the virtual displacements $\delta r_k(k = 1, 2, \ldots, n)$ will be

$$\delta W = \sum_{k=1}^{n} R_k \delta r_k \tag{3.33}$$

For a conservative system, the constraint forces collectively should do no work on the virtual displacements. Therefore, the principle of virtual work states that the system is in equilibrium if and only if the total work done by all the applied forces on virtual displacements is zero, i.e.

$$\delta W = \sum_{k=1}^{n} F_k \delta r_k = 0 \tag{3.34}$$

The virtual work in the equation can also be expressed using the generalized coordinates. For displacement r_k, the virtual displacement is given by:

$$\delta x_k = \sum_{i=1}^{n} \frac{\partial r_k}{\partial q_i} \delta q_i \tag{3.35}$$

Here, time 't' is not involved since virtual displacement does not vary on time. The virtual work can then be expressed as:

$$\delta W = \sum_{k=1}^{n} F_k \delta r_k = \sum_{k=1}^{n} \sum_{i=1}^{n} F_k \frac{\partial r_k}{\partial q_i} \delta q_i \tag{3.36}$$

Interchanging the order of summation leads to:

$$\delta W = \sum_{k=1}^{n} F_k \delta r_k = \sum_{i=1}^{n} \sum_{k=1}^{n} F_k \frac{\partial r_k}{\partial q_i} \delta q_i \qquad (3.37)$$

Define the generalized force as:

$$Q_i = \sum_{k=1}^{n} F_k \frac{\partial r_k}{\partial q_i} \quad (i = 1, 2, \ldots, n) \qquad (3.38)$$

The virtual work principle can then be described analytically as:

$$\delta W = \sum_{i=1}^{n} Q_i \delta q_i = 0 \quad (i = 1, 2, \ldots, n) \qquad (3.39)$$

This means that a system being in static equilibrium is equated so that the total virtual work done by all the generalized forces on chosen virtual displacements is zero. Since the virtual displacements δq_i of coordinate q_i are arbitrary, the above equation implies:

$$Q_i = 0 \quad (i = 1, 2, \ldots, n) \qquad (3.40)$$

This means that a system being in static equilibrium is equated so that all the generalized forces on chosen virtual displacements are zero.

3.7.2 D'Alembert's principle

D'Alembert's principle extends the virtual work for static equilibrium into the realm of *dynamic equilibrium*. It suggests that since the sum of the forces acting on a DoF 'k' results in its acceleration $\ddot{r}_k$, the application of a fictitious force – $m_k \ddot{r}_k$ would produce a state of equilibrium. This is a different interpretation of Newton's second law of motion. The equation for the DoF can then be written as:

$$F_k + f_k - m_k \ddot{r}_k = 0 \qquad (3.41)$$

The application of the virtual work principle yields:

$$\delta W = \sum_{k=1}^{n} (F_k - m_k \ddot{r}_k) \delta r_k = 0 \qquad (3.42)$$

This equation is the analytical expression of D'Alembert's principle. It forms the theoretical basis used to derive Lagrange's equation.

3.7.3 Kinetic energy

It was known from the preceding section that 'n' generalized coordinates can be used to describe the motion of an 'n' DoF system. The displacement of the system at DoF 'k' can be expressed as:

$$r_k = r_k(q_1, q_2, \ldots, q_n, t) \qquad (3.32)$$

The velocity at the same DoF is:

$$\dot{r}_k = \dot{r}_k(\dot{q}_1, \dot{q}_2, \ldots, \dot{q}_n, t) \tag{3.43}$$

Here, $\dot{q}_i (i = 1, 2, \ldots, n)$ is called *generalized velocity*. Velocity $\dot{v}_k$ is a linear function of the generalized velocities. For a mass and spring system with 'n' DoFs, the kinetic energy for DoF 'k' with mass m_k is given as:

$$T_k = \frac{1}{2} m_k r_k r_k = \frac{1}{2} \sum_{k=1}^{n} m_k q_k q_k \tag{3.44}$$

Thus, the system's total kinetic energy will be:

$$T = \frac{1}{2} \sum_{k=1}^{n} m_k \dot{r}_k \dot{r}_k \tag{3.45}$$

By substituting equations (3.43) into (3.45), we find that the total kinetic energy is associated with order zero, order one and order two functions of the generalized velocity.

If the physical constraints of the system are time-invariant, then for DoF 'k' we have,

$$r_k = r_k(q_1, q_2, \ldots, q_n) \tag{3.46}$$

thus

$$\dot{r}_k \dot{r}_k = \sum_{i=1}^{n} \sum_{j=1}^{n} \frac{\partial r_k}{\partial q_i} \frac{\partial r_k}{\partial q_j} \dot{q}_i \dot{q}_j \tag{3.47}$$

The kinetic energy of the system becomes:

$$T = \frac{1}{2} \sum_{k=1}^{n} m_k \sum_{i=1}^{n} \sum_{j=1}^{n} \frac{\partial r_k}{\partial q_i} \frac{\partial r_k}{\partial q_j} \dot{q}_i \dot{q}_j \tag{3.48}$$

Interchanging the summation order leads to:

$$T = \frac{1}{2} \sum_{i=1}^{n} \sum_{j=1}^{n} \left(\sum_{k=1}^{n} m_k \frac{\partial r_k}{\partial q_i} \frac{\partial r_k}{\partial q_j} \right) \dot{q}_i \dot{q}_j \tag{3.49}$$

If the content inside the bracket is defined as '*generalized mass*' and is denoted as m_{ij}, then

$$m_{ij} = \sum_{k} m_k \frac{\partial r_k}{\partial q_i} \frac{\partial r_k}{\partial q_j} \tag{3.50}$$

The generalized mass m_{ij} is a function of the generalized coordinate q_i and it is evident that $m_{ij} = m_{ji}$.

The kinetic energy can then be written as:

$$T = \frac{1}{2} \sum_{i=1}^{n} \sum_{j=1}^{n} m_{ij} \dot{q}_i \dot{q}_j \tag{3.51}$$

Since the generalized mass is a function of only the generalized coordinate q_i, it is obvious that the kinetic energy is a quadratic function of generalized velocity.

If the mass matrix and generalized velocity vector are defined respectively as:

$$[M] = \begin{bmatrix} m_{11} & m_{12} & \dots & m_{1n} \\ m_{21} & m_{22} & \dots & m_{2n} \\ \dots & \dots & \dots & \dots \\ m_{n1} & m_{n2} & \dots & m_{nn} \end{bmatrix} \quad \{\dot{q}\} = \begin{Bmatrix} \dot{q}_1 \\ \dot{q}_2 \\ \dots \\ \dot{q}_n \end{Bmatrix}$$

then equation (3.51) can be written in matrix form as:

$$T = \frac{1}{2}\{\dot{q}\}^T [M]\{\dot{q}\} \tag{3.52}$$

Since kinetic energy T is always a positive quantity (or zero only when all the velocities are zero), the mass matrix has to be positive definite.

3.7.4 Potential energy

In a conservative system, potential energy U is a function of the generalized coordinate q_i.

$$U = U(q_1, q_2, \dots, q_n) \tag{3.53}$$

The work done by the conservative forces is equal to the negative of the potential energy:

$$W = -U(q_1, q_2, \dots, q_n) \tag{3.54}$$

When coordinate q_i varies by ∂q_i, the corresponding potential energy change is:

$$dU = -dW = \sum_{i=1}^{n} \frac{\partial U}{\partial q_i} dq_i \tag{3.55}$$

If ∂q_i is chosen as virtual displacement, then from equation (3.40) we have:

$$\delta W = \sum_{i=1}^{n} Q_i \delta q_i$$

Thus,
$$Q_i = -\frac{\partial U}{\partial q_i} \quad (i = 1, 2, \dots, n) \tag{3.56}$$

This means that the generalized forces corresponding to conservative forces are equal to the partial derivatives of the potential energy of the system with respect to generalized coordinates.

If the system's applied forces are all conservative, then it can be seen by comparing equation (3.56) with equation (3.39) that such a system is in equilibrium when the following equation holds:

$$\frac{\partial U}{\partial q_i} = 0 \quad (i = 1, 2, \dots, n) \tag{3.57}$$

i.e. its potential energy reaches its maximum or minimum.

For a system with 'n' generalized coordinates, expanding U in a Taylor series about its equilibrium position leads to:

$$U = U_0 + \sum_{i=1}^{n} \left(\frac{\partial U}{\partial q_i} \right)_0 q_i + \frac{1}{2} \sum_{i=1}^{n} \sum_{j=1}^{n} \left(\frac{\partial^2 U}{\partial q_i \, \partial q_j} \right)_0 q_i q_j + \ldots \qquad (3.58)$$

In this expression U_0 is an arbitrary constant which can be set to zero. The derivatives of U are evaluated at the equilibrium position '0' and are constants when the coordinates q_i are small quantities and are equal to zero at the equilibrium position. Since U is minimum (or maximum) in the equilibrium position,

$$\left(\frac{\partial U}{\partial q_i} \right)_0 = 0$$

Thus, if terms beyond the second order are ignored, equation (3.58) becomes:

$$U = \frac{1}{2} \sum_{i=1}^{n} \sum_{j=1}^{n} \left(\frac{\partial^2 U}{\partial q_i \, \partial q_j} \right)_0 q_i q_j \qquad (3.59)$$

The second derivative evaluated at '0' position is a constant known as the 'generalized stiffness':

$$k_{ij} = \left(\frac{\partial^2 U}{\partial q_i \, \partial q_j} \right)_0 \qquad (3.60)$$

The potential energy can then be written as:

$$U = \frac{1}{2} \{q\}^T [K] \{q\} \qquad (3.61)$$

where the stiffness matrix $[K]$ and generalized coordinate vector are:

$$[K] = \begin{bmatrix} k_{11} & k_{12} & \ldots & k_{1n} \\ k_{21} & k_{22} & \ldots & k_{2n} \\ \ldots & \ldots & \ldots & \ldots \\ k_{n1} & k_{n2} & \ldots & k_{nn} \end{bmatrix} \qquad \{q\} = \begin{Bmatrix} q_1 \\ q_2 \\ \ldots \\ q_n \end{Bmatrix}$$

3.7.5 Lagrange's equation

Lagrange's equation is based on D'Alembert's principle and equation (3.42):

$$\delta W = \sum_{k=1}^{n} (F_k - m_k \ddot{r}_k) \delta r_k = 0 \qquad (3.42)$$

There are two parts of virtual work in this equation. The first part is the virtual work done by the applied forces which can be estimated by equation (3.37). The second part is the virtual work done by inertial forces. Without proof, this virtual work can be estimated as:

$$\sum_{k} m_k \ddot{r}_k \delta r_k = \sum_{i=1}^{n} \left[\frac{d}{dt} \left(\frac{\partial T}{\partial \dot{q}_i} \right) - \frac{\partial T}{\partial q_i} \right] \delta q_i \qquad (3.62)$$

Inserting equations (3.38) and (3.62) into (3.43) will produce:

$$\sum_{i=1}^{n} \left[\frac{d}{dt} \frac{\partial T}{\partial \dot{q}_i} - \frac{\partial T}{\partial q_i} \right] \delta q_i - \sum_{i=1}^{n} Q_i \delta q_i = 0$$

or
$$\sum_{i=1}^{n} \left[\frac{d}{dt} \frac{\partial T}{\partial \dot{q}_i} - \frac{\partial T}{\partial q_i} - Q_i \right] \delta q_i = 0 \tag{3.63}$$

Since virtual displacements δq_i are arbitrary, they can be chosen such that only one generalized coordinate has non-zero displacement each time, thus equation (3.63) becomes n independent equations:

$$\frac{d}{dt} \frac{\partial T}{\partial \dot{q}_i} - \frac{\partial T}{\partial q_i} - Q_i = 0 \quad (i = 1, 2, \ldots, n) \tag{3.64}$$

This is called Lagrange's equation.

If the system is conservative, then substituting equations (3.56) into (3.64) leads to:

$$\frac{d}{dt} \frac{\partial T}{\partial \dot{q}_i} - \frac{\partial T}{\partial q_i} + \frac{\partial U}{\partial q_i} = 0 \quad (i = 1, 2, \ldots, n) \tag{3.65}$$

Introducing the Lagrangian L function defined as:

$$L = T - U \tag{3.66}$$

and if U is not a function of velocity, then equation (3.65) is rewritten as:

$$\frac{d}{dt} \frac{\partial L}{\partial \dot{q}_i} - \frac{\partial L}{\partial q_i} = 0 \quad (i = 1, 2, \ldots, n) \tag{3.67}$$

This is the compact form of Lagrange's equation for a conservative system.

Using Lagrange's equation to solve problems of conservative systems usually involves the following steps:

(1) Identify the number of degrees of freedom of the system and select generalized coordinates to describe motions of the system.
(2) Estimate the kinetic energy T of the system in terms of the generalized velocities.
(3) Estimate the potential energy U of the system in terms of the generalized coordinates.
(4) Substitute T and U into Lagrange's equation to derive the differential equation of motion for the system.

Example 3.1
Use Lagrange's equation to derive the equation of motion of the SDoF system shown in Figure 3.8.

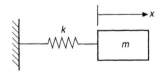

Figure 3.8 An SDoF system

Solution: Choose x as the generalized coordinate for the system, the kinetic and potential energies with respect to the equilibrium position are respectively:

$$T = \frac{1}{2} m\dot{x}^2 \quad U = \frac{1}{2} kx^2$$

Lagrangian:

$$L = T - U = \frac{1}{2} m\dot{x}^2 - \frac{1}{2} kx^2$$

Derivatives:

$$\frac{\partial L}{\partial x} = m\dot{x}, \quad \frac{d}{dt}\frac{\partial L}{\partial \dot{x}} = m\ddot{x}, \quad \frac{\partial L}{\partial x} = -kx$$

Inserting these terms into Lagrange's equation leads to:

$$m\ddot{x}^2 + kx = 0$$

Example 3.2

A double pendulum has lengths of L_1 and L_2, with masses m_1 and m_2 at the end of each massless link. Use Lagrange's equation to derive the equations of motion.

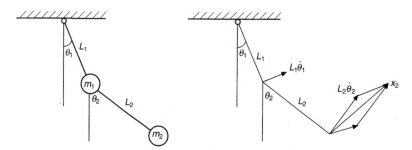

Figure 3.9 A double pendulum

Solution: The kinetic energy of the pendulum is given by:

$$T = \frac{1}{2} m_1 \dot{x}_1^2 + \frac{1}{2} m_2 \dot{x}_2^2$$

where $\dot{x}_1^2 = (L_1\dot{\theta}_1)^2$ and $\dot{x}_2^2 = (L_1\dot{\theta}_1)^2 + (L_2\dot{\theta}_2)^2 + 2L_1\dot{\theta}_1 L_2\dot{\theta}_2 \cos(\theta_2 - \theta_1)$.

$\dot{x}_1$ and $\dot{x}_2$ are the velocities of masses m_1 and m_2 respectively, as shown in Figure 3.9. The potential energy of the double pendulum is given by:

$$U = m_1 g L_1 (1 - \cos \theta_1) + m_2 g[L_1(1 - \cos \theta_1) + L_2(1 - \cos \theta_2)]$$

the Lagrangian is defined as:

$$L = \frac{1}{2} m_1 \dot{x}_1^2 + \frac{1}{2} m_2 \dot{x}_2^2 - m_1 g L_1 (1 - \cos \theta_1)$$
$$- m_2 g \, [L_1 (1 - \cos \theta_1) + L_2 (1 - \cos \theta_2)]$$

So,

$$\frac{\partial L}{\partial \dot{\theta}_1} = (m_1 + m_2)L_1^2\dot{\theta}_1 + m_2 L_1 L_2 \cos(\theta_2 - \theta_1)\dot{\theta}_2$$

$$\frac{d}{dt}\frac{\partial L}{\partial \dot{\theta}_1} = (m_1 + m_2)L_1^2\ddot{\theta}_1 - m_2 L_1 L_2 \sin(\theta_2 - \theta_1)(\dot{\theta}_2 - \dot{\theta}_1)\dot{\theta}_2$$

$$+ m_2 L_1 L_2 \cos(\theta_2 - \theta_1)\ddot{\theta}_2$$

$$\frac{\partial L}{\partial \theta_1} = (m_1 + m_2)gL_1 \sin \theta_1$$

and

$$\frac{\partial L}{\partial \dot{\theta}_2} = m_2 L_2^2 \dot{\theta}_2 + m_2 L_1 L_2 \cos(\theta_2 - \theta_1)\dot{\theta}_1$$

$$\frac{d}{dt}\frac{\partial L}{\partial \dot{\theta}_2} = -m_2 L_1 L_2 \sin(\theta_2 - \theta_1)(\dot{\theta}_2 - \dot{\theta}_1)\dot{\theta}_1$$

$$+ m_2 L_1 L_2 \cos(\theta_2 - \theta_1)\ddot{\theta}_1 + m_2 L_2^2 \ddot{\theta}_2$$

$$\frac{\partial L}{\partial \theta_2} = m_2 gL_2 \sin \theta_2$$

Substituting these derivatives into Lagrange's equation yields:

$$(m_1 + m_2)L_1^2\ddot{\theta}_1 - m_2 L_1 L_2 \sin(\theta_2 - \theta_1)(\dot{\theta}_2 - \theta_1)\dot{\theta}_2 + m_2 L_1 L_2 \cos(\theta_2 - \theta_1)\ddot{\theta}_2$$

$$+ (m_1 + m_2)gL_1 \sin \theta_1 = 0$$

$$- m_2 L_1 L_2 \sin(\theta_2 - \theta_1)(\dot{\theta}_2 - \theta_1)\dot{\theta}_1 + m_2 L_1 L_2 \cos(\theta_2 - \theta_1)\ddot{\theta}_1 + m_2 L_2^2\ddot{\theta}_2 + m_2 gL_2 \sin \theta_2 = 0$$

For small oscillations, $\sin \theta \approx \theta$ and $\cos \theta \approx 1$. Ignoring higher order terms of derivatives such as $(\dot{\theta}_2 - \dot{\theta}_1)\dot{\theta}_1$, the two equations become:

$$(m_1 + m_2)L_1^2\ddot{\theta}_1 + m_2 L_1 L_2\ddot{\theta}_2 + (m_1 + m_2)gL_1\theta_1 = 0$$

$$m_2 L_1 L_2\ddot{\theta}_1 + m_2 L_2^2\ddot{\theta}_2 + m_2 gL_2\theta_2 = 0$$

or in matrix form:

$$\begin{bmatrix} (m_1 + m_2)L_1^2 & m_2 L_1 L_2 \\ m_2 L_1 L_2 & m_2 L_2^2 \end{bmatrix}\begin{Bmatrix} \ddot{\theta}_1 \\ \ddot{\theta}_2 \end{Bmatrix} + \begin{bmatrix} (m_1 + m_2)gL_1 & 0 \\ 0 & m_2 gL_2 \end{bmatrix}\begin{Bmatrix} \theta_1 \\ \theta_2 \end{Bmatrix} = \begin{Bmatrix} 0 \\ 0 \end{Bmatrix}$$

3.8 Vibration of continuous systems

A continuous system here means a system with continuously distributed mass and elasticity. Their bodies are assumed to be homogeneous and isotropic, obeying Hooke's law and their vibration is within the elastic limit. To specify the vibration of every particle in the elastic body of a continuous system, an infinite number of coordinates are necessary, and, therefore, such a system possesses an infinite number of degrees of freedom. Mathematically, functions of both position and time are needed to describe the vibration of a continuous system, resulting in partial differential equations. Physically this not different to a discrete system. In fact, if the continuous mass of a system can be concentrated to a finite number of points connected by elastic elements, then the system will become a discrete system. Contrarily, as the number of degrees of freedom approaches infinity, a discrete system will become a continuous system.

The partial differential equations of simple continuous systems can be solved analytically. Such systems include strings, rod, beams and shells. For systems with more geometrical complexities, we have to use numerical methods to analyse them as discrete systems having a finite number of degrees of freedom, or as a combination of several simple continuous systems coupled together.

3.8.1 The vibrating string

A flexible string of mass ρ per unit length is stretched under tension T. Assume the lateral deflection y of the string is small, the change in tension with deflection will be negligible and ignored. Figure 3.10 shows the free-body diagram of an elementary length dx of the string.

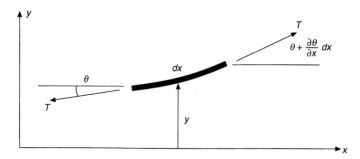

Figure 3.10 Lateral vibration of a string element

Due to a small deflection, the tension T is a constant and $\sin \theta \approx \tan \theta \approx \theta$. The equation of motion in the y-direction can be derived using Newton's second law of motion as:

$$T\left(\theta + \frac{\partial \theta}{\partial x} \, dx\right) - T\theta = \rho \, dx \, \frac{\partial^2 y}{\partial t^2} \tag{3.68}$$

or

$$\frac{\partial \theta}{\partial x} = \frac{\rho}{T} \frac{\partial^2 y}{\partial t^2} \tag{3.69}$$

Because $\dot{\theta} = \dfrac{\partial y}{\partial x}$, equation (3.69) becomes:

$$\frac{\partial^2 y}{\partial x^2} = \frac{1}{c^2} \frac{\partial^2 y}{\partial t^2} \tag{3.70}$$

where $c = \sqrt{T/\rho}$ is the velocity of wave propagation along the string.

Free vibration

Using the parameter separation method introduced in Chapter 2, equation (3.70) can be solved by assuming:

$$y(x, t) = Y(x)G(t) \tag{3.71}$$

then

$$\frac{1}{Y(x)} \frac{d^2 Y(x)}{dx^2} = \frac{1}{c^2} \frac{1}{G(t)} \frac{d^2 G(t)}{dt^2} \tag{3.72}$$

Since the left-hand side of equation (3.72) is independent of time t and the right-hand side of x, both sides must be the same constant. Let the constant be $-\left(\dfrac{\omega}{c}\right)^2$, then

$$\begin{cases} \dfrac{d^2 Y(x)}{dx^2} + \left(\dfrac{\omega}{c}\right)^2 Y(x) = 0 \\[3mm] \dfrac{d^2 G(t)}{dt^2} + \omega^2 G(t) = 0 \end{cases} \tag{3.73}$$

The general solutions for equation (3.73) are:

$$Y(x) = A \sin \frac{\omega}{c} x + B \cos \frac{\omega}{c} x \tag{3.74}$$

$$G(t) = C \sin \omega t + D \cos \omega t \tag{3.75}$$

Here, A, B, C, D are constants. They depend on the boundary conditions (for A and B) and the initial conditions (for C and D).

As an example, we study a string stretched between two fixed points with distance L. The boundary conditions are:

$$y(0, t) = y(L, t) = 0$$

For
$$y(0, t) = 0 \rightarrow B = 0$$

the solution becomes:

$$y(x, t) = C \sin \omega t + D \cos \omega t) \sin \frac{\omega}{c} x \tag{3.76}$$

For
$$y(L, t) = 0 \rightarrow \sin \frac{\omega}{c} L = 0$$

$$\frac{\omega_n}{c} L = \frac{2\pi L}{\lambda} = n\pi \ \text{ or } \ \omega_n = \frac{n\pi c}{L} \quad (n = 1, 2, \ldots, \infty) \tag{3.77}$$

where $\lambda = \dfrac{2\pi c}{\omega_n}$ is the wave length. The natural frequencies of the string will be:

$$f_n = \frac{nc}{2L} = \frac{n}{2L} \sqrt{\frac{T}{\rho}} \tag{3.78}$$

and the mode shapes are

$$Y(x) = A \sin \frac{n\pi x}{L} \tag{3.79}$$

The first and second modes of the string are shown in Figure 3.11.

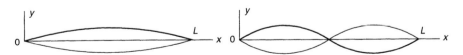

Figure 3.11 The first and second modes of a vibrating string with fixed ends

For general initial conditions, the string will vibrate under all the normal modes, thus,

$$y(x, t) = \sum_{n=1}^{\infty} (C_n \sin \omega_n t + D_n \cos \omega_n t) \sin \frac{n\pi x}{L} \tag{3.80}$$

Both C_n and D_n can be determined by initial conditions $y(x, 0)$ (for D_n) and $\dot{y}(x, 0)$ (for C_n).

Example 3.3
A uniform string of length L is fixed at the ends and stretched under tension T. If the string is displaced into an arbitrary shape $y(x, 0)$ and released, determine the vibration of the string.

Solution: Equation (3.80) shows the string vibration. The coefficients can be determined by the boundary conditions:

$$\dot{y}(x, 0) = \sum_{n=1}^{\infty} C_n \sin \frac{n\pi x}{L} = 0 \rightarrow C_n = 0$$

$$y(x, 0) = \sum_{n=1}^{\infty} D_n \sin \frac{n\pi x}{L}$$

$$D_n = \frac{2}{L} \int_0^L y(x, 0) \sin \frac{n\pi x}{L} \, dx \tag{3.81}$$

3.8.2 Vibrations of membranes

A membrane can be considered as the two-dimensional analogues of strings. Therefore, the theory of a vibrating string is directly applicable here. Consider a homogeneous and perfectly flexible membrane bounded by a plane curve c in the xy-plane, as shown in Figure 3.12.

In order to study the small transverse vibration of the membrane, a few assumptions are made:

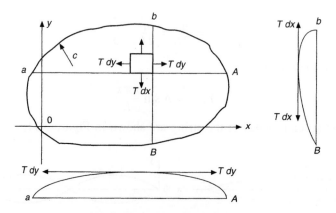

Figure 3.12 A two dimensional membrane

(1) The equilibrium position of the membrane is in the xy-plane. Each point P of the membrane is identified by its xy-coordinates at the equilibrium position.
(2) Each point $P(x, y)$ only moves in the z-direction and its displacement is denoted by $w = w(x, y, t)$. The value w is very small compared with the dimensions of the membrane.
(3) The uniform tension per unit length along the edge T is sufficiently big so that the plane, tangent to the membrane at point P, makes very small angles with the x- and y-axes. It is also not overly big so it can remain a constant during vibration.

The equation of motion

The derivation of the equation of motion of the membrane begins with a small element of area $dxdy$. The area is under tension T. The sum of the projections of the two Tdy forces on the z-axis is $\left(\text{noting that } \dfrac{\partial w}{\partial x} = \theta_x \right)$:

$$Tdy\left(\frac{\partial w}{\partial x} + \frac{\partial}{\partial x}\left(\frac{\partial w}{\partial x} \right)dx \right) - Tdy\,\frac{\partial w}{\partial x} = T\frac{\partial^2 w}{\partial x^2}\,dxdy \qquad (3.82)$$

Similarly, the sum of the projections of the two Tdx forces on the z-axis is $\left(\text{noting that } \dfrac{\partial w}{\partial x} = \theta_x \right)$:

$$Tdx\left(\frac{\partial w}{\partial y} + \frac{\partial}{\partial y}\left(\frac{\partial w}{\partial y} \right)dy \right) - Tdx\,\frac{\partial w}{\partial y} = T\frac{\partial^2 w}{\partial y^2}\,dxdy \qquad (3.83)$$

Let μ be the mass per unit area of the membrane, then the equation of motion becomes:

$$\mu\,dxdy\,\frac{\partial^2 w}{\partial t^2} = T\left(\frac{\partial^2 w}{\partial x^2} + \frac{\partial^2 w}{\partial y^2} \right)dxdy \qquad (3.84)$$

or

$$\frac{\partial^2 w}{\partial t^2} = c^2\left(\frac{\partial^2 w}{\partial x^2} + \frac{\partial^2 w}{\partial y^2} \right) \qquad (3.85)$$

where $c = \sqrt{T/\mu}$.

Using the Laplace operator ∇^2 as

$$\nabla^2 = \frac{\partial^2}{\partial x^2} + \frac{\partial^2}{\partial y^2}$$

equation (3.85) becomes:

$$\frac{\partial^2 w}{\partial t^2} = c^2\nabla^2 w \qquad (3.86)$$

This is the equation of motion of the membrane.

If the membrane is subjected to an external force $p(x, y, t)$ per unit area acting in the z-direction, the equation of motion will become:

$$\mu\frac{\partial^2 w}{\partial t^2} - T\nabla^2 w(x, y, t) = p(x, y, t) \qquad (3.87)$$

Free vibrations of rectangular membranes

To obtain the solution of equation (3.86) for a rectangular membrane with side lengths a and b, as shown in Figure 3.13, the variable separation method is used.

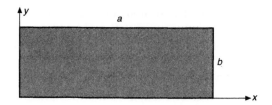

Figure 3.13 A rectangular membrane with sides a and b

Assume the solution (x, y, t) can be written as:

$$w(x, y, t) = X(x)Y(y)f(t) \tag{3.88}$$

Equation (3.86) becomes:

$$c^2\left(\frac{\partial^2 X}{\partial x^2}Y + \frac{\partial^2 Y}{\partial y^2}X\right)f = XY\frac{\partial^2 f}{\partial t^2} \tag{3.89}$$

or

$$c^2\left(\frac{1}{X}\frac{\partial^2 X}{\partial x^2} + \frac{1}{Y}\frac{\partial^2 Y}{\partial y^2}\right)f = \frac{1}{f}\frac{\partial^2 f}{\partial t^2} \tag{3.90}$$

or

$$c^2\left(\frac{X''}{X} + \frac{Y''}{Y}\right) = \frac{f''}{f} \tag{3.91}$$

The right-hand side of equation (3.91) does not expand on x and y and the left-hand side does not expand on t. Hence, both sides should be equal to the same constant, as in the case of the string. Assume the constant is $-\omega^2$, then:

$$\begin{cases} f'' + \omega^2 f = 0 & \text{(a)} \\ \dfrac{Y''}{Y} + \dfrac{\omega^2}{c^2} = -\dfrac{X''}{X} & \text{(b)} \end{cases} \tag{3.92}$$

The left-hand side of equation (3.92b) is not a function of y while the right-hand side is not that of x. Hence, the variable separation method can be applied, yielding:

$$\begin{cases} X'' + \lambda^2 X = 0 & \text{(a)} \\ Y'' + \chi^2 Y = 0 & \text{(b)} \end{cases} \tag{3.93}$$

where $\chi^2 = \dfrac{\omega^2}{c^2} - \lambda^2$ hence,

$$\omega^2 = c^2(\lambda^2 + \chi^2) \tag{3.94}$$

Thus, the problem of finding the solution of equation (3.86) for a membrane becomes the problem of finding solutions of equations (3.92a) and (3.93), each being an ordinary differential equation. If the membrane is fixed at its boundary, then the boundary conditions will be:

$$\begin{cases} w(0, y, t) = 0 \\ w(a, y, t) = 0 \end{cases} \quad 0 \le y \le b, \ 0 \le t \tag{3.94}$$

$$\begin{cases} w(x, 0, t) = 0 \\ w(x, b, t) = 0 \end{cases} \quad 0 \le x \le a, \ 0 \le t \tag{3.95}$$

Inserting these boundary conditions to equation (3.88) yields:

$$\begin{cases} X(0) = X(a) = 0 \\ Y(0) = Y(b) = 0 \end{cases} \tag{3.96}$$

From this equation it is clear that solutions of X and Y will take the forms:

$$\begin{cases} X = \sin \lambda x \\ Y = \sin \chi y \end{cases} \tag{3.97}$$

and also constants λ and χ must satisfy:

$$\begin{cases} \sin \lambda a = 0 \\ \sin \chi b = 0 \end{cases} \tag{3.98}$$

Therefore, λ and χ can be derived from one of the infinite values:

$$\begin{cases} \lambda_m = \dfrac{m\pi}{a} & \text{(a)} \\ \chi_n = \dfrac{n\pi}{b} & \text{(b)} \end{cases} \quad m, n = 1, 2, 3, \ldots \tag{3.99}$$

The natural frequency of the rectangular membrane is given by equation (3.93) as:

$$\omega_{mn}^2 = c^2 \pi^2 \left(\frac{m^2}{a^2} + \frac{n^2}{b^2} \right) \quad m, n = 1, 2, 3, \ldots \tag{3.100}$$

The corresponding mode shape of the membrane is:

$$U(x, y) = X_m Y_n = \sin \frac{m\pi x}{a} \sin \frac{n\pi y}{b} \tag{3.101}$$

and the vibration is given by:

$$w_{mn}(x, y, t) = \sin \frac{m\pi x}{a} \sin \frac{n\pi y}{b} (A_{mn} \cos \omega_{mn} t + B_{mn} \sin \omega_{mn} t) \tag{3.102}$$

This equation describes the free vibration of the rectangular membrane with fixed boundaries. Basically, the membrane performs harmonic motion with circular frequency ω_{mn}. The points of the membrane which remain at rest during a single mode vibration form lines called *nodal lines*. For example, when the membrane vibrates in its first (fundamental) mode, meaning $m = n = 1$, the node lines are the fixed edges of the membrane.

Like the string, the general solution of equation (3.86) is the linear combination of all the w_{mn}, i.e.,

$$w(x, y, t) = \sum_{m=1}^{\infty} \sum_{n=1}^{\infty} \sin \frac{m\pi x}{a} \sin \frac{n\pi y}{b} (A_{mn} \cos \omega_{mn} t + B_{mn} \sin \omega_{mn} t) \quad (3.103)$$

The constants A_{mn} and B_{mn} are to be determined from the initial conditions. Suppose the initial conditions of the membrane are:

$$\begin{cases} w(x, y, 0) = \varphi(x, y) \\ \dfrac{\partial w}{\partial t}(x, y, 0) = \psi(x, y) \end{cases} \quad 0 \le x \le a, 0 \le y \le b \quad (3.104)$$

then, inserting these conditions into equation (3.103) leads to:

$$\begin{cases} \sum_{m=1}^{\infty} \sum_{n=1}^{\infty} A_{mn} \sin \dfrac{m\pi x}{a} \sin \dfrac{n\pi y}{b} = \varphi(x, y) \\ \sum_{m=1}^{\infty} \sum_{n=1}^{\infty} \omega_{mn} B_{mn} \sin \dfrac{m\pi x}{a} \sin \dfrac{n\pi y}{b} = \psi(x, y) \end{cases} \quad (3.105)$$

Using the orthogonality conditions of the trigonometric functions, the two constants can be determined as:

$$A_{mn} = \frac{4}{ab} \int_0^a \int_0^b \varphi(x, y) \sin \frac{m\pi x}{a} \sin \frac{n\pi y}{b} \, dx \, dy$$

$$B_{mn} = \frac{4}{ab\omega_{mn}} \int_0^a \int_0^b \psi(x, y) \sin \frac{m\pi x}{a} \sin \frac{n\pi y}{b} \, dx \, dy$$

Example 3.4

Derive the free vibration of a square membrane and its mode shape nodal lines.

Solution: For a square membrane with length a, the natural frequencies are:

$$\omega_{mn}^2 = \frac{c^2 \pi^2}{a^2} (m^2 + n^2) \quad m, n = 1, 2, 3, \ldots$$

A few natural frequencies can be shown below:

$$\omega_{11} = \frac{c\pi}{a} \sqrt{2} \quad \omega_{12} = \omega_{21} = \frac{c\pi}{a} \sqrt{5} \quad \omega_{22} = \frac{c\pi}{a} \sqrt{8}$$

$$\omega_{13} = \frac{c\pi}{a} \sqrt{10} \quad \omega_{23} = \omega_{32} = \frac{c\pi}{a} \sqrt{13} \quad \ldots$$

The mode shape corresponding to the first natural frequency ω_{11} is:

$$U_{11} = \sin \frac{\pi x}{a} \sin \frac{\pi y}{a}$$

The free vibration of the square membrane is given by:

$$w_{11} = \sin \frac{\pi x}{a} \sin \frac{\pi y}{a} \left(A_{11} \cos \frac{c\pi}{a} \sqrt{2} t + B_{11} \sin \frac{c\pi}{a} \sqrt{2} t \right)$$

The nodal lines are at the edges $x = 0$, $x = a$, $y = 0$, and $y = a$.

The modes corresponding to the natural frequency $\dfrac{c\pi}{a} \sqrt{5}$ (for ω_{12} and ω_{21}) are:

$$U_{12} = \sin \frac{\pi x}{a} \sin \frac{2\pi y}{a}$$

$$U_{21} = \sin \frac{2\pi x}{a} \sin \frac{\pi y}{a}$$

and the vibrations are respectively:

$$w_{12} = \sin \frac{\pi x}{a} \sin \frac{2\pi y}{a} \left(A_{12} \cos \frac{c\pi}{a} \sqrt{5}t + B_{12} \sin \frac{c\pi}{a} \sqrt{5}t \right)$$

$$w_{21} = \sin \frac{2\pi x}{a} \sin \frac{\pi y}{a} \left(A_{21} \cos \frac{c\pi}{a} \sqrt{5}t + B_{21} \sin \frac{c\pi}{a} \sqrt{5}t \right)$$

To simplify the examination of the nodal lines, assume $B_{12} = B_{21} = 0$, then the sum of the two vibrations (of the same natural frequencies) is given by:

$$w_{12} + w_{21} = \cos \frac{c\pi}{a} \sqrt{5}t \left[A_{12} \sin \frac{\pi x}{a} \sin \frac{2\pi y}{a} + A_{21} \sin \frac{2\pi x}{a} \sin \frac{\pi y}{a} \right]$$

$$= 2 \cos \frac{c\pi}{a} \sqrt{5}t \sin \frac{\pi x}{a} \sin \frac{\pi y}{a} \left[A_{12} \cos \frac{\pi y}{a} + A_{21} \cos \frac{\pi x}{a} \right]$$

Therefore, the nodal lines are determined by equation:

$$A_{12} \cos \frac{\pi y}{a} + A_{21} \cos \frac{\pi x}{a} = 0$$

or

$$\cos \frac{\pi y}{a} = - \frac{A_{12}}{A_{21}} \cos \frac{\pi x}{a}$$

Thus, for different initial conditions resulting in different A_{12} and A_{21} values, the nodal lines will be different. However, all these nodal lines should unexceptionally pass through the centre point $x = y = a/2$ – a point called 'pole'.

3.8.3 Longitudinal vibration of a bar

Like the previous continuous systems, in order to measure the vibration of a bar (or rod), we need to make assumptions which are generally satisfied. The bar is homogeneous and slender. During the vibration, the cross-sections normal to the axis of the bar remain plane and normal to the axis. Consider an elastic bar of length L with varying cross-sectional area $A(x)$, as shown in Figure 3.14.

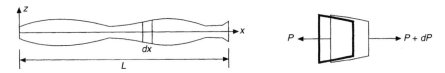

Figure 3.14 Longitudinal vibration of a bar

Suppose the mass density of the bar is ρ and the Young's modulus E. The forces acting on the cross-sections of a small bar element dx are P and $P + dP$ and the axial stress is given by:

$$P = \sigma A = EA \frac{\partial u}{\partial x} \tag{3.106}$$

and
$$dP = \frac{\partial P}{\partial x} dx \tag{3.107}$$

Here, $\frac{\partial u}{\partial x}$ is the axial strain. If the external axial force per unit length is denoted as $f(x, t)$, then the equation of motion of the small bar element becomes:

$$(P + dP) + f(x, t) \, dx - P = \rho A(x) \, dx \frac{\partial^2 u}{\partial t^2} \tag{3.108}$$

or
$$\frac{\partial}{\partial x} EA(x) \frac{\partial u(x, t)}{\partial x} + f(x, t) = \rho(x) A(x) \frac{\partial^2 u}{\partial t^2} \tag{3.109}$$

For a uniform bar, this equation is reduced to:

$$c^2 \frac{\partial^2 u(x, t)}{\partial x^2} = \frac{\partial^2 u(x, t)}{\partial t^2} \tag{3.110}$$

where $c = \sqrt{E/\rho}$. \tag{3.111}

Using the variable separation approach, equation (3.110) has the following solution:

$$u(x, t) = U(x)T(t) = \left(A \cos \frac{\omega x}{c} + B \sin \frac{\omega x}{c} \right) (C \cos \omega t + D \sin \omega t) \tag{3.112}$$

Here, constants A and B are determined by the boundary conditions while C and D are determined by the initial conditions.

Example 3.5
Find the natural frequencies, mode shapes and the free vibration solution of a fixed–free bar.

Solution: The boundary conditions are:

$$u(0, t) = 0 \quad \frac{\partial u}{\partial x} (L, t) = 0$$

Insetting these conditions into equation (3.112) leads to:

$$A = 0 \quad \cos \frac{\omega L}{c} = 0$$

Thus, the natural frequencies are given by:

$$\frac{\omega_n L}{c} = (2n + 1) \frac{\pi}{2} \quad n = 0, 1, 2, \ldots$$

or
$$\omega_n = \frac{(2n + 1)\pi c}{2L} \quad n = 0, 1, 2, \ldots$$

and the mode shapes are:

$$U_n(x) = B_n \sin \frac{(2n + 1)\pi x}{2L} \quad n = 1, 2, \ldots$$

The free vibration solution of the bar becomes:

$$u(x, t) = \int_{n=1}^{\infty} u_n(x, t)$$

or
$$u(x, t) = \int_{n=1}^{\infty} \sin \frac{(2n+)\pi x}{2L} \, C_n \cos \frac{(2n+)\pi ct}{2L} + D_n \sin \frac{(2n+)\pi ct}{2L}$$

where constants can be determined by:

$$C_n = \frac{2}{L} \int_0^L u(x, 0) \sin \frac{(2n+1)\pi x}{2L} \, dx, \quad D_n = \frac{4}{(2n+1)\pi c} \int_0^L \dot{u}(x, 0) \sin \frac{(2n+1)\pi x}{2L} \, dx$$

3.8.4 Transverse vibration of a beam

When a slender beam vibrates in the direction normal to its longitudinal axis, its main deformation is due to bending. Such a vibration is called transverse, or lateral, vibration.

The equation of motion

Consider the free body diagram of a small element of a beam, as shown in Figure 3.15.

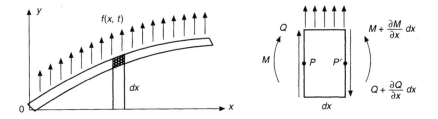

Figure 3.15 A small element of the beam and free body diagram

In Chapter 2, the equation of motion at the 'y' direction has been derived using Newton's law as:

$$\frac{\partial^2}{\partial x^2}\left(EI(x)\frac{\partial^2 y}{\partial x^2}\right) + \rho A(x)\frac{\partial^2 y}{\partial t^2} = f(x, t) \qquad (2.147)$$

If the external force $f(x, t)$ is made zero, the free vibration of a uniform beam can then be described by:

$$c^2 \frac{\partial^2}{\partial x^2} \frac{\partial^2 y}{\partial x^2} + \frac{\partial^2 y}{\partial t^2} = 0 \quad c^2 = \frac{EI}{\rho A} \qquad (2.148)$$

The free vibration

The free vibration solution of the beam has been given in Chapter 2 as:

$$\frac{d^4 Y(x)}{dx^4} - \beta^4 Y(x) = 0 \quad \beta^4 = \frac{\omega^2}{c^2} = \frac{\rho A \omega^2}{EI} \tag{2.152}$$

$$\frac{d^2 T(t)}{dt^2} + \omega^2 T(t) = 0 \tag{2.153}$$

The solution to equation (2.153) is in the form of:

$$T(t) = A \cos \omega t + B \sin \omega t \tag{3.113}$$

where A and B are constants to be determined by the initial conditions. The solution of equation (2.152) can be assumed to take the form of:

$$Y(x) = C e^{sx} \tag{3.114}$$

where s and C are constants. Therefore,

$$s^4 - \beta^4 = 0 \tag{3.115}$$

Thus, $\quad\quad\quad\quad\quad\quad s_{1,2} = \pm \beta \quad s_{3,4} = \pm i\beta$

Hence, the solution to equation (2.152) is:

$$Y(x) = C_1 e^{\beta x} + C_2 e^{-\beta x} + C_3 e^{i\beta x} + C_4 e^{i\beta x} \tag{3.116}$$

where C_1 to C_4 are constants to be determined by the boundary conditions of the beam.

This solution can also be expressed as:

$$Y(x) = C_1 \cos \beta x + C_2 \sin \beta x + C_3 \cosh \beta x + C_4 \sinh \beta x \tag{3.117}$$

C_1 to C_4 are now different constants.

Boundary conditions

From equation (3.117) it can be seen that there are four constants to be determined from the boundary conditions. For a beam in general, each end will contain two boundary conditions, thus enabling the solution of the four constants.

The boundary conditions of some common ends of a beam can be summarized below:

(1) Free end:

Bending moment $= EI(x)\dfrac{\partial^2 y}{\partial x^2} = 0$, and Shear force $= \dfrac{\partial}{\partial x}\left[EI(x)\dfrac{\partial^2 y}{\partial x^2}\right] = 0$

(2) Simply supported end:

Deflection $= y = 0$ and Bending moment $= EI(x)\dfrac{\partial^2 y}{\partial x^2} = 0$

(3) Fixed end:

Deflection $= y = 0$ and Slope $= \dfrac{\partial y}{\partial x} = 0$

Natural frequencies and mode shapes

The natural frequencies of the beam are obtained from equation (2.152),

$$w = (\beta L)^2 \sqrt{\frac{EI}{\rho A L^4}} \tag{3.118}$$

The mode shapes of the beam are given by equation (3.117).

The normal modes $Y(x)$ will satisfy equation (2.152). Thus, for mode 'i' and mode 'j',

$$c^2 \frac{d^4 Y_i}{dx^4} - \omega_i^2 Y_i = 0 \tag{3.119}$$

and

$$c^2 \frac{d^4 Y_j}{dx^4} - \omega_j^2 Y_j = 0 \tag{3.120}$$

Multiplying equation (3.119) by Y_j and equation (3.120) by Y_i, subtracting the two resultant equations and integrating on the full length of the beam, gives:

$$\int_0^L \left[c^2 \frac{d^4 Y_i}{dx^4} Y_j - \omega_i^2 Y_i Y_j \right] dx - \int_0^L \left[c^2 \frac{d^4 Y_j}{dx^4} Y_j - \omega_j^2 Y_j Y_i \right] dx = 0 \tag{3.121}$$

or

$$\int_0^L Y_i Y_j \, dx = -\frac{c^2}{\omega_i^2 - \omega_j^2} \int_0^L (Y_i'''' Y_j - Y_j'''' Y_i) \, dx$$

or

$$\int_0^L Y_i Y_j \, dx = -\frac{c^2}{\omega_i^2 - \omega_j^2} [Y_i Y_j''' - Y_i''' Y_j + Y_i'' Y_j' - Y_i' Y_j'']\Big|_0^L \tag{3.122}$$

The right-hand side of this equation is zero for any combination of free, fixed or simply supported ends of the beam. This is because:

for a free end: $Y'' = 0$ (bending moment) $Y''' = 0$ (shear force)
for a fixed end: $Y = 0$ (deflection) $Y' = 0$ (slope)
for a simple end: $Y'' = 0$ (bending moment) $Y = 0$ (deflection)

Thus, it has been shown that:

$$\int_0^L Y_i Y_j \, dx = 0$$

This shows the orthogonality of normal modes of a beam for transverse vibration.

Example 3.6
Find the natural frequencies and mode shapes of a simply supported beam.

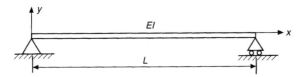

Figure 3.16 A simply supported beam

Solution: The boundary conditions are:

$$Y(0) = 0 \quad Y''(0) = 0$$

$$Y(L) = 0 \quad Y''(L) = 0$$

Substituting these conditions into equation (3.117) leads to:

$$C_1 = 0 \quad C_3 = 0$$

and

$$C_2 \sin \beta L + C_4 \sinh \beta L = 0$$

$$-\beta^2 C_2 \sin \beta L + \beta^2 C_4 \sinh \beta L = 0$$

Therefore,

$$C_4 = 0, \quad \text{and} \quad \sin \beta L = 0$$

$$\sqrt{\omega/c}\, L = i\pi \quad \text{where } i = 1, 2, 3, \ldots$$

The natural frequencies are thus given by:

$$\omega_i = \frac{i^2 \pi^2 c}{L^2}$$

and the normal mode is given by:

$$Y_i = \sin \frac{i\pi x}{L}$$

Literature

1. Dimarogonas, A. 1996: *Vibration for Engineers*. Prentice-Hall, Inc.
2. Inman, D.J. 1994: *Engineering Vibration*. Prentice-Hall, Inc.
3. Norton, M. 1989: *Fundamentals of Noise and Vibration Analysis for Engineers*. Cambridge University Press, Cambridge, UK.
4. Rao, S.S. 1990: *Mechanical Vibrations*. Addison-Wesley, USA.
5. Timoshenko, S., Young, D.H. and Weaver, W. Jr. 1974: *Vibration Problems in Engineering* (4th edn). John Wiley, New York.

Modal analysis theory of an SDoF dynamic system

Unlike classical vibration theory which is primarily engaged with the response of a dynamic system, modal analysis is concerned with its intrinsic properties. A most effective way of investigation for modal analysis is using the frequency response function. This chapter will define the frequency response function of an SDoF system, and study different methods of presenting the function and its intrinsic properties. These properties will form the basis for experimental modal analysis methods. The approach in this chapter will be extended to a system with more than one degree of freedom.

Modal analysis can also be carried out using the time domain impulse response or the response due to ambient vibration input. This topic will be dealt with in later chapters.

4.1 Frequency response functions of an SDoF system

Some mechanical and structural systems can be idealized as SDoF systems. The theory for an SDoF system forms the basis for the analysis of a system with more than one DoF. It also provides physical insight into the vibration of a structural system.

We will use the SDoF system shown in Figure 4.1 that has a mass, a spring and a damper with either viscous or structural (hysteretic) damping.

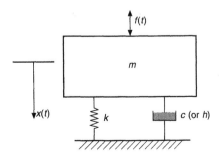

Figure 4.1 An SDoF system with a harmonic excitation

For a harmonic force $f(t) = F(\omega)e^{j\omega t}$, the response of the system is another harmonic function $x(t) = X(\omega)e^{j\omega t}$ where $X(\omega)$ is a complex amplitude. Substituting them into the equations of motion for different damping models in Chapter 3, we can derive the ratio of the displace response and the force input as:

$$\text{For viscous damping: } \frac{X(\omega)}{F(\omega)} = \frac{1}{k - \omega^2 m + j\omega c} \tag{4.1}$$

$$\text{For structural damping: } \frac{X(\omega)}{F(\omega)} = \frac{1}{k - \omega^2 m + jh} \tag{4.2}$$

This ratio, often denoted as $\alpha(\omega)$, is defined as the *frequency response function* (FRF) of the system. Although defined as the ratio of the force and response, the FRF is independent of them. When damping is zero, the complex FRF function is relegated to a real function.

The FRF is the main function on which modal analysis will depend. Although in theory the FRF is dictated only by the system, in reality the accuracy of measured FRF data is critical to the success of modal analysis. In the following three chapters, as we study the modal analysis theory, the accuracy of FRF data is not an issue.

The FRF defined in equations (4.1) and (4.2) can take different forms. For the viscous damping case, for example,

$$\alpha(\omega) = \frac{1/k}{1 - \dfrac{\omega^2}{\omega_0^2} + j2\dfrac{\omega}{\omega_0}} \tag{4.3}$$

or

$$\alpha(\omega) = \frac{1/m}{\omega_0^2 - \omega^2 + j2\omega\omega_0\xi} \tag{4.4}$$

For the structural damping case, the expressions are similar:

$$\alpha(\omega) = \frac{1/k}{1 - \dfrac{\omega^2}{\omega_0^2} + j\eta^2} \tag{4.5}$$

or

$$\alpha(\omega) = \frac{1/m}{\omega_0^2 - \omega^2 + j\omega_0^2\eta} \tag{4.6}$$

The defined FRF uses displacement as the response. It is known as receptance FRF. The vibration response can also be velocity or acceleration. By replacing the displacement response $X(\omega)$ with velocity $\dot{X}(\omega)$ and acceleration $\ddot{X}(\omega)$, two different types of FRFs can be defined as:

$$\text{Mobility FRF for viscous damping: } Y(\omega) = \frac{\dot{X}(\omega)}{F(\omega)} = \frac{j\omega}{k - \omega^2 m + j\omega c} \tag{4.7}$$

$$\text{Mobility FRF for hysteretic damping: } Y(\omega) = \frac{\dot{X}(\omega)}{F(\omega)} = \frac{j\omega}{k - \omega^2 m + jh} \tag{4.8}$$

$$\text{Accelerance FRF for viscous damping: } A(\omega) = \frac{\ddot{X}(\omega)}{F(\omega)} = \frac{-\omega^2}{k - \omega^2 m + j\omega c} \tag{4.9}$$

Accelerance FRF for hysteretic damping: $A(\omega) = \dfrac{\ddot{X}(\omega)}{F(\omega)} = \dfrac{-\omega^2}{k - \omega^2 m + jh}$ (4.10)

It is evident that the three types of FRFs, $\alpha(\omega)$, $Y(\omega)$, and $A(\omega)$, are easily interchangeable. All three are complex functions of frequency. Their amplitudes follow:

$$|A(\omega)| = \omega\,|Y(\omega)| = \omega^2 |\alpha(\omega)|$$ (4.11)

The phase difference among them remains constant at any frequency:

$$\theta_{A(\omega)} = \theta_{Y(\omega)} + \frac{\pi}{2} = \theta_{\alpha(\omega)} + \pi$$ (4.12)

The reciprocals of the three FRFs of an SDoF system also bear useful physical significance and are sometimes used in modal analysis. They are respectively:

$$\text{Dynamic stiffness} = \frac{1}{\alpha(\omega)} = \frac{\text{force}}{\text{displacement}}$$ (4.13)

$$\text{Mechanical impedance} = \frac{1}{Y(\omega)} = \frac{\text{force}}{\text{velocity}}$$ (4.14)

$$\text{Apparent mass} = \frac{1}{A(\omega)} = \frac{\text{force}}{\text{acceleration}}$$ (4.15)

The FRF of an SDoF system can be presented in different forms to those in previous equations. For the case of viscous damping, the receptance FRF can be factorized to become:

$$\alpha(\omega) = \frac{R}{j\omega - \lambda} + \frac{R^*}{j\omega - \lambda^*}$$ (4.16)

where

$$R = \frac{1}{2m\omega_0 j}$$ (4.17)

$$\lambda = (-\zeta + \sqrt{1 - \zeta^2}\, j)\omega_0$$ (4.18)

Conjugate coefficients R and R^* are called *residues* of the receptance. λ and λ^* are the complex poles of the SDoF system. If the same receptance FRF is seen as the transfer function with the real part of the Laplace variable nullified, then it can be expressed as:

$$\alpha(j\omega) = \frac{1}{(j\omega)^2 m + (j\omega)c + k}$$ (4.19)

The FRF can also be seen as the inverse Fourier transform of the impulse response of the system, hence:

$$\alpha(\omega) = \mathscr{F}^{-1}(h(t)) = \int_{-\infty}^{\infty} h(t)e^{j\omega t}\, dt$$ (4.20)

4.2 Graphical display of a frequency response function

Graphical display of an FRF plays a vital role in modal analysis. A different graphical display highlights different information an FRF carries. Since experimental modal analysis often relies on curve fitting of FRF data, sound understanding of FRFs in graphical forms is imperative. In the following, we will show that even for the FRF of an SDoF system which seems to be analytically simple and trivial, much insight about the function can be gained by studying it in various forms of graphical display.

We will use first the receptance FRF to begin our exploration of the graphical display. Since the receptance FRF is a complex function of frequency, it is impossible to fully display it using merely one two-dimensional plot. A three-dimensional plot of a receptance FRF of a typical SDoF system is shown in Figure 4.2.

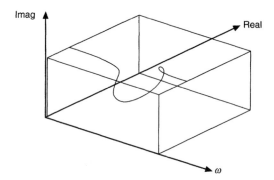

Figure 4.2 A three-dimensional plot of an FRF

Such a three-dimensional plot is complete because it shows the true face of an FRF. It is, however, difficult to be used especially for modal analysis where characteristics such as resonance need to be visually available. From Figure 4.2, we can see that the 3-D plot, when projected to the frequency–real plane, becomes the real part of the FRF. Likewise, its projection to the frequency–imaginary plane gives the imaginary part of the FRF and that to the real–imaginary plane is the Nyquist plot. These plots (and their variations) highlight different aspects of the FRF.

The need to investigate an FRF from a 2-D graphical manifestation gives rise to a number of different graphical presentations of it. Those often used in modal analysis are discussed below.

4.2.1 Amplitude–phase plot and log–log plot

The amplitude–phase plot consists of two parts: the magnitude of the FRF versus frequency and the phase versus frequency. The phase plot does not have much variety since the information of phase cannot be processed numerically in the same way magnitude data can. Therefore, the main focus will be on the magnitude plot of an FRF. This plot can be plotted on a linear scale for both frequency and magnitude axes (linear–linear plot). Figures 4.3, 4.4 and 4.5 show the receptance, mobility and accelerance FRFs of the same SDoF system.

A distinct feature of these figures is the prominence of the resonance. But because of that, it is difficult to appreciate the whole FRF curve, since the high resonance peak dwarfs the rest of the curve. To overcome this, it is customary in modal analysis

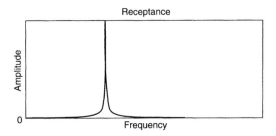

Figure 4.3 Receptance FRF of an SDoF system

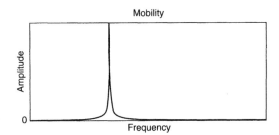

Figure 4.4 Mobility FRF of an SDoF system

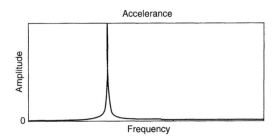

Figure 4.5 Accelerance FRF of an SDoF system

to plot FRF data using logarithmic scales. This can be done in two different ways: (1) logarithmic scale for modulus axis only (linear–log plot) and (2) logarithmic scales for both modulus and frequency axes (log–log plot). In both cases, the magnitude of the FRF is converted into its decibel scale defined as:

$$FRF_{dB} = 20 \log_{10} \frac{\text{linear magnitude}}{\text{unit FRF}} \qquad (4.21)$$

For example, the receptance FRF in dB is estimated as:

$$\alpha(\omega)_{dB} = 20 \log_{10} \frac{|\alpha(\omega)|}{1 \dfrac{\text{metre}}{\text{newton}}} \qquad (4.22)$$

Figure 4.6 shows the linear–log plot of an FRF in receptance, mobility and accelerance forms.

Receptance, mobility and inertance FRF of an SDOF system

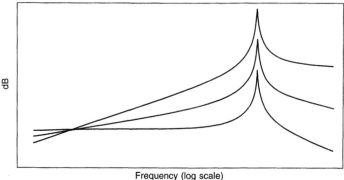

Figure 4.6 Linear–log plot of an FRF in receptance, mobility and accelerance forms

Since the analytical expression of an FRF is determined by the system parameters (mass, stiffness and damping) as well as the variable ω, it is reasonable to expect that these parameters can be derived from the FRF plot easily. However, the linear–linear plot of an FRF shown in Figures 4.3 to 4.5 is overwhelmed by the resonance peak. It is not that obvious how the physical parameters can be derived. From the SDoF vibration theory we know that at low frequency, the FRF is dominated by the stiffness characteristic of the system. At high frequency, the mass characteristic prescribes the FRF. At the vicinity of resonance, the damping characteristic commands the function. The linear–linear plot does not help. But we can use log–log plot to study the mass and stiffness characteristics of the system since the plot enhances off-resonance regions. This is discussed in Section 4.3.

4.2.2 Real and imaginary plots

The real and imaginary plots consist of two parts: the real part of the FRF versus frequency and the imaginary part of the FRF versus frequency. Using the structural damping model, the real and imaginary parts of the FRF are:

$$\text{Re}(\alpha(\omega)) = \frac{k - \omega^2 m}{(k - \omega^2 m)^2 + h^2} \tag{4.23}$$

$$\text{Im}(\alpha(\omega)) = \frac{-h}{(k - \omega^2 m)^2 + h^2} \tag{4.24}$$

Figures 4.7 and 4.8 show these two functions. Obviously the natural frequency occurs when the real part becomes zero. This observation is less useful than it appears to be since experimental FRF data may not have sufficient frequency resolution to pinpoint the location of the zero real part.

Likewise, the real and imaginary parts of the FRF with viscous damping are derived as:

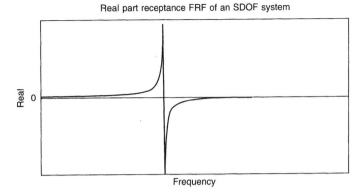

Figure 4.7 Real part of an SDoF FRF with structural damping

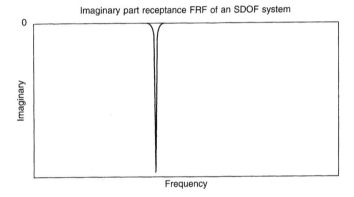

Figure 4.8 Imaginary part of an SDoF FRF with structural damping

$$\mathrm{Re}(\alpha(\omega)) = \frac{k - \omega^2 m}{(k - \omega^2 m)^2 + (\omega c)^2} \tag{4.25}$$

$$\mathrm{Im}(\alpha(\omega)) = \frac{-\omega c}{(k - \omega^2 m)^2 + (\omega c)^2} \tag{4.26}$$

these functions are shown in Figures 4.9 and 4.10.

4.2.3 Nyquist plot

A Nyquist plot shows on the complex plane the real part of an FRF against its imaginary part with frequency as an implicit variable. The benefit of using Nyquist plot comes from the circularity of an FRF on the complex plane. This will be shown in Section 4.3. For an SDoF system with structural damping, we can draw a Nyquist plot for its receptance, mobility and accelerance FRFs in Figure 4.11. These three plots are not drawn to scale.

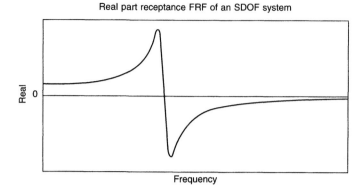

Figure 4.9 Real part of an SDoF FRF with viscous damping

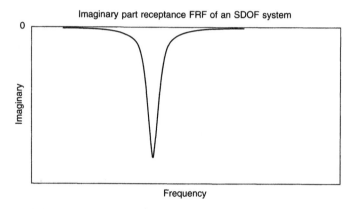

Figure 4.10 Imaginary part of an SDoF FRF with viscous damping

Although all three plots in Figure 4.11 appear to be circles, only the receptance FRF is a real one. From equations (4.23) and (4.24) we can see that the receptance FRF begins from point $\left(\dfrac{k}{k^2 + h^2}, \dfrac{-h}{k^2 + h^2} \right)$ while both mobility and accelerance FRFs begin from the origin. All three end at the origin. For measured FRF data, only a finite frequency range is covered and a limited number of data points are available so that we always have a fraction of the complete Nyquist plot.

For an SDoF system with viscous damping, we can also draw a Nyquist plot for its receptance, mobility and accelerance FRFs. All three plots begin and end at the origin. Although all three plots in Figure 4.12 appear to be circles, only the mobility FRF is a real one. This is explained in section 4.3.

4.2.4 Dynamic stiffness plot

Dynamic stiffness is the inverse of receptance FRF. Analytically, it shows a remarkable

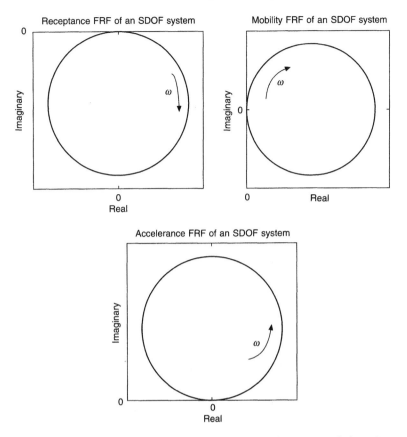

Figure 4.11 Nyquist plot of the three FRFs with structural damping

simplicity compared with the receptance itself. Using the FRF in equation (4.12), the inverse of the FRF becomes:

$$\frac{1}{\alpha(\omega)} = k - \omega^2 m + jh \qquad (4.27)$$

The real part of the dynamic stiffness is a linear function of ω^2. If plotted against ω^2, the straight line intercepts with the frequency axis when the frequency is equal to the natural frequency. The imaginary part is a constant. When plotted against ω^2, the height of the horizontal line tells the extent of structural damping, as shown in Figure 4.13.

For the FRF with viscous damping model, the same simplicity exists, except that the imaginary part of the dynamic stiffness is now a linear function of frequency, resulting in a tilted straight line. The slope of the line derives the amount of viscous damping, as shown in Figure 4.14.

$$\frac{1}{\alpha(\omega)} = k - \omega^2 m + j\omega c \qquad (4.28)$$

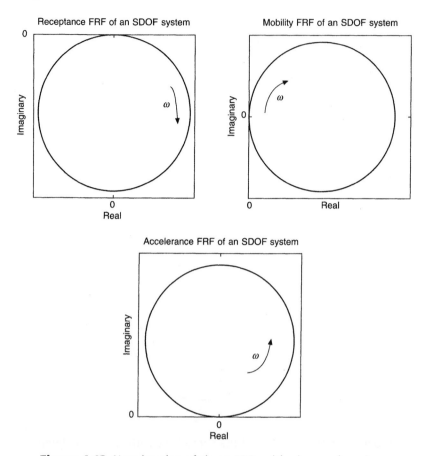

Figure 4.12 Nyquist plot of three FRFs with viscous damping

4.3 Properties of the FRF of an SDoF system

The frequency response function of an SDoF system can be displayed in a number of different ways, as shown in the preceding section. Each display method is able to highlight a specific aspect of the FRF. For example, linear modulus versus frequency highlights the resonance while other parts of the function are disfigured. Through different ways of FRF display, some interesting properties of the FRF can be revealed. In this section, we will try to exploit further the display of an FRF and identify its properties which will form bases on which several modal analysis methods will be developed.

4.3.1 Asymptoticity of log–log plots

As we have seen, the linear–linear plot of an FRF is dominated by the resonance. It is unlikely to expose the beginning and the far end of the FRF where stiffness and mass characteristics of the system are portrayed. The alternative is the log–log plot

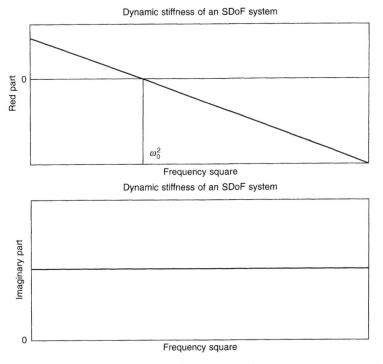

Figure 4.13 Real and imaginary plots of dynamic stiffness with structural damping

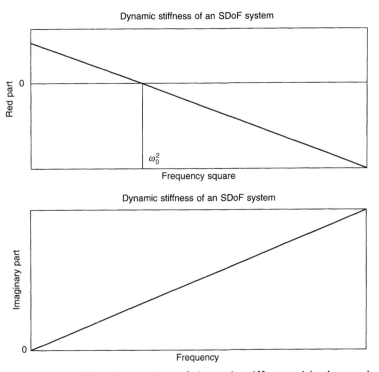

Figure 4.14 Real and imaginary plots of dynamic stiffness with viscous damping

which is able to compensate this deficiency. Using receptance FRF as an example, we know that:

$$\alpha(\omega)|_{\omega<<\omega_0} \approx \frac{1}{k} \qquad (4.29)$$

or

$$\alpha(\omega)|_{\omega<<\omega_0} \approx -20 \log k \quad \text{dB} \qquad (4.30)$$

Therefore, the beginning of the log–log plot of the receptance FRF follows a horizontal asymptote line that shows an intercept at $-20 \log k$, as shown in Figure 4.15. This asymptote line is also called the stiffness line for the FRF.

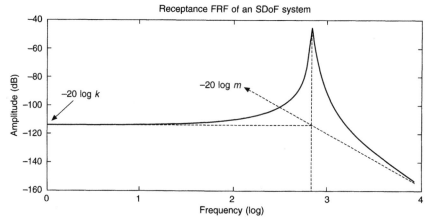

Figure 4.15 Asymptote lines of a receptance FRF

Likewise, the mass line can be found from the log–log plot of the receptance FRF. At high frequency, we have:

$$\alpha(\omega)|_{\omega>>\omega_0} \approx \frac{1}{m\omega^2} \qquad (4.31)$$

or

$$\alpha(\omega)|_{\omega>>\omega_0} \approx -20 \log m - 40 \log \omega \quad \text{dB} \qquad (4.32)$$

This shows that towards the end the log–log plot the receptance FRF conforms to an asymptote line that intercepts the vertical axis at $-20 \log m$. This asymptote line is called the mass line for the FRF.

From the mass and stiffness lines and their intercepts, we are able to estimate the mass and stiffness quantities of the SDoF system. This is an outcome a simple linear plot of the FRF does not offer. We can also see that the damping amount in the system should not vary the mass and stiffness lines since damping contributes little away from resonance.

4.3.2 Circularity of the Nyquist plot

Another useful property of the FRF of an SDoF system is the circularity of its Nyquist plot. We have seen that the Nyquist plots of all three forms of FRFs appear to be circles. Here, we have the analytical explanation. For an SDoF system with viscous damping, the mobility FRF is given as:

$$Y(\omega) = \frac{j\omega}{k - m\omega^2 + j\omega c} \tag{4.17}$$

Its real and imaginary parts can be derived as:

$$\text{Re}(Y(\omega)) = \frac{\omega^2 c}{(k - m\omega^2)^2 + (\omega c)^2} \tag{4.33a}$$

and

$$\text{Im}(Y(\omega)) = \frac{\omega(k - m\omega^2)}{(k - m\omega^2)^2 + (\omega c)^2} \tag{4.33b}$$

Mathematically, it is easy to verify the following equation:

$$\left(\text{Re}(Y(\omega)) - \frac{1}{2c}\right)^2 + (\text{Im}(Y(\omega)))^2 = \left(\frac{1}{2c}\right)^2 \tag{4.34}$$

According to analytical geometry, this equation depicts a circle centred at point $\left(\frac{1}{2c}, 0\right)$ with a radius of $\frac{1}{2c}$. Neither the receptance nor accelerance shares this circularity. Figure 4.16 shows the circle of the mobility FRF data. The FRF commences from and ends at the origin of the complex plane.

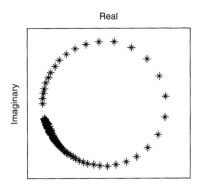

Figure 4.16 Nyquist plot of mobility FRF with viscous damping

For comparison, the same FRF in receptance form is plotted in Figure 4.17. Due to the significant presence of damping, the receptance Nyquist plot is clearly skewed from a perfect circle.

For an SDoF system with structural damping, the circularity also exists except that it lies with receptance FRF. The real and imaginary parts of the receptance FRF are found to be:

$$\text{Re}(\alpha(\omega)) = \frac{(k - m\omega^2)}{(k - m\omega^2)^2 + h^2} \tag{4.23}$$

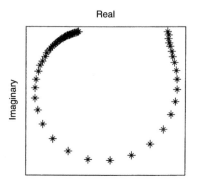

Figure 4.17 Nyquist plot of receptance FRF with viscous damping

and
$$\text{Im}(\alpha(\omega)) = \frac{-h}{(k - m\omega^2)^2 + h^2} \qquad (4.24)$$

It can be shown that the following equation satisfies:

$$(\text{Re}(\alpha(\omega)))^2 + \left(\text{Im}(\alpha(\omega)) + \frac{1}{2h}\right)^2 = \left(\frac{1}{2h}\right)^2 \qquad (4.35)$$

This equation represents a circle with its centre at point $\left(0, -\dfrac{1}{2h}\right)$ and radius equal to $\dfrac{1}{2h}$. Neither the mobility nor accelerance share the circularity. The receptance FRF does not commence from the origin of the complex plane. Figure 4.18 shows the circle of the receptance FRF.

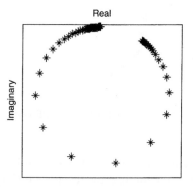

Figure 4.18 Nyquist plot of receptance FRF with structural damping

The circularity of the receptance FRF gives rise to a classical modal analysis method called the circle fit method. This method is numerically simple. It is also possible to interpolate data points after circle fit to estimate more precisely the natural frequency, and, therefore, other parameters. The modal analysis methods related to the circularity property of an FRF will be discussed in Chapter 8.

4.3.3 Linearity of dynamic stiffness

The remarkable linearity of the dynamic stiffness of an SDoF system has been shown by the dynamic stiffness plots in Figures 4.13 and 4.14. For an SDoF receptance FRF with structural damping, the real part of the inverse FRF is a linear function of ω^2 and the imaginary part is a constant. This simplicity is not repeated on the inverse of mobility or accelerance FRFs. For the FRF with viscous damping model, the same simplicity has been observed. The real part of the dynamic stiffness has not changed but the imaginary part becomes a linear function of frequency ω.

The dynamic stiffness separates damping property of the system from the mass and stiffness properties. This is a characteristic no other form of SDoF FRFs share. The simplicity of the real and imaginary plots of dynamic stiffness suggests that line fitting can be used to estimate parameters of the system. Relevant methods will be presented in Chapter 8.

Literature

1. Bishop, R.E.D. and Gladwell, G.M.L. 1963: An investigation into the theory of resonance testing. *Philosophical Transactions of the Royal Society of London*, **255 A 1055**, 241–280.
2. Dobson, B.J. 1984: Modal analysis using dynamic stiffness data. *Royal Naval Engineering College (RNEC)*, TR–84015.
3. Ewins, D.J. 1984: *Modal Testing – Theory and Practice*. Research Studies Press, England.
4. Goydar, H.G.D. 1980: Methods and application of structural modelling from measured structural frequency response data. *Journal of Sound and Vibration*, **68**(2), 209–230.
5. Kennedy, C.C. and Pancu, C.D.P. 1947: Use of vectors in vibration measurement and analysis. *Journal of the Aeronautical Sciences*, **14**(11), 603–625.
6. Maia, N.M.M. *et al.* 1997: *Theoretical and Experimental Modal Analysis*. Research Studies Press, UK and John Wiley & Sons, USA.
7. Randall, R. 1977: *Frequency Analysis*. Bruel & Kjaer.
8. Silva, J.M.M. and Maia, N.M.M. (eds) 1998: *Modal Analysis and Testing*. NATO Science Series E: Applied Sciences, Vol. 363.

5

Modal analysis of an undamped MDoF system

The analysis of a multi-degree-of-freedom (MDoF) dynamic system is a natural extension of that of an SDoF system. For many mechanical and structural systems, more than one coordinate is needed to describe its motion and vibration sufficiently. The result is an MDoF model. Such a model characterizes a system in terms of mass and stiffness matrices. As a result, the matrix method becomes an essential part of the analysis. This chapter is focused on the modal analysis theory of an undamped MDoF system. Like an SDoF system, the analysis extends from the traditional vibration analysis for an MDoF system where response was the main concern. Damping will be introduced in the next chapter.

5.1 Normal modes and orthogonality of an undamped MDoF system

5.1.1 Normal modes of an undamped MDoF system

From Chapter 3 we know that the equation of motion for free vibration of an MDoF system leads to the following eigenvalue problem:

$$([K] - \omega^2[M])\{\Psi\} = \{0\} \tag{5.1}$$

Here, both the mass matrix $[M]$ and the stiffness matrix $[K]$ are symmetric. The mass matrix is usually positive definite while the stiffness matrix may become semi-positive definite if the system possesses rigid body vibration modes.

The solution to equation (5.1) is composed of 'n' eigenvalues ω_r^2 and 'n' eigenvectors $\{\Psi\}_r$, ($r = 1, 2, \ldots, n$). The square roots of these eigenvalues are the natural frequencies of the system and the eigenvectors its mode shapes. It is evident that mode shapes are not unique since multiples of $\{\Psi\}_r$ also satisfy equation (5.1). Mode shapes $\{\Psi\}_r$ are referred to as *principal modes*, or *normal modes* of the system. They are also called the *undamped modes* for the obvious reason.

From linear algebra we know that, if the natural frequencies are non-zero and distinct, then all the mode shapes are independent. Therefore, 'n' mode shapes collectively form a basis for 'n' vector space. If some natural frequencies are the same, the independence of the corresponding mode shapes is usually lost.

Example 5.1
Shown in Figure 5.1 is a 4 degree-of-freedom mass–spring system.

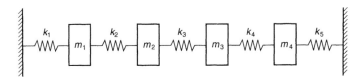

Figure 5.1 A 4 degree-of-freedom mass–spring system

The stiffness and mass elements are assigned as: $k_1 = k_2 = k_3 = k_4 = k_5 = 1000$ N/m and $m_1 = m_2 = m_3 = m_4 = 1$ kg. Therefore, the mass and stiffness matrices of the system are respectively:

$$[M] = \begin{bmatrix} 1 & 0 & 0 & 0 \\ 0 & 1 & 0 & 0 \\ 0 & 0 & 1 & 0 \\ 0 & 0 & 0 & 1 \end{bmatrix} \text{kg} \quad \text{and} \quad [K] = \begin{bmatrix} 2000 & -1000 & 0 & 0 \\ -1000 & 2000 & -1000 & 0 \\ 0 & -1000 & 2000 & -1000 \\ 0 & 0 & -1000 & 2000 \end{bmatrix} \text{N/m}$$

The eigenvalue solution of equation (5.1) produces the following eigenvalues and eigenvectors which show the natural frequencies and mode shapes of the system:

$$\omega_1^2 = 381.967 \text{ (rad/s)}^2 \quad \omega_2^2 = 1381.97 \text{ (rad/s)}^2$$

$$\omega_3^2 = 2618.03 \text{ (rad/s)}^2 \quad \omega_4^2 = 3618.03 \text{ (rad/s)}^2$$

and
$$\{\Psi\}_1 = \begin{Bmatrix} 0.37363 \\ 0.60455 \\ 0.60455 \\ 0.37363 \end{Bmatrix} \quad \{\Psi\}_2 = \begin{Bmatrix} -0.36180 \\ -0.22361 \\ 0.22361 \\ 0.36180 \end{Bmatrix} \quad \{\Psi\}_3$$

$$= \begin{Bmatrix} -0.44721 \\ 0.27639 \\ 0.27639 \\ -0.44721 \end{Bmatrix} \quad \{\Psi\}_4 = \begin{Bmatrix} -0.82706 \\ 1.33821 \\ -1.33821 \\ 0.82706 \end{Bmatrix}$$

It is customary in modal analysis to collate all the eigenvalues and eigenvectors together in matrix forms. This will greatly convenience the analysis. If eigenvalues are arranged in an ascending order and the mode shapes are accordingly allocated in columns, then a natural frequency matrix and a mode shape matrix (also called modal matrix) can be formed as:

Natural frequency matrix $[\,^{\cdot}\omega_r^2\,.] = \begin{bmatrix} \omega_1^2 & 0 & \cdots & 0 \\ 0 & \omega_2^2 & \cdots & 0 \\ \cdots & \cdots & \cdots & \cdots \\ 0 & 0 & \cdots & \omega_n^2 \end{bmatrix}$ (5.2)

Mode shape matrix $[\Psi] = [\{\Psi\}_1, \{\Psi\}_2, \ldots, \{\Psi\}_n]$ (5.3)

For instance, for the system in Example 5.1, these matrices are:

$$[\,^{\cdot}\omega_r^2\,.] = \begin{bmatrix} 381.966 & & & \\ & 1381.97 & & \\ & & 2618.03 & \\ & & & 3618.03 \end{bmatrix} (\text{rad/s})^2$$

and $[\Psi] = \begin{bmatrix} 0.37363 & -0.36180 & -0.44721 & -0.82706 \\ 0.60455 & -0.22361 & 0.27639 & 1.33821 \\ 0.60455 & 0.22361 & 0.27639 & -1.33821 \\ 0.37363 & 0.36180 & -0.44721 & 0.82706 \end{bmatrix}$

With (5.2) and (5.3), equation (5.1) can now be recast in matrix form as:

$$[K][\Psi] = [M][\Psi][\,^{\cdot}\omega_r^2\,.] \tag{5.4}$$

Similarly,

$$[\Psi]^T[K] = [\,^{\cdot}\omega_r^2\,.][\Psi]^T[M] \tag{5.5}$$

It is easy to verify numerically that for Example 5.1, the following holds:

$$\begin{bmatrix} 2000 & -1000 & 0 & 0 \\ -1000 & 2000 & -1000 & 0 \\ 0 & -1000 & 2000 & -1000 \\ 0 & 0 & -1000 & 2000 \end{bmatrix}$$

$$\times \begin{bmatrix} 0.37363 & -0.36180 & -0.44721 & -0.82706 \\ 0.60455 & -0.22361 & 0.27639 & 1.33821 \\ 0.60455 & 0.22361 & 0.27639 & -1.33821 \\ 0.37363 & 0.36180 & -0.44721 & 0.82706 \end{bmatrix}$$

$$
= \begin{bmatrix} 1 & 0 & 0 & 0 \\ 0 & 1 & 0 & 0 \\ 0 & 0 & 1 & 0 \\ 0 & 0 & 0 & 1 \end{bmatrix} \begin{bmatrix} 0.37363 & -0.36180 & -0.44721 & -0.82706 \\ 0.60455 & -0.22361 & 0.27639 & 1.33821 \\ 0.60455 & 0.22361 & 0.27639 & -1.33821 \\ 0.37363 & 0.36180 & -0.44721 & 0.82706 \end{bmatrix}
$$

$$
\times \begin{bmatrix} 381.966 & & & \\ & 1381.97 & & \\ & & 2618.03 & \\ & & & 3618.03 \end{bmatrix}
$$

5.1.2 Orthogonality properties of an undamped MDoF system

The orthogonality properties of an undamped MDoF system are manifested in the relationship between its spatial model and the modal model. Consider the rth and sth modes of the system, equation (5.1) becomes:

$$
([K] - \omega_r^2 [M])\{\Psi\}_r = \{0\} \tag{5.6}
$$

and

$$
([K] - \omega_s^2 [M])\{\Psi\}_s = \{0\} \tag{5.7}
$$

Pre-multiplying equation (5.7) by $\{\Psi\}_r^T$ yields:

$$
\{\Psi\}_r^T ([K] - \omega_s^2 [M])\{\Psi\}_s = 0 \tag{5.8}
$$

Meanwhile, transposing equation (5.6) and post-multiplying it by $\{\Psi\}_s$ gives

$$
\{\Psi\}_r^T ([K] - \omega_r^2 [M])\{\Psi\}_s = 0 \tag{5.9}
$$

Subtracting equations (5.8) and (5.9) yields:

$$
(\omega_s^2 - \omega_r^2)\{\Psi\}_r^T [M]\{\Psi\}_s = 0 \tag{5.10}
$$

Since $\omega_s^2 \neq \omega_r^2$, equation (5.10) suggests that:

$$
\{\Psi\}_r^T [M]\{\Psi\}_s = 0 \quad \text{for} \quad r \neq s \tag{5.11}
$$

Substituting equations (5.11) into (5.9) yields:

$$
\{\Psi\}_r^T [K]\{\Psi\}_s = 0 \quad \text{for} \quad r \neq s \tag{5.12}
$$

These two equations mean that mode shapes are orthogonal to each other with respect to matrices $[M]$ and $[K]$. Pre-multiplying equation (5.6) by $\{\Psi\}_r^T$ yields:

$$
\{\Psi\}_r^T [M]\{\Psi\}_s = \omega_r^2 \{\Psi\}_r^T [M]\{\Psi\}_r \quad (r = 1, 2, \ldots, n) \tag{5.13}
$$

Let
$$\{\Psi\}_r^T [M]\{\Psi\}_r = m_r \tag{5.14}$$

and
$$\{\Psi\}_r^T [K]\{\Psi\}_r = k_r \tag{5.15}$$

then
$$\omega_r^2 = \frac{k_r}{m_r} \quad (r = 1, 2, \ldots, n) \tag{5.16}$$

Here m_r and k_r are called *modal mass* and *modal stiffness*, or *generalized mass* and *generalized stiffness*, of the rth mode respectively, but they do not have the same units as mass and stiffness.

The results of equations (5.6) to (5.15) pronounce that the mode shapes of an undamped MDoF system with distinct natural frequencies are orthogonal to each other with respect to system mass and stiffness matrices. This is known as the principle of orthogonality. In matrix forms, this principle can be expressed in a more succinct form:

$$[\Psi]^T [M][\Psi] = [\cdot m_i \cdot] = \begin{bmatrix} m_1 & 0 & \cdots & 0 \\ 0 & m_2 & \cdots & 0 \\ \cdots & \cdots & \cdots & \cdots \\ 0 & 0 & \cdots & m_n \end{bmatrix} \tag{5.17}$$

$$[\Psi]^T [K][\Psi] = [\cdot k_i \cdot] = \begin{bmatrix} k_1 & 0 & \cdots & 0 \\ 0 & k_2 & \cdots & 0 \\ \cdots & \cdots & \cdots & \cdots \\ 0 & 0 & \cdots & k_n \end{bmatrix} \tag{5.18}$$

and
$$[\cdot \omega_r^2 \cdot] = [\cdot k_i \cdot][\cdot m_i \cdot]^{-1} \tag{5.19}$$

Matrix $[\cdot m_i \cdot]$ is referred to as the modal mass matrix, and $[\cdot k_i \cdot]$ the modal stiffness matrix. In them, diagonal element m_i is the modal mass of the ith mode and k_i the modal stiffness of it.

Example 5.2
Verify the orthogonality properties of the 4DoF system given in Example 5.1.

Solution: From the known natural frequency and mode shape matrices of the system, the orthogonality properties are verified below.

$$[\Psi]^T [M][\Psi] = \begin{bmatrix} 1.01017 & 0 & 0 & 0 \\ 0 & 0.36180 & 0 & 0 \\ 0 & 0 & 0.55279 & 0 \\ 0 & 0 & 0 & 0.94966 \end{bmatrix}$$

and
$$[\Psi]^T [K][\Psi] = \begin{bmatrix} 385.851 & & & \\ & 499.995 & & \\ & & 1447.22 & \\ & & & 3435.90 \end{bmatrix}$$

It is also easy to see that:

$$
\begin{bmatrix} 385.851 & & & \\ & 499.995 & & \\ & & 1447.22 & \\ & & & 3435.90 \end{bmatrix}
\begin{bmatrix} 1.01017 & & & \\ & 0.36180 & & \\ & & 0.55279 & \\ & & & 0.94966 \end{bmatrix}^{-1}
$$

$$
= \begin{bmatrix} 381.966 & & & \\ & 1381.97 & & \\ & & 2618.03 & \\ & & & 3618.03 \end{bmatrix} = [\,{}^{\cdot}\omega_r^2\,\cdot]
$$

Since 'n' mode shapes of an MDoF system collectively form a basis for 'n' vector space, it can be said that the free vibration of the system consists of a linear combination of all modes. The equation of motion of the system has been converted into an eigenvalue problem as:

$$([K] - \omega^2[M])\{X\} = \{0\} \tag{3.21}$$

As we have:

$$\{X\} = [\Psi]\{Y\} \tag{5.20}$$

here $\{Y\}$ consists of the principal coordinates, this equation becomes:

$$([K] - \omega^2[M])[\Psi]\{Y\} = \{0\} \tag{5.21}$$

Pre-multiplying the equation with matrix $[\Psi]^T$ and using the principle of orthogonality, we can transform this equation into:

$$([\,{}^{\cdot}k_i\,\cdot] - \omega^2[\,{}^{\cdot}m_i\,\cdot])\{Y\} = \{0\} \tag{5.22}$$

Therefore, we found that the mode shapes are able to 'diagonalize' the matrix equation of motion and decouple the 'n' intertwined equations into 'n' independent ones. This means an MDoF system now effectively becomes a collection of separate SDoF systems. For example, the 2DoF system shown in Figure 5.2 can be converted into two SDoF systems.

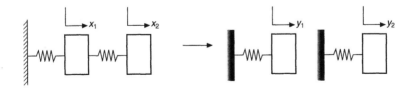

Figure 5.2 Modal decoupling of a 2DoF system

The decoupling not only provides a convenient numerical analysis for an MDoF system but also offers a unique physical interpretation of its modal behaviour.

5.2 Frequency response functions of an undamped MDoF system

5.2.1 Dynamic stiffness matrix and receptance matrix

In Chapter 3, it has been shown that, using Lagrange's equation or Newton's laws of motion, the matrix form of the equations of motion of an MDoF conservative system becomes:

$$[M]\{\ddot{x}(t)\} + [K]\{x(t)\} = \{f(t)\} \tag{5.23}$$

Here, $\{f(t)\}$ is an $n \times 1$ force vector of 'n' external forces. If these forces are harmonic with the same frequency and phase (assume the phase is zero), then:

$$\{f(t)\} = \left\{ \begin{array}{c} F_1 \\ F_2 \\ \ldots \\ \ldots \\ F_n \end{array} \right\} \sin \omega t = \{F\} \sin \omega t \tag{5.24}$$

Here F_r ($r = 1, 2, \ldots, n$) are the amplitudes of the harmonic forces. They are real quantities. Figure 5.3 shows a cantilever beam excited by a group of external forces.

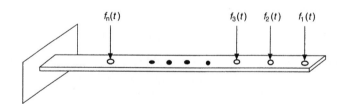

Figure 5.3 A cantilever excited by a number of external forces

As described in Chapter 2, the system will vibrate harmonically. The displacement and acceleration vectors can then be expressed respectively as:

$$\{x(t)\} = \left\{ \begin{array}{c} X_1 \\ X_2 \\ \ldots \\ \ldots \\ X_n \end{array} \right\} \sin \omega t = \{X\} \sin \omega t \tag{5.25}$$

$$\{\ddot{x}(t)\} = -\omega^2 \begin{Bmatrix} X_1 \\ X_2 \\ \cdots \\ \cdots \\ X_n \end{Bmatrix} \sin \omega t = -\omega^2 \{X\} \sin \omega t \qquad (5.26)$$

Substituting equations (5.25) and (5.26) into (5.23) will yield:

$$-\omega^2 [M]\{X\} \sin \omega t + [K]\{X\} \sin \omega t = \{F\} \sin \omega t \qquad (5.27)$$

or
$$([K] - \omega^2[M])\{X\} = \{F\} \qquad (5.28)$$

Matrix $([K] - \omega^2[M])$ is known as the *dynamic stiffness matrix* of an MDoF system (it has the unit of stiffness) and is denoted as $[Z(\omega)]$, i.e.:

$$[Z(\omega)] = [K] - \omega^2[M] \qquad (5.29)$$

and $z_{ij}(\omega) = k_{ij} - \omega^2 m_{ij}$. Thus, (5.28) can be written as:

$$[Z(\omega)]\{X\} = \{F\} \qquad (5.30)$$

If matrix $[Z(\omega)]$ is non-singular (which is true unless frequency ω is equal to one of the natural frequencies), then the amplitude responses of the system can be obtained by:

$$\{X\} = [Z(\omega)]^{-1}\{F\} \qquad (5.30)$$

The inverse of matrix $[Z(\omega)]$ is defined as the *receptance FRF matrix* of the system and is denoted as $[\alpha(\omega)]$, i.e.:

$$[\alpha(\omega)] = ([K] - \omega^2[M])^{-1} \qquad (5.31)$$

$$= \begin{bmatrix} \alpha_{11}(\omega) & \alpha_{12}(\omega) & \cdots & \alpha_{1n}(\omega) \\ \alpha_{21}(\omega) & \alpha_{22}(\omega) & \cdots & \alpha_{2n}(\omega) \\ \cdots & \cdots & \cdots & \cdots \\ \alpha_{n1}(\omega) & \alpha_{n2}(\omega) & \cdots & \alpha_{nn}(\omega) \end{bmatrix}$$

Equation (5.30) can now be written as:

$$\{X\} = [\alpha(\omega)]\{F\} \qquad (5.32)$$

The receptance matrix is symmetrical because the dynamic stiffness matrix is symmetrical. This symmetry property manifests the *reciprocity nature* of a linear MDoF system regarding its vibration responses. Thus, the response at coordinate 'i' due to a single force applied at coordinate 'j' is the same as the response at coordinate 'j' due to the same force applied at coordinate 'i'. When the response and excitation coordinates coincide ($i = j$), the FRF is referred to as a *point FRF*. Otherwise it is called a *transfer FRF*.

It is important to know that, although derived from forced vibration, the receptance FRFs reflect the properties of a linear vibrating system, similar to the natural frequencies and mode shapes of the system. Therefore, they do not depend upon external forces. The dependency can only occur if the system's dynamic behaviour is nonlinear.

Example 5.3
Show the receptance $\alpha_{11}(\omega)$ of the 4DoF system given in Example 5.1.

Solution: For the system in Example 5.1, its receptance $\alpha_{11}(\omega)$ can be calculated using equation (5.31) and shown in Figure 5.4 in a linear scale.

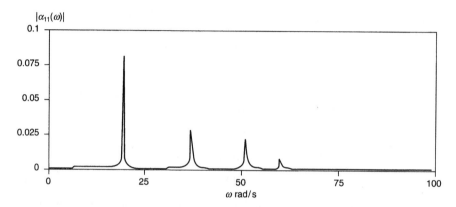

Figure 5.4 Receptance $\alpha_{11}(\omega)$ of the 4DoF system in Figure 5.1

5.2.2 Physical interpretation of a receptance FRF

The physical meaning of a receptance FRF in the MDoF case is not as apparent as that in an SDoF case. However, it does exist for each receptance function in matrix $[\alpha(\omega)]$. We begin with the response amplitude at coordinate 'i', equation (5.32) can be reduced into:

$$X_i = \alpha(\omega)_{i1}F_1 + \alpha(\omega)_{i2}F_2 + \ldots + \alpha(\omega)_{in}F_n \qquad (5.33)$$

If only one force is applied to the system, say F_j, then (5.33) will reduce to:

$$\alpha(\omega)_{ij} = \frac{X_i}{F_j} \quad (F_r = 0,\ r = 1, 2, \ldots, n \text{ and } r \neq j) \qquad (5.34)$$

This suggests that the ijth element in matrix $[\alpha(\omega)]$ is the frequency response function when the system only has one input force applied at coordinate 'j' and the response is measured at coordinate 'i'. This is the physical interpretation of a receptance FRF for an MDoF system.

If there is an additional force (say F_r and $r \neq j$) applied to the system, then the ratio $\dfrac{X_i}{F_j}$ will not yield the receptance FRF $\alpha(\omega)_{ij}$ since this time:

$$\frac{X_i}{F_j} = \alpha(\omega)_{ij} + \alpha(\omega)_{ir}\frac{F_r}{F_j} \qquad (5.35)$$

Like the SDoF case, the mobility and accelerance FRFs of an MDoF system can be derived from the receptance counterparts. Thus, if the accelerance FRF matrix is denoted as $[A(\omega)]$ and mobility as $[Y(\omega)]$, then we have:

$$[Y(\omega)] = -j\omega[\alpha(\omega)] \qquad (5.36)$$

and
$$[A(\omega)] = -\omega^2[\alpha(\omega)] \qquad (5.37)$$

5.2.3 Display of an FRF of an undamped MDoF system

Like an SDoF case, an FRF of an MDoF system can be displayed in different ways. An appropriate selection of display can help to highlight particular properties of the FRF. This is important given that an MDoF FRF contains more than one resonance and therefore has special features such as anti-resonances and minima. Magnitude plot, log–log magnitude plots and the inverse FRF are several customary ways of displaying an FRF. Other methods such as the Nyquist plot and real–imaginary plot are not theoretically applicable to the FRFs of an *undamped* MDoF system. Those methods, which are also vitally important in modal analysis, will be covered in the next chapter where a *damped* MDoF system will be investigated.

A magnitude plot of an FRF can display the magnitude of the FRF against frequency in linear scale. For instance, Figure 5.3 shows $\alpha_{11}(\omega)$ of the 4DoF system given in Example 5.1. The magnitude plot clearly exhibits the resonances. However, details of the FRF curve are swamped because of the prominence of resonance peaks. In particular, characteristics such as anti-resonances become invisible.

Like for an SDoF case, a very effective remedy to show the details of an FRF curve is to replace the linear amplitude scale with logarithmic amplitude scale. Thus, details of the whole FRF curve can be exposed. In practice, it is most convenient to utilize decibel scale which is referred to a unit quantity of the FRF. The dB scale has been defined in Chapter 4. Thus, the receptance of the 4DoF system given in Example 5.1, if plotted in dB scale, becomes the curve shown in Figure 5.5.

The comparison of Figures 5.5 with 5.4 clearly demonstrates the advantages of using the dB scale, the whole receptance curve – both the resonance and anti-resonances – is clearly visible. Figure 5.6 shows the log–log plot of the same FRF. This plot

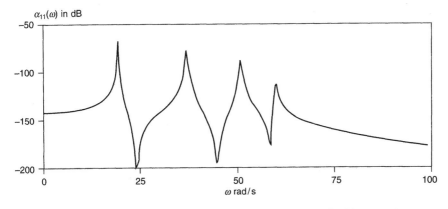

Figure 5.5 Receptance $\alpha_{11}(\omega)$ of the 4DoF system in Figure 5.1

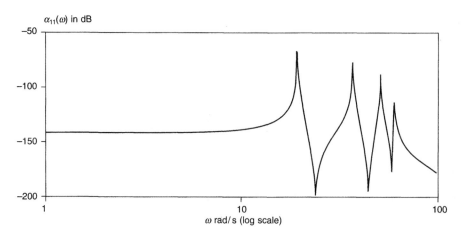

Figure 5.6 Receptance $\alpha_{11}(\omega)$ of the 4DoF system in Figure 5.1

reveals the asymptotic properties of the FRF which will be discussed later in this chapter.

An FRF can also be presented in its inverse form in linear or dB scale. Figure 5.7 shows the inverse receptance $\alpha_{11}(\omega)$ of the 4DoF system in Figure 5.1 and Figure 5.8 the same inverse receptance FRF in dB scale.

Apparently, the inverse FRF helps to expose the vicinity of anti-resonances and minima of the FRF. However, its real significance lies in the modal data extraction to be discussed in Chapter 8.

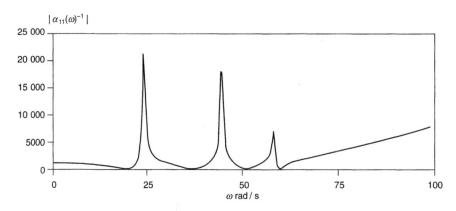

Figure 5.7 Inverse of receptance $\alpha_{11}(\omega)$ of the 4DoF system in Figure 5.1 (linear scale)

5.3 Mass-normalized modes and modal model of an undamped MDoF system

Mode shapes of a system are not unique since multiples of them are equally valid. Mass-normalized mode shapes are unique presentations of the mode shapes. They are

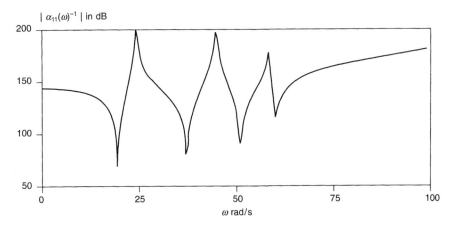

Figure 5.8 Inverse of receptance $\alpha_{11}(\omega)$ of the 4DoF system in Figure 5.1 (dB scale)

important in the further development of modal analysis theory, especially when applied to experimental modal analysis.

A mass-normalized mode shape is a mode shape normalized using the modal mass. Since $\{\Psi\}_r^T [M]\{\Psi\}_r = m_r$ $(r = 1, 2, \ldots, n)$, mode shape $\{\Psi\}_r$ can be normalized in the following way:

$$\{\phi\}_r = \frac{1}{\sqrt{m_r}} \{\Psi\}_r \quad (r = 1, 2, \ldots, n) \tag{5.38}$$

Here, $\{\phi\}_r$ is called the *mass-normalized mode shape* of the system. Equation (5.38) can also be written in matrix form:

$$[\Phi] = [\Psi][\,^{\cdot}m_i\,]^{-1/2} \tag{5.39}$$

Therefore, it can be seen that, using mass-normalized mode shapes, equations (5.7 and 5.8) become:

$$[\Phi]^T[M][\Phi] = [I] \tag{5.40}$$

$$[\Phi]^T [K][\Phi] = [\,^{\cdot}\omega_r^2\,] \tag{5.41}$$

Mass-normalized mode shape matrix $[\Phi]$ is *unique* to an MDoF system. The importance of the mass-normalized mode shapes, however, lies in the fact that the frequency response functions of an undamped MDoF system can be expressed in terms of mass-normalized mode shapes and natural frequencies. Such an expression of an FRF manifests the essence of modal analysis. This will be discussed in the next section in this chapter.

Matrices $[\,^{\cdot}\omega_r^2\,]$ and $[\,^{\cdot}\Phi\,]$ of an MDoF system constitute its modal model of the system. This model can be derived from the spatial model of the system which comprises matrices $[M]$ and $[K]$. The response model of the system contains its FRFs.

Example 5.4
The mass and stiffness matrices of the 4DoF system in Example 5.1 are respectively:

$$[M] = \begin{bmatrix} 1 & 0 & 0 & 0 \\ 0 & 1 & 0 & 0 \\ 0 & 0 & 1 & 0 \\ 0 & 0 & 0 & 1 \end{bmatrix} \text{kg} \quad \text{and} \quad [K] = \begin{bmatrix} 2000 & -1000 & 0 & 0 \\ -1000 & 2000 & -1000 & 0 \\ 0 & -1000 & 2000 & -1000 \\ 0 & 0 & -1000 & 2000 \end{bmatrix} \text{N/m}$$

and the mode shape matrix is:

$$[\Psi] = \begin{bmatrix} 0.37363 & -0.36180 & -0.44721 & -0.82706 \\ 0.60455 & -0.22361 & 0.27639 & 1.33821 \\ 0.60455 & 0.22361 & 0.27639 & -1.33821 \\ 0.37363 & 0.36180 & -0.44721 & 0.82706 \end{bmatrix}$$

Therefore, the modal mass matrix can be calculated as:

$$[\,m_i\,] = [\Psi]^T [M][\Psi] = \begin{bmatrix} 1.01017 & 0 & 0 & 0 \\ 0 & 0.36180 & 0 & 0 \\ 0 & 0 & 0.55279 & 0 \\ 0 & 0 & 0 & 0.94966 \end{bmatrix}$$

According to equation (5.7), the mass-normalized mode shape matrix of the 4DoF system becomes:

$$[\Phi] = \begin{bmatrix} 0.37363 & -0.36180 & -0.44721 & -0.82706 \\ 0.60455 & -0.22361 & 0.27639 & 1.33821 \\ 0.60455 & 0.22361 & 0.27639 & -1.33821 \\ 0.37363 & 0.36180 & -0.44721 & 0.82706 \end{bmatrix} \begin{bmatrix} 1.01017 & 0 & 0 & 0 \\ 0 & 0.36180 & 0 & 0 \\ 0 & 0 & 0.55279 & 0 \\ 0 & 0 & 0 & 0.94966 \end{bmatrix}^{-1/2}$$

$$= \begin{bmatrix} 0.37175 & -0.60150 & -0.60150 & -0.37175 \\ 0.60150 & -0.37175 & 0.37175 & 0.60150 \\ 0.60150 & 0.37175 & 0.37175 & -0.60150 \\ 0.37175 & 0.60150 & -0.60150 & 0.37175 \end{bmatrix}$$

This mass-normalized mode shape matrix can be used to verify equations (5.40 and 41).

5.4 Frequency response functions and the modal model

5.4.1 Decomposition of an FRF using modal data

Though the receptance FRF matrix is defined in equation (5.31), the actual derivation of the matrix by that equation can be very time consuming since, for every frequency given, an $n \times n$ dynamic stiffness matrix has to be inverted. However, using the orthogonality properties of an MDoF system, the receptance matrix can be derived much more easily from the natural frequency and mode shape matrices. We begin at equation (5.31) which can lead to:

$$[\Phi]^T[[K] - \omega^2[M]][\Phi] = [\Phi]^T[\alpha(\omega)]^{-1}[\Phi] \qquad (5.42)$$

or
$$[\cdot(\omega_r^2 - \omega^2).] = [\Phi]^T[\alpha(\omega)]^{-1}[\Phi] \qquad (5.43)$$

i.e.
$$[\alpha(\omega)] = [\Phi][\cdot(\omega_r^2 - \omega^2).]^{-1}[\Phi]^T \qquad (5.44)$$

It can be seen that $[\alpha(\omega)]$ is symmetric, which is indicative of the principle of reciprocity.

For a single receptance FRF $\alpha_{jk}(\omega)$, equation (5.44) can also be written as:

$$\alpha_{jk}(\omega) = \frac{\phi_{j1}\phi_{k1}}{\omega_1^2 - \omega^2} + \frac{\phi_{j2}\phi_{k2}}{\omega_2^2 - \omega^2} + \ldots + \frac{\phi_{jn}\phi_{kn}}{\omega_n^2 - \omega^2} \qquad (5.45)$$

$$= \{\phi_{j1}\phi_{k1} \quad \phi_{j2}\phi_{k2} \quad \ldots \quad \phi_{jn}\phi_{kn}\} \left\{ \begin{array}{c} \dfrac{1}{\omega_1^2 - \omega^2} \\ \dfrac{1}{\omega_2^2 - \omega^2} \\ \ldots \\ \dfrac{1}{\omega_n^2 - \omega^2} \end{array} \right\} \qquad (5.46)$$

Equation (5.45) is the essence of modal analysis. By expressing an FRF in terms of modal data, it becomes clear that the FRF comprises the contributions of all individual modes.

Example 5.5
For the 4DoF system shown in Figure 5.1, the receptance $\alpha_{11}(\omega)$ can be decomposed into the contributions of four individual modes. Figure 5.9 shows the $\alpha_{11}(\omega)$ due to each individual mode in non-solid lines and the complete $\alpha_{11}(\omega)$ in solid line.

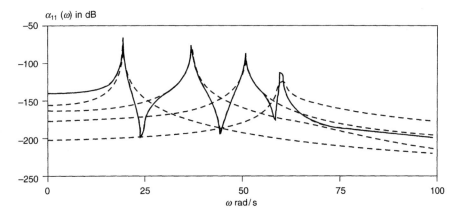

Figure 5.9 Composition of a receptance FRF

Equation (5.45) can also be written in a different form as:

$$\alpha_{jk}(\omega) = \{_rA_{jk}\}\left\{\frac{1}{\omega_r^2 - \omega^2}\right\} \qquad (5.47)$$

Here, $_rA_{jk} = \phi_{jr}\phi_{kr}$ is the product of the jth and kth elements in the rth mode shape $\{\phi\}_r$. It is known as the '*modal constant*'. They are related but not equal to *residue*. Thus, the pth column of matrix $[\alpha(\omega)]$ can be written as:

$$\{\alpha(\omega)\}_p = \begin{Bmatrix} \alpha(\omega)_{p1} \\ \alpha(\omega)_{p2} \\ \dots \\ \alpha(\omega)_{pn} \end{Bmatrix} = [A_p]\left\{\frac{1}{\omega_r^2 - \omega^2}\right\} \tag{5.48}$$

Here,

$$[A_p] = \begin{bmatrix} _1A_{p1} & _2A_{p1} & \cdots & _nA_{p1} \\ _1A_{p2} & _2A_{p2} & \cdots & _nA_{p2} \\ \cdots & \cdots & \cdots & \cdots \\ _1A_{pn} & _2A_{pn} & \cdots & _nA_{pn} \end{bmatrix} \tag{5.49}$$

Matrix $[A_p]$ is called modal constant matrix for the pth column of receptance matrix $[\alpha(\omega)]$. Each column of $[\alpha(\omega)]$ has its own modal constant matrix. The mass-normalized mode shapes of a system can be derived from one of its modal constant matrices.

5.4.2 Other forms of an FRF

A receptance FRF can be presented in forms other than that in equations (5.45) and (5.46). One useful form is a ratio of two polynomials:

$$\alpha_{jk}(\omega) = \frac{c_0 + c_1\omega^2 + c_2\omega^4 + \dots + c_{n-1}\omega^{2(n-1)}}{(\omega_1^2 - \omega^2)(\omega_2^2 - \omega^2)\dots(\omega_n^2 - \omega^2)} \tag{5.50}$$

Here, coefficients c_r $(r = 0, 1, 2, \dots, n - 1)$ are real constants or zero. If the receptance has 'm' frequencies which makes the numerator equal to zero, then $\alpha_{jk}(\omega)$ can be factorized to become:

$$\alpha_{jk}(\omega) = \frac{C(\Omega_1^2 - \omega^2)(\Omega_2^2 - \omega^2)\dots(\Omega_m^2 - \omega^2)}{(\omega_1^2 - \omega^2)(\omega_2^2 - \omega^2)\dots(\omega_n^2 - \omega^2)} \tag{5.51}$$

Here, Ω_r $(r = 1, 2, \dots, m)$ are the anti-resonance frequencies. The reciprocal of constant C will be the effective stiffness between coordinates j and k when the system is 'grounded', or it will be the effective mass between j and k when the system is in a 'free–free' condition.

5.5 Asymptote properties of FRFs of an undamped MDoF system

The asymptote (also known as skeleton) properties discussed in Chapter 3 for the FRF of an SDoF system can be extended to the MDoF system. The asymptotes of an

MDoF system provide insights into the system and are useful in developing an understanding of the system's dynamic behaviour. However, explanation of mass and stiffness lines now becomes inconspicuous.

It is convenient to use a 2DoF system to discuss the asymptote properties of MDoF systems, since the outcome is generally extendable. Figure 5.10 shows a 2DoF mass–stiffness system and its $\alpha(\omega)_{11}$ receptance FRF in log–log scale.

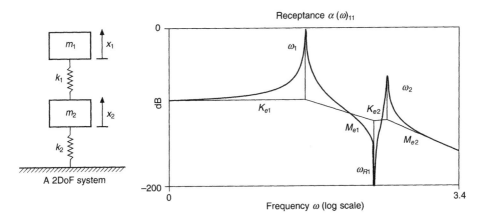

Figure 5.10 A 2DoF system and its $\alpha_{11}(\omega)$ receptance FRF

From the results derived in Chapter 3 for an SDoF system, we know that the following equivalent mass and stiffness quantities can be identified:

$$K_{e1} = \frac{k_1 k_2}{k_1 + k_2} \tag{5.52}$$

$$M_{e1} = \frac{K_{e1}}{\omega_{R1}^2} \tag{5.53}$$

$$K_{e2} = \frac{M_{e1}}{\omega_1^2} \tag{5.54}$$

$$M_{e2} = \frac{K_{e2}}{\omega_2^2} = m_2 \tag{5.55}$$

In addition, the receptance $\alpha(\omega)_{11}$ of the system can be expressed as:

$$\alpha(\omega)_{11} = C \frac{(\Omega_1^2 - \omega^2)}{(\omega_1^2 - \omega^2)(\omega_2^2 - \omega^2)} \tag{5.56}$$

The static stiffness of the receptance can be derived as:

$$K_{static} = K_{e1} = \frac{1}{\alpha(\omega)_{11}}\bigg|_{\omega=0} = \frac{\omega_1^2 \omega_2^2}{C\Omega_1^2} \tag{5.57}$$

Thus, the four asymptote lines in Figure 5.10 can also be represented by the following linear functions of log ω.

Line 1 $y = 20 \log\left(\dfrac{C\Omega_1^2}{\omega_1^2\omega_2^2}\right)$ $(0 < \omega < \omega_1)$ $\qquad$ (5.58)

Line 2 $y = 20 \log\left(\dfrac{C\Omega_1^2}{\omega_2^2\omega^2}\right)$ $(\omega_1 < \omega < \Omega_1)$ $\qquad$ (5.59)

Line 3 $y = 20 \log\left(\dfrac{C}{\omega_2^2}\right)$ $(\Omega_1 < \omega < \omega_2)$ $\qquad$ (5.60)

Line 4 $y = 20 \log\left(\dfrac{C}{\omega^2}\right)$ $(\omega_2 < \omega < \infty)$ $\qquad$ (5.61)

These four asymptote lines are connected, as shown in Figure 5.10. It is also noted that equations (5.58) to (5.61) are effectively equivalent to equations (5.52) to (5.55).

The conclusion of the asymptotes for a 2DoF system can be extended to an MDoF system. For example, a shaft–disc propulsion system is modelled by an NDoF system (Figure 5.11) to study their dynamics.

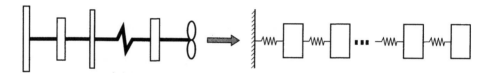

Figure 5.11 A shaft-disc propulsion system and its model

Take the point receptance FRF at the far right end of the model. The function can be written as:

$$\alpha_{NN}(\omega) = \frac{m_1 m_2 \ldots m_{N-1}(\Omega_1^2 - \omega^2)(\Omega_2^2 - \omega^2)\ldots(\Omega_{N-1}^2 - \omega^2)}{m_1 m_2 \ldots m_N(\omega_1^2 - \omega^2)(\omega_2^2 - \omega^2)\ldots(\omega_N^2 - \omega^2)} \qquad (5.62)$$

or $\qquad$ $$\alpha_{NN}(\omega) = \frac{1}{m_N}\frac{\Omega_1^2 \ldots \Omega_{n-1}^2}{\omega_1^2 \ldots \omega_N^2}\frac{\left(1 - \dfrac{\omega^2}{\Omega_1^2}\right)\ldots\left(1 - \dfrac{\omega^2}{\Omega_{N-1}^2}\right)}{\left(1 - \dfrac{\omega^2}{\omega_1^2}\right)\ldots\left(1 - \dfrac{\omega^2}{\omega_N^2}\right)} \qquad (5.63)$$

It is obvious that the DC component of the FRF is:

$$\alpha_{NN}(0) = \frac{1}{k_{NN}} = \frac{1}{m_N}\frac{\Omega_1^2 \ldots \Omega_{N-1}^2}{\omega_1^2 \ldots \omega_N^2} \qquad (5.64)$$

$\alpha_{NN}(0)$ is the point flexibility of the system. It indicates the local deflection due to a static unity load. Therefore, the first stiffness line for the FRF is located at:

$$y = -20 \log k_{NN} \qquad (5.65)$$

Between the first resonance frequency ω_1 and the first anti-resonance frequency Ω_1, lies the first mass line. From equation (5.65) we know this mass line is located at:

$$y = 20 \log \frac{1}{k_{NN}} \frac{1}{\dfrac{\omega^2}{\omega_1^2}} = -20 \log \frac{k_{NN}}{\omega_1^2} - 20 \log \omega^2 \tag{5.66}$$

The first equivalent mass is given in the equation as:

$$m_{e1} = \frac{k_{NN}}{\omega_1^2} \tag{5.67}$$

Between the first anti-resonance frequency Ω_1 and the second resonance frequency ω_2, lies the second stiffness line for the FRF. From equation (5.65) we also know this stiffness line lies in:

$$y = 20 \log \frac{1}{k_{NN}} \frac{1}{\dfrac{\Omega_1^2}{\omega_1^2}} = -20 \log \left(k_{NN} \frac{\Omega_1^2}{\omega_1^2} \right) \tag{5.68}$$

It is easy to verify that the first mass line intercepts the second stiffness line at frequency Ω_1.

The second equivalent mass is given in the equation as:

$$k_{e2} = k_{NN} \frac{\Omega_1^2}{\omega_1^2} \tag{5.69}$$

From here we can derive generally the ith equivalent stiffness and mass of the point FRF as:

$$k_{ei} = k_{NN} \frac{\Omega_1^2}{\omega_1^2} \cdots \frac{\Omega_{i-1}^2}{\omega_{i-1}^2} \tag{5.70}$$

and

$$m_{ei} = k_{NN} \frac{1}{\omega_i^2} \frac{\Omega_1^2}{\omega_1^2} \cdots \frac{\Omega_{i-1}^2}{\omega_{i-1}^2} \tag{5.71}$$

As the FRF passes its last resonance, the mass line lies in:

$$y = 20 \log \frac{1}{m_N} \frac{1}{\omega^2} = -20 \log m_N - 20 \log \omega^2 \tag{5.72}$$

The equivalent mass becomes:

$$m_{eN} = m_N \tag{5.73}$$

The asymptote properties of an MDoF system help in understanding the dynamic behaviour of the system from its FRFs. In Chapter 8, the asymptote properties will be used in modal analysis.

5.6 Other forms of orthogonality properties of an undamped MDoF system

The orthogonality properties of an MDoF system presented in Section 5.3 are of a most familiar form. There are also other forms of orthogonality properties which are less well known but nonetheless find certain applications in modal analysis.

5.6.1 Orthogonality between spatial model and response model

The dynamic stiffness and receptance matrix of an MDoF system are a matrix inverse pair. This suggests that column vectors of matrix $[Z(\omega)]$ and matrix $[\alpha(\omega)]$ are orthogonal (in fact, orthonormal) to each other. For the first column of both matrices, for instance, we have:

$$\{Z(\omega)\}_1^T \{\alpha(\omega)\}_1 = \sum_{i=1}^{n} z_{1i}(\omega)\alpha_{i1}(\omega) = 1 \tag{5.74}$$

This equation reflects the orthogonality property of the system in the frequency domain. Further expansion will show that it is equivalent to an orthogonality of the system in the modal domain. Using equation (5.45), (5.74) can be expanded as:

$$\{Z(\omega)\}_1^T \{\alpha(\omega)\}_1$$

$$= z_{11} \sum_{r=1}^{n} \frac{\phi_{1r}\phi_{1r}}{\omega_r^2 - \omega^2} + z_{12} \sum_{r=1}^{n} \frac{\phi_{2r}\phi_{1r}}{\omega_r^2 - \omega^2} + \ldots + z_{1n} \sum_{r=1}^{n} \frac{\phi_{nr}\phi_{1r}}{\omega_r^2 - \omega^2} \tag{5.75}$$

$$= \sum_{k=1}^{n} z_{1k} \frac{\phi_{k1}\phi_{11}}{\omega_1^2 - \omega^2} + \sum_{k=1}^{n} z_{1k} \frac{\phi_{k2}\phi_{12}}{\omega_2^2 - \omega^2} + \ldots + \sum_{k=1}^{n} z_{1k} \frac{\phi_{kn}\phi_{1n}}{\omega_n^2 - \omega^2}$$

$$= \frac{\phi_{11}}{\omega_1^2 - \omega^2} \sum_{i=1}^{n} z_{1i}\phi_{i1} + \frac{\phi_{12}}{\omega_2^2 - \omega^2} \sum_{i=1}^{n} z_{1i}\phi_{i2} + \ldots + \frac{\phi_{1n}}{\omega_n^2 - \omega^2} \sum_{i=1}^{n} z_{1i}\phi_{in} \tag{5.76}$$

$$= \sum_{r=1}^{n} \left(\frac{\phi_{1r}}{\omega_r^2 - \omega^2} \sum_{i=1}^{n} z_{1i}\phi_{ir} \right) = 1 \tag{5.77}$$

Using the first column of the receptance matrix for the whole dynamic stiffness matrix, we have the following:

$$[Z(\omega)]\{\alpha(\omega)\}_1 = [\{Z(\omega)\}_1, \{Z(\omega)\}_2, \ldots, \{Z(\omega)\}_n]^T \{\alpha(\omega)\}$$

$$= \{\delta_{11}, \delta_{21}, \delta_{31}, \ldots, \delta_{n1}\}^T \tag{5.78}$$

$$= \left(\frac{\phi_{11}}{\omega_1^2 - \omega^2} \right)[Z(\omega)]\{\phi\}_1 + \left(\frac{\phi_{12}}{\omega_2^2 - \omega^2} \right)[Z(\omega)]\{\phi\}_2$$

$$+ \ldots + \left(\frac{\phi_{1n}}{\omega_n^2 - \omega^2} \right)[Z(\omega)]\{\phi\}_n \tag{5.79}$$

Here, δ_{ij} is a Kronecker factor whose value is zero but becomes one if i and j are equal.

Since $[Z(\omega)]\{\phi\}_r = ([K] - \omega^2[M])\{\phi\}_r$ (5.80)

$$= ([K] - \omega_r^2[M])\{\phi\}_r + (\omega_r^2 - \omega^2)[M]\{\phi\}_r \tag{5.81}$$

$$= (\omega_r^2 - \omega^2)[M]\{\phi\}_r \tag{5.82}$$

$$[Z(\omega)]\{\alpha(\omega)\}_1 = \sum_{k=1}^{n} \phi_{1k}[M]\{\phi\}_k = \{\delta\}_1 \tag{5.83}$$

Here, $\{\delta\}_1$ is a Kronecker vector whose elements are zeros except the first element is equal to one. Denote $\{M_r\}$ as the rth row of matrix $[M]$. For the whole receptance matrix, equation (5.83) can be expanded as:

$$[Z(\omega)][\alpha(\omega)]$$

$$= \begin{bmatrix} \sum\limits_{k=1}^{n} \phi_{1k}\{M_1\}^T\{\phi\}_k & \sum\limits_{k=1}^{n} \phi_{2k}\{M_1\}^T\{\phi\}_k & \cdots & \sum\limits_{k=1}^{n} \phi_{nk}\{M_1\}^T\{\phi\}_k \\ \sum\limits_{k=1}^{n} \phi_{1k}\{M_2\}^T\{\phi\}_k & \sum\limits_{k=1}^{n} \phi_{2k}\{M_2\}^T\{\phi\}_k & \cdots & \sum\limits_{k=1}^{n} \phi_{nk}\{M_2\}^T\{\phi\}_k \\ \cdots & \cdots & \cdots & \cdots \\ \sum\limits_{k=1}^{n} \phi_{1k}\{M_n\}^T\{\phi\}_k & \sum\limits_{k=1}^{n} \phi_{2k}\{M_n\}^T\{\phi\}_k & \cdots & \sum\limits_{k=1}^{n} \phi_{nk}\{M_n\}^T\{\phi\}_k \end{bmatrix}$$

$$= \lfloor \cdot I. \rfloor \tag{5.84}$$

or: $$\sum_{k=1}^{n} \phi_{rk}\{M_s\}^T\{\phi\}_k = \delta_{rs} \tag{5.85}$$

Therefore, the frequency domain orthogonality properties between the FRF and dynamic stiffness matrices are equivalent to a new form of modal domain orthogonality properties which involve all the mode shapes at the one time. This is unlike the modal domain orthogonality properties given by equation (5.40) which can be rewritten as:

$$\sum_{k=1}^{n} \phi_{rk}\{M_k\}^T\{\phi\}_s = \delta_{rs} \tag{5.86}$$

Comparison of equations (5.85) and (5.86) reveals the two different modal domain orthogonality properties.

5.6.2 Orthogonality between measured modes and their reciprocal modal vectors

The orthogonality of an MDoF system between the reciprocal modal vector and its mode shape is used primarily to check the accuracy of measured vibration modes in the absence of a reliable analytical mass matrix. A reciprocal modal vector is defined as:

$$\{\chi\}_r = [M]\{\phi\}_r \quad (r = 1, 2, \ldots, n) \tag{5.87}$$

It is apparent from equation (5.40) that orthogonality exists between the reciprocal modal vector $\{\chi\}$ and the measured mode shapes $\{\phi\}$ such that:

$$\{\chi\}_p^T \{\phi\}_q = \{\phi\}_p^T \{\chi\}_q = \delta_{pq} \tag{5.88}$$

or: $$[\chi]^T [\Phi] = [\Phi]^T [\chi] = [\cdot I \cdot] \tag{5.89}$$

In practice, the reciprocal modal vector $\{\chi\}$ cannot be calculated using equation (5.87) since matrix $[M]$ is not available. Therefore, measured receptance FRF data are used to derive vector $\{\chi\}$. Since the pth column of the experimental receptance matrix can be expressed as:

$$\{\alpha(\omega)\}_p = \left\{ \sum_{r=1}^{n} \frac{\phi_{pr}\phi_{ir}}{\omega_r^2 - \omega^2} \cdots \sum_{r=1}^{n} \frac{\phi_{pr}\phi_{nr}}{\omega_r^2 - \omega^2} \right\}^T \tag{5.90}$$

$$= \sum_{r=1}^{n} \frac{\phi_{pr}}{\omega_r^2 - \omega^2} \{\phi\}_r \tag{5.91}$$

it can be seen that

$$\{\chi\}_q^T \{\alpha(\omega)\}_p = \frac{\phi_{pq}}{\omega_p^2 - \omega^2} \tag{5.92}$$

For a given frequency range, the reciprocal modal vector $\{\chi\}$ can be determined from equation (5.92) by taking a number of different frequency points and formulating a set of linear equations for the elements in vector $\{\chi\}$. For points $\bar{\omega}_1, \bar{\omega}_2, \ldots, \bar{\omega}_n$ within a frequency range of interest, equation (5.92) can be expanded as:

$$[\{\alpha(\omega_1)\}_p, \ldots, \{\alpha(\omega_n)\}_p]^T \{\chi\}_q = \left\{ \begin{array}{c} \dfrac{\phi_{pq}}{\omega_q^2 - \bar{\omega}_1^2} \\ \cdots \\ \cdots \\ \dfrac{\phi_{pq}}{\omega_q^2 - \bar{\omega}_n^2} \end{array} \right\} \tag{5.93}$$

The thus determined vector $\{\chi\}$ can then be used for the orthogonality of measured modes $\{\phi\}$ from outside the frequency range using equation (5.88).

5.6.3 Orthogonality between submatrices

A submatrix of an MDoF system is a matrix that describes the quantity and connectivity of a physical component, such as a spring or a mass, in the system matrix. For instance, for a simple 'n' DoF mass–spring system, submatrix $[k]_r$ exists for the spring connecting coordinates 'i' and 'j' with stiffness $k_{(r)}$:

$$[k]_r^{(ij)} = k_{(r)}(\{\delta\}_i + \{\delta\}_j)(\{\delta\}_i + \{\delta\}_j)^T = k_{(r)}[\Gamma]_r^{(ij)} \tag{5.94}$$

Here, vector $\{\delta_i\}$ is an $n \times 1$ Kronecker vector. Matrix $[\Gamma]_r^{(ij)}$ is the participation matrix for the rth stiffness component. Most of elements in matrix $[\Gamma]_r^{(ij)}$ are zeros. The stiffness matrix of the system becomes a straight summation of all the stiffness submatrices, namely:

$$[K] = \sum_r [k]_r = \sum_r k_{(r)}[\Gamma]_r \tag{5.95}$$

The same can be applied on the system mass matrix by defining the submatrices for the masses of the system. If two spring components are not physically connected so that no energy path exists between them, then

$$[k]_r^{(ij)} [k]_s^{(pq)} = k_{(r)}(\{\delta\}_i + \{\delta\}_j)(\{\delta\}_i + \{\delta\}_j)^T k_{(s)}(\{\delta\}_p$$

$$+ \{\delta\}_q)(\{\delta\}_p + \{\delta\}_q)^T \tag{5.96}$$

$$= [0] \quad (\text{for } i \neq j \neq p \neq q) \tag{5.97}$$

This is because:

$$(\{\delta\}_i + \{\delta\}_j)^T(\{\delta\}_p + \{\delta\}_q) = 0 \tag{5.98}$$

On the other hand, if spring components r and s are connected both to coordinate 'p', then

$$[k]_r^{(ip)} [k]_s^{(pq)} = k_{(r)}(\{\delta\}_i + \{\delta\}_p)(\{\delta\}_i + \{\delta\}_p)^T k_{(s)}(\{\delta\}_p + \{\delta\}_q)(\{\delta\}_p + \{\delta\}_q)^T$$

$$\tag{5.99}$$

$$= k_{(r)}k_{(s)}([\sigma_{ii}] + [\sigma_{ip}] + [\sigma_{pi}] + [\sigma_{pp}])([\sigma_{pp}] + [\sigma_{pq}] + [\sigma_{qp}] + [\sigma_{qq}]) \tag{5.100}$$

$$= k_{(r)}k_{(s)}[\sigma_{ip} + \sigma_{iq} + \sigma_{pp} + \sigma_{pq}] \tag{5.101}$$

where $[\sigma_{ip}]$ is a matrix which has zero elements everywhere except element ip is one.

Using the submatrix approach, the equation of motion of an MDoF system can be written as:

$$\left(\sum_r [m]_r\right)\{\ddot{x}\} + \left(\sum_r [k]_r\right)\{x\} = \{0\} \tag{5.102}$$

Pre-multiplying both sides by the participation matrix of the rth stiffness component yields:

$$[\Gamma]_r^{(ij)} \sum_r [m]_r \{\ddot{x}\} + [\Gamma]_r^{(ij)} \sum_r [k]_r \{x\} = \{0\} \tag{5.103}$$

Equation (5.103) contains a subset of equations given in the general equation of motion of the system. This subset of equations is concerned only with coordinates i and j – the coordinate's stiffness component k_r is connected to. Therefore, only masses and spring components connected to coordinates i and j will appear in equation (5.103).

Likewise, stiffness submatrices can be used for the case of frequency domain orthogonality. In this case, equation (5.74) can be rewritten as:

$$\left(\sum_r [k]_r - \omega^2 [M]\right)\{\alpha(\omega)\}_k = \{\delta\}_k \tag{5.104}$$

Pre-multiplying both sides by the participation matrix of the rth stiffness component yields:

$$[\Gamma]_r^{(ij)}\left(\sum_r [k]_r - \omega^2 [M]\right)\{\alpha(\omega)\}_k = \{0\} \quad (i \neq j \neq k) \tag{5.105}$$

This equation involves only the mass and spring components connected to coordinates i and j, and also only the ith and jth receptance FRFs in vector $\{\alpha(\omega)\}_k$. In the event of using modal and FRF data to update an analytical model or to locate structural damage, a set of equations can be derived from equation (5.105) to determine the constants $k_{(r)}$ of those submatrices which contribute to the stiffness changes.

5.6.4 Orthogonality with incomplete data

A dynamic stiffness vector $\{Z(\omega)\}$ or receptance vector $\{\alpha(\omega)\}$ is incomplete if certain elements of the vector are missing. The orthogonality among incomplete data of an MDoF system is of interest because the dynamic stiffness matrix is usually sparse or banded. The orthogonality presented in equation (5.74) appears to involve all the elements in the receptance and the dynamic stiffness vectors. However, because of the sparseness of the dynamic stiffness matrix $[Z(\omega)]$, normally fewer than all the elements in the receptance vectors are involved. Therefore, equation (5.74) can be reduced to:

$$\{Z(\omega)\}_i^T \{\alpha_j(\omega)\} = \sum_{r=e1}^{e2} z_{ir}(\omega)\alpha_{rj}(\omega) = \delta_{ij} \tag{5.106}$$

where only elements between and including $e1$ and $e2$ of vector $\{Z(\omega)\}_i$ are non-zero. Here, it is assumed for the sake of simplicity that non-zero elements in vector $\{Z(\omega)\}_i$ are all between elements $e1$ and $e2$ consecutively. As a result, receptance FRFs falling outside this range become irrelevant to this orthogonality equality.

5.7 Harmonic response of an undamped MDoF system using FRFs

Having defined and discussed the receptance FRF of an MDoF system, it is convenient to derive its forced response when the system is subjected to a series of harmonic excitations. Assume the harmonic excitation forces have the same frequency and zero phase:

$$\{f(t)\} = \{F\} \sin \omega t \tag{5.107}$$

then the responses of a system will also be harmonic such that:

$$\{x(t)\} = \{X\} \sin \omega t \qquad (5.108)$$

From the discussion leading to equation (5.28), we know that vector $\{X\}$ can be derived using the receptance FRFs or the modal data of the system as:

$$\{X\} = [\alpha(\omega)]\{F\} \qquad (5.109)$$

$$= [\Phi][\cdot\omega_r^2 - \omega^2\cdot]^{-1}[\Phi]^T\{F\} \qquad (5.110)$$

Therefore, the responses of the system subjected to harmonic excitation forces are:

$$\{x(t)\} = [\Phi][\cdot\omega_r^2 - \omega^2\cdot]^{-1}[\Phi]^T\{F\}\sin \omega t \qquad (5.111)$$

Each response can be written as:

$$x_i(t) = \left(\sum_{r=1}^{n} \frac{\phi_{ir}}{\omega_r^2 - \omega^2} \sum_{k=1}^{n} \phi_{kr} F_k \right) \sin \omega t \qquad (5.112)$$

Equation (5.112) shows that at any given moment, the response $x_i(t)$ enlists the contribution of all the modes of the system. In fact, the participation factor of the rth mode is $\dfrac{\phi_{ir}}{\omega_r^2 - \omega^2}$. The equation also shows that different harmonic forces are effectively 'weighted' by a mode shape element. Namely, the contribution of the force $F_k \sin \omega t$ for mode 'r' is effectively weighted by the element ϕ_{kr} in the mode.

Example 5.6
To determine the displacement responses of the 2DoF system if the driving forces are harmonic and in phase (Figure 5.12),

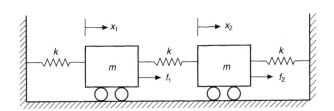

Figure 5.12 A 2DoF system excited by external forces

the equation of motion of the system is derived as:

$$\begin{bmatrix} m & 0 \\ 0 & m \end{bmatrix}\begin{Bmatrix} \ddot{x}_1 \\ \ddot{x}_2 \end{Bmatrix} + \begin{bmatrix} 3k & -k \\ -k & 3k \end{bmatrix}\begin{Bmatrix} x_1 \\ x_2 \end{Bmatrix} = \begin{Bmatrix} F_1 \\ F_2 \end{Bmatrix}\sin \omega t$$

The dynamic stiffness matrix will be:

$$[Z(\omega)] = \begin{bmatrix} 3k - \omega^2 m & -k \\ -k & 3k - \omega^2 m \end{bmatrix}$$

$$[\alpha(\omega)] = \begin{bmatrix} 3k - \omega^2 m & -k \\ -k & 3k - \omega^2 m \end{bmatrix}^{-1}$$

$$= \frac{1}{|Z(\omega)|} \begin{bmatrix} 3k - \omega^2 m & k \\ k & 3k - \omega^2 m \end{bmatrix} (|Z(\omega)| = |(3k - \omega^2 m)^2 - k^2|)$$

$$\begin{bmatrix} m & 0 \\ 0 & m \end{bmatrix} \begin{Bmatrix} \ddot{x}_1 \\ \ddot{x}_2 \end{Bmatrix} + \begin{bmatrix} 3k & -k \\ -k & 3k \end{bmatrix} \begin{Bmatrix} x_1 \\ x_2 \end{Bmatrix} = \begin{Bmatrix} F_1 \\ F_2 \end{Bmatrix} \sin \omega t$$

Or: $\quad x_1(t) = \dfrac{1}{|Z(\omega)|} \{(3k - \omega^2 m) F_1 + k F_2\} \sin \omega t$

and $\quad x_2(t) = \dfrac{1}{|Z(\omega)|} \{k F_1 + (3k - \omega^2 m) F_2\} \sin \omega t$

5.8 Anti-resonances and minima of an FRF

The anti-resonance of an MDoF FRF receive less attention than the resonance because they are not associated with severe vibration. It is also because many modal analysis algorithms identify modal parameters only from the FRF data near resonances. But anti-resonance and minimum are sometimes as important as the resonance.

Anti-resonance is another manifestation of the dynamic characteristics of a system. While resonances of a system are 'global' parameters (i.e. the same resonances are expected to appear on any FRF data of the system), anti-resonances are 'local' parameters. Different FRFs have different anti-resonances (or no anti-resonances at all). Therefore, the discussion of anti-resonances has to be based upon a given FRF.

5.8.1 Anti-resonances and spatial data

The anti-resonance of a system can be derived analytically if the spatial model of a system is available. The receptance FRF between coordinates 'i' and 'j' of the system is derived from equations (5.31) and (5.34) as:

$$\alpha(\omega)_{ij} = \left. \frac{X_i}{F_j} \right|_{F_r = 0 \ (r=1,2,\ldots,n, \, r \neq j)} \tag{5.113}$$

or $\qquad \alpha(\omega)_{ij} = (-1)^{i+j} \dfrac{\det([K]_{ij} - \omega^2 [M]_{ij})}{\det([K] - \omega^2 [M])} \tag{5.114}$

where $[K]_{ij}$ is obtained from $[K]$ after its ith row and jth column are deleted. Equation (5.114) is derived from matrix inversion as the FRF matrix is the inverse of the system dynamic stiffness matrix. $[M]_{ij}$ and $[K]_{ij}$ are symmetric only if $i = j$. When an anti-resonance of $\alpha(\omega)_{ij}$ occurs, $\alpha(\omega)_{ij} = 0$. Therefore, the anti-resonance of $\alpha(\omega)_{ij}$ can be determined by the positive roots of ω in the following equation:

$$\det([K]_{ij} - \omega^2 [M]_{ij}) = 0 \tag{5.115}$$

Equation (5.115) is equivalent to the following eigenvalue problem where the positive eigenvalues are the squares of the anti-resonances for $\alpha(\omega)_{ij}$:

$$([K]_{ij} - \Omega_a^2 [M]_{ij})\{\mu\} = \{0\} \qquad (5.116)$$

The eigenvector in this equation $\{\mu\}$ does not offer apparent physical interpretation. Since matrices $[M]_{ij}$ and $[K]_{ij}$ are normally non-symmetrical, eigenvalues of equation (5.116) may become negative or even complex. However, only those positive eigenvalues represent anti-resonances.

5.8.2 Anti-resonances and modal data

Although the anti-resonances of a system can be estimated from its spatial data, the understanding of their formation will be best achieved from the viewpoint of modal data. For the sake of simplicity, FRF $\alpha_{11}(\omega)$ of the 2DoF system shown in Figure 5.13 is used as an example.

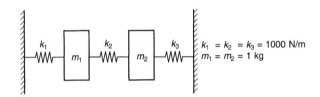

Figure 5.13 A 2DoF mass–spring system

The natural frequency and mode shape matrices of the system are respectively:

$$[\mathop{}_{\cdot}\omega_r^2.] = \begin{bmatrix} 1000 & 0 \\ 0 & 3000 \end{bmatrix} (\text{rad/s})^2 \quad [\phi] = \begin{bmatrix} 0.707 & 0.707 \\ -0.707 & 0.707 \end{bmatrix}$$

Using equation (5.45), the receptance $\alpha_{11}(\omega)$ can then be expressed as a summation of two ratios as:

$$\alpha_{11}(\omega) = \frac{\phi_{11}\phi_{11}}{\omega_1^2 - \omega^2} + \frac{\phi_{12}\phi_{12}}{\omega_2^2 - \omega^2} = \frac{0.5}{1000 - \omega^2} + \frac{0.5}{3000 - \omega^2} \qquad (5.117)$$

The trough between the two resonances ω_1 and ω_2 is either an anti-resonance or a minimum. Since both modal constants, the numerators of two ratios, in equation (5.117) are positive, it is clear that between ω_1 and ω_2, the first ratio is negative while the second positive. Therefore, there is a frequency in the range at which receptance $\alpha_{11}(\omega)$ becomes zero. When the function is plotted on a dB graph, this zero receptance will signify the anti-resonance. This is illustrated in Figure 5.14.

The positive and negative signs in Figure 5.14 are only explanatory. They do not actually apply to the log plot of an FRF. Nevertheless, an anti-resonance is displayed only on a log plot.

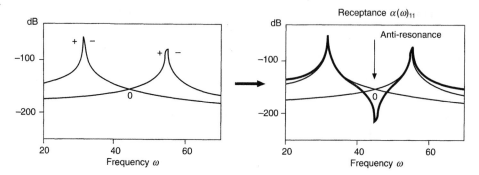

Figure 5.14 Anti-resonance of an FRF of a 2DoF mass–spring system

Likewise, receptance $\alpha_{21}(\omega)$ of the 2DoF system can be studied. The FRF is expressed as:

$$\alpha_{21}(\omega) = \frac{0.5}{1000 - \omega^2} + \frac{-0.5}{3000 - \omega^2} \tag{5.118}$$

Since the numerator of the second ratio is negative, the signs of both ratios before and after their own resonances can be marked in Figure 5.15. It is evident that there cannot be a zero between two resonances. Indeed, there exists a minimum rather than an anti-resonance, as illustrated by Figure 5.15.

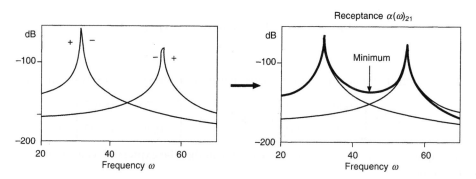

Figure 5.15 Minimum of an FRF of a 2DoF mass–spring system

The study of this 2DoF system reveals that there exists an intrinsic relationship between the anti-resonances of a structural system and its mode shapes. The formation of anti-resonances/minima in an FRF is generally dictated by the signs of modal constants which are determined by the signs of the elements in the mode shape matrix. This conclusion can be extended to a general MDoF system. Thus, generally for two consecutive resonances, the same signs of modal constants indicate the existence of an anti-resonance and different signs a minimum. Care must be taken when applying this conclusion to any FRFs with close resonances since values of FRF in between may no longer be dominated by the contribution of these resonances.

As a derivation of this conclusion, it can be seen that any point FRF ($\alpha_{ii}(\omega)$) will possess an anti-resonance between two consecutive resonances. This is because for such a function, all the numerators in its expression given in equation (5.45) will be positive. For a transfer FRF, anti-resonance depends on the mode shapes. When the mode shape matrix of a system is given, one should be able to sketch any FRF of the system.

5.8.3 Minima of an FRF

The minima of an FRF are the lowest points between two consecutive resonances but they are not anti-resonances. They are not as significant as resonances and anti-resonances. The existence and locations of the minima merit mainly theoretical interests in the study of modal analysis. Similar to anti-resonances, minima are also 'non-global' properties of a system. Therefore, different FRFs have minima of different frequencies.

Unlike resonances and anti-resonances, the exact frequency of a minimum cannot be worked out easily from the spatial data of a system. Its approximate location, however, can be determined most conveniently by using the modal data. By taking the derivative of receptance in equation (5.43) with respect to frequency ω, the minimum frequency can be derived.

$$\frac{\partial \alpha_{jk}(\omega)}{\partial \omega} = \frac{\partial}{\partial \omega}\left(\frac{\phi_{j1}\phi_{k1}}{\omega_1^2 - \omega^2} + \frac{\phi_{j2}\phi_{k2}}{\omega_2^2 - \omega^2} + \ldots + \frac{\phi_{jn}\phi_{kn}}{\omega_n^2 - \omega^2}\right) \tag{5.119}$$

$$= (-2\omega)\left(\frac{\phi_{j1}\phi_{k1}}{(\omega_1^2 - \omega^2)^2} + \frac{\phi_{j2}\phi_{k2}}{(\omega_2^2 - \omega^2)^2} + \ldots + \frac{\phi_{jn}\phi_{kn}}{(\omega_n^2 - \omega^2)^2}\right) \tag{5.120}$$

When the derivative in equation (5.120) becomes zero, the roots will be those frequencies corresponding to the minima of the receptance FRF. Obviously, the task of finding these roots is unrealistic. An alternative is to find the approximate minimum between two consecutive modes by only taking these two modes into account in the calculation. Therefore, the minimum between mode 1 and mode 2 can be determined approximately from:

$$\frac{\phi_{j1}\phi_{k1}}{(\omega_1^2 - \omega^2)^2} + \frac{\phi_{j2}\phi_{k2}}{(\omega_2^2 - \omega^2)^2} = 0 \tag{5.121}$$

or:

$$\omega^4(\phi_{j1}\phi_{k1} + \phi_{j2}\phi_{k2}) - \omega^2(2\phi_{j1}\phi_{k1}\omega_2^2 + 2\phi_{j2}\phi_{k2}\omega_1^2)$$

$$+ (\phi_{j1}\phi_{k1}\omega_2^4 + \phi_{j2}\phi_{k2}\omega_1^4) = 0 \tag{5.122}$$

For such a quadratic equation of ω^2, the condition of having real roots will be:

$$(2\phi_{j1}\phi_{k1}\omega_2^2 + 2\phi_{j2}\phi_{k2}\omega_1^2) - 4(\phi_{j1}\phi_{k1} + \phi_{j2}\phi_{k2})(\phi_{j1}\phi_{k1}\omega_2^4 + \phi_{j2}\phi_{k2}\omega_1^4) > 0 \tag{5.123}$$

or:

$$-4\phi_{j1}\phi_{k1}\phi_{j2}\phi_{k2}(\omega_2^2 - \omega_1^2)^2 > 0 \tag{5.124}$$

Therefore, if the product $\phi_{j1}\phi_{k1}\phi_{j2}\phi_{k2}$ is negative, there will be a minimum between mode 1 and mode 2. If the product is positive, however, it has been shown before that there will be an anti-resonance between the two modes.

Literature

1. Brock, J.E. 1968: Optimal Matrices Describing Linear Systems. *AIAA Journal*, **6**(7), 1292–1296.
2. Ewins, D.J. 1984: *Modal Testing – Theory and Practice*. Research Studies Press: England.
3. Gladwell, G.M.L. 1986: The Inverse Mode Problem for Lumped-mass Systems. *Quarterly Journal of Mechanics and Applied Mathematics*, **39**, 297–307.
4. He, J. and Imregun, M. 1995: Different Forms of Orthogonality for Multi-Degree-of-Freedom Vibrating Systems. *The International Journal of Analytical and Experimental Modal Analysis*, **10**(3), 131–141.
5. Kidder, R.L. 1980: Reduction of Structural Frequency Equations. *AIAA Journal*, **11**(6), 892.
6. Maia, N.M.M. *et al.* 1997: *Theoretical and Experimental Modal Analysis*. Research Studies Press, UK and John Wiley & Sons, USA.
7. Mottershead, J.E. 1999: On the natural frequencies and anti-resonances of modified structures. *Proceedings of the 17th International Modal Analysis Conference*, Kissimmee, FL, USA, 648.
8. Newland, D.E. 1989: *Mechanical Vibration Analysis and Computation*. Longman Scientific & Technical, UK.
9. Richardson, M.H. 1997: Is it a mode shape, or an operating deflection shape? *Sound and Vibration*, **31**, 54–61.
10. Salter, J. 1969: *Steady-State Vibration*. Kenneth Mason Press.
11. Zhang, Q. Shih, C.Y. and Allemang, R.J. 1989: Orthogonality criterion for experimental modal analysis. *Vibration Analysis – Techniques and Application*, DE-Vol. 18-4, 251–258.
12. Zaveri, K. 1984: *Modal Analysis of Large Structures – Multiple Exciter Systems*. Bruel & Kjaer Publications.

6

Modal analysis of a damped MDoF system

The modal analysis theory for an undamped MDoF system is applicable for dynamic structures when damping is negligible. The presence of damping does not change every aspect of the theory presented in Chapter 5. However, more mathematical treatment is needed in order to extend the modal analysis theory into the case for a damped MDoF system.

The two main damping models used in MDoF modal analysis are the viscous damping model and the structural damping model. These are the same models used for SDoF systems except that they are now applied to an MDoF system. Like mass and stiffness properties, now the distribution of damping is an important property as well as its amount.

The coupled equations of motion for an undamped MDoF system can be uncoupled using the principle of orthogonality. Therefore, analysis of individual modes becomes convenient. However, once damping is present, it is generally difficult or not possible to uncouple the equations of motion. Therefore, damped MDoF systems demand extra theoretical treatment.

In dealing with a damped MDoF dynamic system, it always helps to draw a comparison between its treatment and that of an SDoF damped system.

6.1 Proportional damping models

A proportional damping model is the first analytical model used to study damping for an MDoF system. Unlike mass and stiffness properties, damping cannot usually be modelled. This became a stumbling block to the analysis of a damped MDoF system. The proposition of proportional damping enabled the analysis to proceed.

Proportional damping has found significant applications in finite element analysis where damping needs to be incorporated in order to carry out meaningful response analysis and prediction. In modal analysis theory, the significance of proportional damping will become evident when we realize that a system with proportional damping would have the mode shapes identical to that of its undamped counterpart.

We begin with the analysis of free vibration. If the damping distribution of the system of n DoFs with viscous damping is denoted as a matrix $[C]$, the matrix equation of motion of the system is given by:

$$[M]\{\ddot{x}\} + [C]\{\dot{x}\} + [K]\{x\} = \{0\} \tag{6.1}$$

Here, matrix $[C]$ is positive definite or positive semi-definite. Unlike the undamped case, there generally does not exist a set of principal coordinates, which uncouple equation (6.1). In particular, if we use the mode shape matrix $[\Phi]$, then both the mass matrix $[M]$ and the stiffness matrix $[K]$ can be diagonalized, as discussed in Chapter 5. However, damping matrix $[C]$ cannot be, leaving the equations still coupled. The exceptional damping distribution which does allow the diagonalization of matrix $[C]$ as well as $[K]$ and $[M]$ is called proportional damping.

The proportional damping model made a significant contribution at the early development of modal analysis. Without needing sophisticated mathematical treatment, a structure assumed with proportional damping can be analysed using the theory for an undamped MDoF system. The initial proposal of proportional damping came long before the study of modal analysis. Rayleigh indicated in his work *The Theory of Sound*, first published in 1845, that if the viscous damping matrix $[C]$ is proportional to mass and stiffness matrices (or that if the damping forces are proportional to the kinetic and potential energies of the system), then it can be written as:

$$[C] = \alpha[M] + \beta[K] \tag{6.2}$$

where α and β are real positive constants. Equation (6.1) can then be uncoupled like the matrix equation for an undamped system. The substitution of equation (6.2) into (6.1) leads to:

$$[M]\{\ddot{x}\} + (\alpha[M] + \beta[K])\{\dot{x}\} + [K]\{x\} = \{0\} \tag{6.3}$$

Repeating the uncoupling process for the undamped case using the undamped mode shape matrix $[\Phi]$ (obtained by assuming $[C] = [0]$ from equation (6.1)) will lead to the uncoupled equations:

$$[\cdot m_r \cdot]\{\ddot{x}_p\} + (\alpha[\cdot m_r \cdot] + \beta[\cdot k_r \cdot])\{\dot{x}_p\} + [\cdot k_r \cdot]\{x_p\} = \{0\} \tag{6.4}$$

or

$$[\cdot m_r \cdot]\{\ddot{x}_p\} + [\cdot c_r \cdot]\{\dot{x}_p\} + [\cdot k_r \cdot]\{x_p\} = \{0\} \tag{6.5}$$

where diagonal matrix $[\cdot c_r \cdot]$ is called the *modal damping matrix* or *generalized damping matrix* of the system. Obviously, the undamped mode shape matrix $[\Phi]$ can diagonalize the proportional damping matrix as well as the mass and stiffness matrices. Therefore, $[\Phi]$ (which is a real matrix) is also the mode shape matrix for the system having proportional viscous damping model. For modal analysis, this is the most important characteristic from a proportional damping model.

Equation (6.5) consists of n uncoupled equations. Using the theory of an SDoF system, the damped natural frequencies of the rth mode, $\underline{\omega}_r$, can be estimated by:

$$\underline{\omega}_r = \omega_r \sqrt{1 - \zeta_r^2} \tag{6.6}$$

$$\zeta_r = \frac{\alpha}{2\omega_r} + \frac{\beta\omega_r}{2} \tag{6.7}$$

Like an SDoF system, ζ_r is defined as the damping ratio. The difference is that this time the damping ratio is for the rth mode. Equation (6.7) shows that the damping ratio for a system with proportional viscous damping is different for each mode.

The proportional damping defined in equation (6.2) is not the only damping model that facilitates decoupling of system equations. A more general expression of a viscous damping model is given as:

$$[K][M]^{-1}[C] = [C][M]^{-1}[K] \tag{6.8}$$

This can be derived by assuming that $[\Psi]^T[C][\Psi]$ is a diagonal matrix. Then it follows that:

$$[\Psi]^T[C][\Psi]\,[\cdot m_r \cdot]^{-1}[\cdot k_r \cdot] = [\cdot k_r \cdot][\cdot m_r \cdot]^{-1}[\Psi]^T\,[C][\Psi]$$

Using equations (5.17) and (5.18), this equation can be transformed into equation (6.8).

The structural damping model is also used in the analysis of an MDoF system. The equation of motion of a system of 'n' DoFs with structural damping is given by:

$$[M]\{\ddot{x}\} + [K]\{x\} + j[H]\{x\} = \{0\} \tag{6.9}$$

Here $[H]$ is the structural damping matrix and 'j' the imaginary unit. This representation is similar to that of an SDoF system. If damping matrix $[H]$ is proportional to mass and stiffness, then it can be written as:

$$[H] = \mu[M] + v[K] \tag{6.10}$$

Here, both μ and v are real positive constants. It can be found that like the proportional viscous damping case, the mode shape matrix of the system with proportional structural damping will be identical to that of the undamped system. The natural frequencies of the system are given as:

$$\lambda_r^2 = \omega_r^2(1 + j\eta_r) \tag{6.11}$$

Also,

$$\eta_r = v + \frac{\mu}{\omega_r^2} \tag{6.12}$$

Here, η_r is defined as the damping loss factor of the rth vibration mode. Equation (6.12) shows that the damping loss factor for a system with proportional structural damping is different for each mode. If the damping matrix is only proportional to the stiffness matrix ($\mu = 0$), then the damping loss factor for all the modes will be equal to the constant v.

6.2 Non-proportional viscous damping model

When the viscous damping of an 'n' DoF system is non-proportional, the solution of equation (6.1) is in the form:

$$\{x(t)\} = \{X\}e^{st} \tag{6.13}$$

Here, s is the Laplace operator and $\{X\}$ a complex vector for displacement amplitudes. Then equation (6.1) becomes:

$$(s^2[M] + s[C] + [K])\{X\} = \{0\} \tag{6.14}$$

This is a complex and higher order eigenvalue problem. The solution to this problem relies on the state–space approach introduced in Chapter 1. This approach invents a new displacement vector defined as:

$$\{y\} = \begin{Bmatrix} x \\ \dot{x} \end{Bmatrix}_{2n \times 1} \tag{6.15}$$

With this new vector, equation (6.1) can be transformed into a new matrix equation which is twice as big in size:

$$[[C] : [M]]\{\dot{y}\} + [[K] : [0]]\{y\} = \{0\} \tag{6.16}$$

Together with the following identity:

$$[[M] : [0]]\{\dot{y}\} + [[0] : [-M]]\{y\} = \{0\} \tag{6.17}$$

equation (6.1) is transformed into:

$$\begin{bmatrix} C & M \\ M & 0 \end{bmatrix}_{2n \times 2n} \{\dot{y}\} + \begin{bmatrix} K & 0 \\ 0 & -M \end{bmatrix}_{2n \times 2n} \{y\} = \{0\} \tag{6.18}$$

or

$$[A]\{\dot{y}\} + [B]\{y\} = \{0\} \tag{6.19}$$

Equation (6.19) is a normal eigenvalue problem and its solution consists of $2N$ complex eigenvalues λ_r (in complex conjugate pairs) and $2n$ corresponding complex eigenvectors $\{\theta\}_r$ (also in complex conjugate pairs). They satisfy:

$$(\lambda_r[A] + [B])\{\theta\}_r = \{0\} \quad (r = 1, 2, \dots, 2n) \tag{6.20}$$

The complex eigenvectors $\{\theta\}_r$ can be grouped in the ascending order of their eigenvalues to form the eigenvector matrix $[\theta]$. The orthogonality properties for this expanded eigenvalue problem can be expressed as:

$$[\theta]^T[A][\theta] = [\cdot a_i \cdot] \tag{6.21}$$

$$[\theta]^T[B][\theta] = [\cdot b_i \cdot] \tag{6.22}$$

and

$$[\cdot b_i \cdot][\cdot a_i \cdot]^{-1} = [\cdot \lambda_i \cdot] \tag{6.23}$$

These solutions indicate that there exist damped natural modes. However, these modes are not the same as the undamped natural modes whose elements are either in phase or 180° out of phase with each other. For the non-proportionally damped system, there are phase differences between various parts of the system, resulting in complex mode shapes. This difference is manifested by the fact that, for undamped modes all points on the structure pass through their equilibrium positions simultaneously, and for complex modes this is not true. Thus, undamped modes have well-defined nodal points or lines while complex modes do not have stationary nodal lines.

The design of equation (6.17) has ensured that the combined equation in (6.18) has two symmetrical matrices $[A]$ and $[B]$. Any other choices such as:

$$[[K] : [0]]\{\dot{y}\} + [[0] : [-K]]\{y\} = \{0\} \tag{6.24}$$

will result in a new eigenvalue problem as:

$$\begin{bmatrix} C & M \\ K & 0 \end{bmatrix} \{\dot{y}\} + \begin{bmatrix} C & 0 \\ K & -K \end{bmatrix} \{y\} = \{0\} \tag{6.25}$$

Here, the non-symmetry of the first matrix may cause some unnecessary numerical difficulties.

6.3 Non-proportional structural damping model

The equation of motion of an MDoF system with non-proportional structural damping is already given by equation (6.9):

$$[M]\{\ddot{x}\} + [K]\{x\} + j[H]\{x\} = \{0\} \tag{6.9}$$

Like the SDoF case, a structural damping matrix can be seen as the imaginary part of a complex stiffness matrix defined as:

$$[K]_c = [K] + j[H] \tag{6.26}$$

The solution to equation (6.9) can be assumed mathematically as:

$$\{x(t)\} = \{X\}e^{j\lambda t} \tag{6.27}$$

Here, λ is the complex frequency accommodating both oscillation and free decay of the vibration; $\{X\}$ is a complex vector for displacement amplitudes. This form of solution, once substituted into equation (6.9), leads to a complex eigenvalue problem:

$$([K]_c - \lambda^2[M])\{X\} = \{0\} \tag{6.28}$$

The solution to equation (6.28) will yield a diagonal eigenvalue matrix $[\cdot\lambda_r^2\cdot]$ and an eigenvector matrix $[\Psi]$. The eigenvalue λ_r^2 is related to the natural frequency ω_r and damping loss factor η_r of the system such that:

$$\lambda_r^2 = \omega_r^2(1 + j\eta_r) \quad (r = 1, 2, \ldots, N) \tag{6.29}$$

Here, λ_r is also known as the complex natural frequency of the system. Matrix $[\cdot\lambda_r^2\cdot]$ is the natural frequency matrix of the system. The corresponding eigenvector $\{\Psi\}_r$ is a complex vector. When all the eigenvectors are grouped together in the ascending order of ω_r, the resultant complex matrix is the mode shape matrix $\{\Psi\}$ of the system. The complex mode shape $\{\Psi\}_r$ is not unique as any multiples of it still satisfy equation (6.28).

The orthogonality properties of the system reveal:

$$[\Psi]^T [M][\Psi] = [\cdot m_r\cdot] \tag{6.30}$$

$$[\Psi]^T [M]_c[\Psi] = [\cdot k_r\cdot] \tag{6.31}$$

$$[\cdot k_r\cdot][\cdot m_r\cdot]^{-1} = [\cdot\lambda_r^2\cdot] \tag{6.32}$$

Both the modal mass m_r and modal stiffness k_r are complex quantities.

It is evident from the analysis in both Sections 6.2 and 6.3 that an MDoF system with a structural damping model requires much simpler analytical treatment than that with a viscous damping model. When there is no evidence that using different damping models will cause significant variance in the output, it is much more convenient to assume the structural damping model in modal analysis.

6.4 Mass-normalized modes of a damped MDoF system

The same reasons for wanting to have mass-normalized mode shapes for an undamped MDoF system exist for a damped system. For a system with structural damping, the procedure of deriving the mode shapes is identical to that for the undamped system except that we are now dealing with complex quantities. Equation (6.30) determines the complex modal mass matrix. The mass-normalized mode shape matrix $[\Phi]$ can be found from:

$$[\Phi] = [\cdot m_r \cdot]^{-\frac{1}{2}} [\Psi] \tag{6.33}$$

It can be shown that the complex matrix $[\Phi]$ will be unique to the system regardless of different multiples of the mode shape matrix $[\Psi]$ used. Using the mass-normalized mode shapes, the orthogonality of an MDoF system can be written as:

$$[\Phi]^T [M][\Phi] = [I] \tag{6.34}$$

$$[\Phi]^T [K_c][\Phi] = [\cdot \lambda_r^2 \cdot] \tag{6.35}$$

In equation (6.34), the mass matrix is seen as a complex matrix with a zero imaginary part.

With the mass-normalized mode shapes, we can diagonalize the FRF matrix of the MDoF system with structural damping:

$$[\Phi]^T ([K] + j[H] - \lambda^2 [M])[\Phi] = [\cdot (\lambda_r^2 - \omega^2) \cdot] \tag{6.36}$$

This result will be useful in deriving the modal composition of a complex FRF.

6.5 Frequency response functions of a damped MDoF system

6.5.1 Dynamic stiffness matrix and receptance matrix

The dynamic stiffness matrix $[Z(\omega)]$ and receptance FRF matrix $[\alpha(\omega)]$ of a damped MDoF system can be defined in exactly the same way as that for an undamped system:

$$[Z(\omega)] = [K]_c - \omega^2[M] = [K] - \omega^2[M] + j[H] \tag{6.37}$$

$$[\alpha(\omega)] = ([K]_c - \omega^2[M])^{-1} \tag{6.38}$$

The physical meaning of the individual receptance FRF $\alpha(\omega)_{ij}$ has been given in Chapter 5. In brief, $\alpha(\omega)_{ij}$ represents the ratio of the displacement response at DoF 'i' and the only force input of the system acting at DoF 'j'. This is inherited by a damped MDoF system. Owing to the existence of damping, the FRF is now a complex function of frequency ω with a non-zero imaginary part. It does not become infinity at a natural frequency of the system.

6.5.2 Composition of a receptance FRF using vibration modes

The derivation of receptance FRF matrix from the inverse of the dynamic stiffness matrix is not a practical numerical treatment. More than that, the approach does not lend any insight into the modal composition of an FRF. Like the undamped case discussed in Chapter 5, the receptance matrix of an MDoF system with structural damping can be determined from the modal data of the system as:

$$[\alpha(\omega)] = [\Phi][\cdot(\lambda_r^2 - \omega^2).]^{-1}[\Phi]^T \tag{6.39}$$

It can be seen that matrix $[\alpha(\omega)]$ is complex, symmetric and always of full rank. The same discussion can also reveal that for a single receptance FRF, for instance $\alpha_{jk}(\omega)$, equation (6.39) can be written as:

$$\alpha_{jk}(\omega) = \frac{\phi_{j1}\phi_{k1}}{\lambda_1^2 - \omega^2} + \frac{\phi_{j2}\phi_{k2}}{\lambda_2^2 - \omega^2} + \ldots + \frac{\phi_{jn}\phi_{kn}}{\lambda_n^2 - \omega^2} \tag{6.40}$$

This equation appears to be the same as the derivation for the undamped MDoF system. However, the numerators on the right-hand side, which are elements of the mode shape matrix, are complex quantities. So are the natural frequencies in the denominators.

Compared with equation (5.45), it becomes obvious that by using the structural damping model, the theory of modal analysis can almost be the same between damped and undamped cases. Using the modal constant, we can rewrite equation (6.40) as:

$$\alpha_{jk}(\omega) = \sum_{r=1}^{n} \frac{{}_r A_{jk}}{\lambda_r^2 - \omega^2} \tag{6.41}$$

$$= \{{}_1 A_{jk} \quad {}_2 A_{jk} \quad {}_n A_{jk}\} \begin{Bmatrix} \dfrac{1}{\lambda_1^2 - \omega^2} \\ \dfrac{1}{\lambda_2^2 - \omega^2} \\ \ldots \\ \dfrac{1}{\lambda_n^2 - \omega^2} \end{Bmatrix} \tag{6.42}$$

$$= \{A_{jk}\}^T \left\{ \frac{1}{\lambda_r^2 - \omega^2} \right\} \tag{6.43}$$

Thus, the pth column of matrix $[\alpha(\omega)]$ can be written as:

$$\{\alpha(\omega)\}_p = \begin{Bmatrix} \alpha(\omega)_{p1} \\ \alpha(\omega)_{p2} \\ \ldots \\ \alpha(\omega)_{pn} \end{Bmatrix} = [A_p] \left\{ \frac{1}{\lambda_r^2 - \omega^2} \right\} \tag{6.44}$$

$$\text{Here, } [A_p] = \begin{bmatrix} \phi_{p1}\phi_{11} & \phi_{p2}\phi_{12} & \cdots & \phi_{pn}\phi_{1n} \\ \phi_{p1}\phi_{21} & \phi_{p2}\phi_{22} & \cdots & \phi_{pn}\phi_{2n} \\ \cdots & \cdots & \cdots & \cdots \\ \phi_{p1}\phi_{n1} & \phi_{p2}\phi_{n2} & \cdots & \phi_{pn}\phi_{nn} \end{bmatrix} = \begin{bmatrix} {}_1A_{p1} & {}_2A_{p1} & \cdots & {}_nA_{p1} \\ {}_1A_{p2} & {}_2A_{p2} & \cdots & {}_nA_{p2} \\ \cdots & \cdots & \cdots & \cdots \\ {}_1A_{pn} & {}_2A_{pn} & \cdots & {}_nA_{pn} \end{bmatrix}$$

$$(6.45)$$

Matrix $[A_p]$ is called the modal constant matrix for the pth column of receptance matrix $[\alpha(\omega)]$. Each column of $[\alpha(\omega)]$ has its own modal constant matrix. The mass-normalized mode shapes of a system can be derived from one of its modal constant matrices.

6.5.3 Display and properties of an FRF of a damped MDoF system

Like an SDoF system, the FRF of an MDoF system merits a variety of different ways of graphical displays in order to disclose its full characteristics. The FRF of an undamped MDoF system has been deprived from a number of useful displays such as Nyquist plot because of its zero imaginary part. The FRF for a damped MDoF system does not have such a deficiency and should therefore be able to have a larger variety of display methods.

In this part, we will use a 4DoF system with non-proportional structural damping. The system matrices are given below.

$$[M] = \begin{bmatrix} 1 & 0 & 0 & 0 \\ 0 & 1 & 0 & 0 \\ 0 & 0 & 1 & 0 \\ 0 & 0 & 0 & 1 \end{bmatrix} \text{kg} \quad [K] = \begin{bmatrix} 2000 & -1000 & 0 & 0 \\ -1000 & 2000 & -1000 & 0 \\ 0 & -1000 & 2000 & -1000 \\ 0 & 0 & -1000 & 2000 \end{bmatrix} \text{N/m}$$

$$[H] = \begin{bmatrix} 20 & 0 & 0 & 0 \\ 0 & 5 & 0 & 0 \\ 0 & 0 & 0 & 0 \\ 0 & 0 & 0 & 0 \end{bmatrix} \text{N/m}$$

(1) Amplitude–phase plot and log–log plot

The amplitude–phase plot of the FRF for a damped MDoF system consists of the plot of its magnitude versus frequency and that of its phase versus frequency. Figure 6.1 shows the amplitude and phase of the mobility $Y_{11}(\omega)$ of the 4DoF system in linear scale. As expected, the amplitude plot is dominated by resonance. The phase plot reflects the existence of a vibration mode from its visible phase change.

Like the undamped case, a better use of the amplitude information of an FRF is to plot it in decibel scale. The receptance $\alpha_{11}(\omega)$ of the 4DoF system can be plotted in dB scale in Figure 6.2. From there, not only the four resonances are obvious, but also

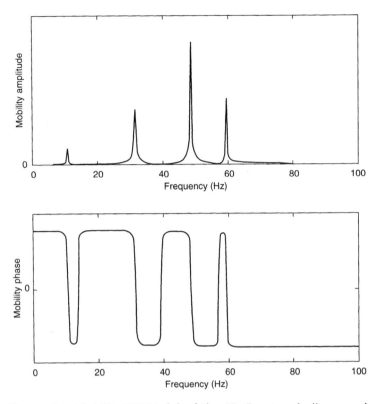

Figure 6.1 Mobility FRF $Y_{11}(\omega)$ of the 4DoF system in linear scale

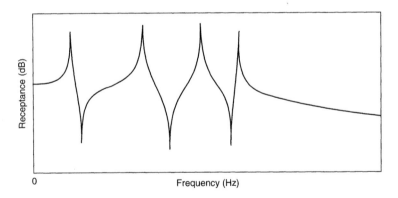

Figure 6.2 Receptance FRF $\alpha_{11}(\omega)$ of the 4DoF system in dB scale

the anti-resonances can be immediately identified. Figure 6.3 plots the four receptance FRFs in dB scale which are related to DoF 1. The resonances, anti-resonances and minima of these FRFs are readily visible. This is an outcome the FRF plot in linear scale does not offer.

To study the asymptote properties, an MDoF FRF needs to be displayed in a log–log plot. The asymptote properties of the MDoF FRF described in Chapter 5 are still

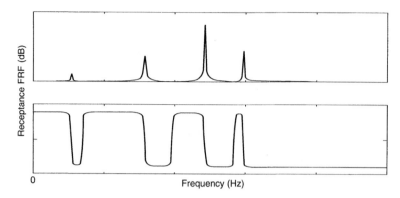

Figure 6.3 Four FRFs of the 4DoF system in dB scale

applicable here. This is because damping mainly changes the amplitudes of FRF data near resonances, anti-resonances and minima. The resonance and anti-resonance frequencies do not alter significantly. In theory, this does not affect the asymptote properties. When damping causes significant complexity to the FRF, it becomes unreliable to derive the asymptote properties. Figure 6.4 shows the log–log plot of the first point FRF of the 4DoF system in receptance, mobility and acceleration forms.

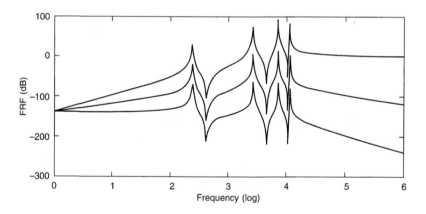

Figure 6.4 Log–log plot of the first point FRF of the 4DoF system in receptance, mobility and accelerance forms

2. Real and imaginary plots

The real and imaginary plots consist of two parts: the real part of the FRF versus frequency and its imaginary part versus frequency. Real and imaginary plots are retracted to be its first part without damping. For an MDoF system with structural damping, the real and imaginary parts can be derived analytically as:

$$\mathrm{Re}(\alpha_{jk}(\omega)) = \mathrm{Real}\left(\sum_{r=1}^{n}\frac{\phi_{jr}\phi_{kr}}{\lambda_r^2 - \omega^2}\right) \tag{6.46}$$

$$\mathrm{Im}(\alpha_{jk}(\omega)) = \mathrm{Imag}\left(\sum_{r=1}^{n}\frac{\phi_{jr}\phi_{kr}}{\lambda_r^2 - \omega^2}\right) \tag{6.47}$$

Figures (6.5) and (6.6) show the real and imaginary plots of an FRF for the 4DoF system.

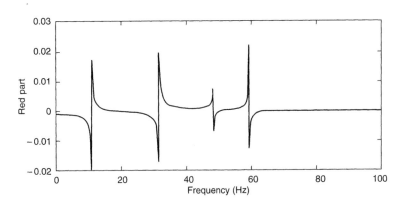

Figure 6.5 Real part of a receptance FRF of the 4DoF system

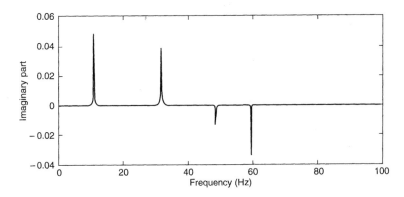

Figure 6.6 Imaginary part of a receptance FRF of the 4DoF system

3. Nyquist plot

The main benefit of using the Nyquist plot for an SDoF FRF comes from its circularity property in the complex plane. This is still valid for a damped MDoF system. The circularity property does not exactly apply here since any vibration mode will be influenced by other modes of the system, thus compromising the simplicity form of an SDoF FRF. However, at the vicinity of a prominent vibration mode, we can assume that the FRF is dominated only by that mode. Thus, the Nyquist plot is still one of the most useful plots for a damped MDoF FRF. Figure 6.7 shows the Nyquist plot of an FRF of the 4DoF system. The data points do not connect full circles because of frequency resolution.

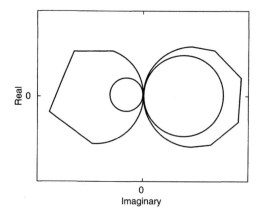

Figure 6.7 Nyquist plot of an FRF of the 4DoF system

   The local dominance of an individual mode and the near circularity of the Nyquist plot of an MDoF FRF means that the SDoF theory can be used in the analysis. This opens an avenue for expanding the SDoF modal analysis method into the MDoF domain. This will be discussed more in Chapter 9.

4. Dynamic stiffness plot

For an SDoF FRF, this plot has shown a remarkable simplicity and easiness to relate the modal parameters with the spatial parameters of the system. For an MDoF FRF, these advantages are heavily discounted. As we can see, the following inequality indicates that the dynamic stiffness plot of an MDoF FRF will not resemble that of an SDoF one.

$$\frac{1}{\alpha_{jk}(\omega)} \neq \sum_{r=1}^{n} \frac{\lambda_r^2 - \omega^2}{\phi_{jr}\phi_{kr}} \tag{6.48}$$

The difficulty of obtaining the same simplicity for the MDoF FRF is compounded by the complexity of the numerators (modal constants). This complexity skews the real and imaginary plots of the FRF so that the plots at a locally dominated mode cannot achieve separate real and imaginary parts.

   Different methods of displaying the same MDoF FRF aim to highlight particular aspects of the function. Although for the same FRF, the amount of information should be identical, it is a matter of making some information more visible and, therefore, useful to modal analysis. Later, we will see that different modal analysis methods rely on these FRF plots.

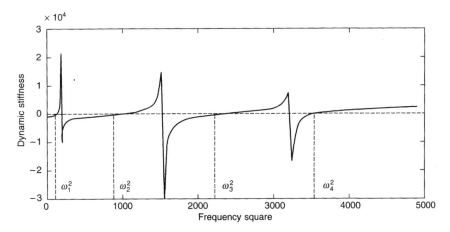

Figure 6.8 The real part of the dynamic stiffness $1/\alpha_{11}(\omega)$ for the 4DoF system

6.6 Time response of a damped MDoF system

Like the FRF of the SDoF system in Chapter 4, an FRF of a damped MDoF system can be written in the form of residues and complex roots:

$$\alpha_{ik}(\omega) = \sum_r \left(\frac{_rA_{ik}}{j\omega - \lambda_r} + \frac{_rA_{ik}^*}{j\omega - \lambda_r^*} \right) \tag{6.49}$$

The complex roots λ_r and λ_r^* are a conjugate pair. So are the complex residues $_rA_{ik}$ and $_rA_{ik}^*$. A more general expression is to use the Laplace operator as the variable so that the transfer function between the same coordinates i and k becomes:

$$\alpha_{ik}(s) = \sum_r \left(\frac{_rA_{ik}}{s - s_r} + \frac{_rA_{ik}^*}{s - s_r^*} \right) \tag{6.50}$$

where

$$s_r, s_r^* = -\zeta_r\omega_r \pm \sqrt{1 - \zeta_r^2}\,\omega_r j \tag{6.51}$$

The inverse Laplace transform of this transfer function is the impulse response of the MDoF system between coordinates i and k:

$$h_{ik}(t) = \sum_{r=1}^{n} {}_rA_{ik}e^{s_r t} + \sum_{r=1}^{n} {}_rA_{ik}^*e^{s_r^* t} = \sum_{r=1}^{2n} {}_rA_{ik}e^{s_r t} \tag{6.52}$$

Here,

$$_rA_{ik}\big|_{r>n} = {}_{r-n}A_{ik}^* \quad \text{and} \quad \lambda_r\big|_{r>n} = \lambda_{r-n}^*$$

In general, the free vibration response of an MDoF system at coordinate i is a combination of the time response contributed by each vibration mode. This is akin to the modal contribution of vibration modes in the composition of an FRF. The response can be written as:

$$x_i(t) = \sum_{r=1}^{2n} \varphi_{ir}e^{\lambda_r t} \tag{6.53}$$

Here, elements φ_{ir} $(r = 1, 2, \ldots, 2n)$ are the ith elements on the rth column of the $2n \times 2n$ mode shape matrix.

The forced response of a damped MDoF system can be derived using Laplace transform and its inverse. The equation of motion of the system under arbitrary external loads is given as:

$$[M]\{\ddot{x}\} + [C]\{\dot{x}\} + [K]\{x(t)\} = \{f(t)\} \tag{6.54}$$

The Laplace transform with zero initial conditions produces:

$$(s^2[M] + s[C] + [K])\{X(s)\} = \{F(s)\} \tag{6.55}$$

Therefore, the force response in the s-domain is given as:

$$\{X(s)\} = (s^2[M] + s[C] + [K])^{-1} \{F(s)\} \tag{6.56}$$

The inverse Laplace transform generates the response in the time domain as:

$$\{x(t)\} = L^{-1}\{X(s)\} = L^{-1} ((s^2[M] + s[C] + [K])^{-1}\{F(s)\}) \tag{6.57}$$

This transform requires the information of the initial conditions of the system.

6.7 Forced normal modes of a damped MDoF system

The mode shape of a damped system is in general complex, and therefore is different from the undamped mode shape of the same system if damping is removed. The only exception is when damping is proportional.

For an undamped system, the equations of motion can be uncoupled under the principal coordinates. As a result, one mode is usually independent of other modes. These modes have been known as the normal mode. For normal mode vibration, every point of the system undergoes harmonic motion and passes through the equilibrium position simultaneously.

The existence of non-proportional damping invalidates this property since the uncoupling of equations becomes impossible. However, by using a number of harmonic forces to counterbalance the damping forces, it is possible to excite the normal mode vibration for a damped MDoF system. Consider a viscously damped 'n' DoF system excited by harmonic forces of frequency ω. The matrix equation of motion of the system is:

$$[M]\{\ddot{x}\} + [C]\{\dot{x}\} + [K]\{x\} = \{F\}\sin \omega t \tag{6.58}$$

Assume the solution to equation (6.58) is in the form of:

$$\{x\} = \{X\} \sin (\omega t - \theta) \tag{6.59}$$

For any given excitation frequency ω, there exist 'n' solutions of the type given by equation (6.59), where each of the modes $\{\psi\}_i$ is associated with a definite phase θ_i and a corresponding distribution of forces $\{F\}$ which is required for its excitation. The response under these conditions is called the *forced normal modes*. Forced normal modes are also referred to as the characteristic phase lag modes.

Now equation (6.58) can be divided into its real and imaginary parts:

$$[([K] - [M]\omega^2) \sin\theta - [C]\omega \cos \theta]\{X\} = \{0\} \qquad (6.60)$$

$$[([K] - [M]\omega^2) \cos\theta + [C]\omega \sin \theta]\{X\} = \{F\} \qquad (6.61)$$

Equation (6.60) is the same as:

$$[[I] \tan \theta - ([K] - [M]\omega^2)^{-1} [C]\omega]\{X\} = \{0\} \qquad (6.62)$$

This is an eigenvalue problem. For each given frequency ω, there exists n eigenvalues $\tan \theta_i$ and eigenvectors $\{\Psi\}_i$ such that:

$$[([K] - [M]\omega^2) \tan\theta_i - [C]\omega]\{\Psi\}_i = \{0\} \qquad (6.63)$$

Compared with the eigenvalue solution and the orthogonality of an undamped MDoF system, we find that:

$$\tan \theta_i = \frac{\omega\{\Psi\}_i^T [C]\{\Psi\}_i}{\{\Psi\}_i^T [[K] - \omega^2[M]]\{\Psi\}_i} \qquad (6.64)$$

Equation (6.64) shows that each eigenvalue $\tan \theta_i$ is a continuous function of frequency ω. When ω is small, so is $\tan \theta_i$. As ω approaches one of the undamped natural frequencies of the system ω_i, $\tan \theta_i$ approaches infinity and therefore θ_i becomes $\pi/2$. The eigenvector defined by equation (6.63) becomes the mode shape of the damped system.

Equations (6.63) and (6.64) also suggest that at any frequency ω, the mode shapes depend only on the shape or distribution of matrix $[C]$ and not on its intensity. If every element in $[C]$ is multiplied by a constant factor, then equation (6.64) shows that the eigenvalue $\tan \theta_i$ will all be increased by the same ratio. Thus, equation (6.63), which determines the mode shapes, will be multiplied throughout by the same factor and the mode shape will be unchanged.

Equation (6.63) can be rewritten as:

$$\left[([K] - [M]\omega^2) - \frac{[C]\omega}{\tan \theta_i} \right]\{\Psi\}_i = \{0\} \qquad (6.65)$$

Substituting $\theta_i = \frac{\pi}{2}$ into equation (6.65) reveals:

$$([K] - [M]\omega_i^2)\{\Psi\}_i = \{0\} \qquad (6.66)$$

This equation lends the theory to the modal testing procedures using multi-excitation forces in order to excite out the undamped modes from a damped structure. It is of interest to note that the damping types are irrelevant when the undamped modes are thus excited.

We can see that when the frequency ω is equal to one of the undamped natural frequencies ω_i, the mode shape for one of the eigenvalues $\tan \theta_i$ (which is now equal to infinity, or the angle θ_i is $\pi/2$) is identical to the undamped mode shape. In this case, the forces required can be obtained from equation (6.61) to be:

$$\omega_i[C]\{\Psi\}_i = \{F\}_i \qquad (6.67)$$

6.8 Remarks on complex modes

For modal analysis, perhaps the most significant outcome of having non-proportional damping are complex vibration modes. The concept of mode shapes was well interpreted in the case of no damping with clearly defined physical meanings. When mode shapes become complex, points on a structure no longer move in a clear pattern of either in or out of phase.

Mathematically, we are able to interpret the origins of complex modes. If a dynamic structure is modelled as an MDoF system with mass, stiffness and damping matrices, then it is easy to see that the eigenvectors of the system, which are the mode shapes, only become complex if (a) one or more of the matrices are not symmetric, or (b) the damping matrix is not diagonalizable using the undamped mode shapes. In the first case, the mass and stiffness matrices are symmetric. The damping matrix can be skew-symmetric if the gyroscopic forces of a rotating machine are present. In this case, even if no damping terms exist in the matrix, the eigenvalue problem will lead to complex mode shapes. In the second case, if damping is not proportional, the eigenvectors will be complex. If the state–space formulation is used, it will result in a set of eigenvectors which diagonalize two system matrices, but not the damping matrix (and many stiffness matrices) of the system.

Complex modes share some characteristics with real modes. When using structural damping model, complex modes are orthogonal to each other with respect to the system mass matrix and complex stiffness matrix. When using a viscous damping model, complex modes are orthogonal to each other with respect to matrices $[A]$ and $[B]$ in equation (6.20). Because of these orthogonality properties, complex modes can uncouple the equations of motion (for structural damping) or the reconstructed equations of motion (for viscous damping).

Unlike real modes, elements in a complex mode have different phase angles valued between 0 and 180 degrees. This means that points on a vibrating structure do not pass through the equilibrium position at the same time, although they still share the same oscillation frequency. The nodal point of the structure is not fixed. It moves with a certain period dictated by the complexity of the mode.

The real difficulty for the complex mode in modal analysis is not the ability to simulate and study it, but to identify them using the concept and theory in the derivation of modal data. In this case, usually there are no correct answers to compare with. As a result, different interpretations and outcomes may occur from the same measured FRF data. The main question is to ascertain the authenticity of the derived complex modes. There are several factors in modal analysis which may result in computational complex modes. For example, if two close real modes are not analysed correctly or accurately, then both may become 'complex' modes. The deviation of the phase for modal constants also results in untrue complexity of modes. The presence of nonlinearity may lead to the identification of false complex modes.

The amount of complexity of an authentic complex mode is difficult to quantify, but there is a need to derive a parameter for that purpose so that complex modes can be analysed. The most obvious feature for complexity of a mode shape is the phase angles of its elements. The more each angle deviates from 0 or 180 degrees, the more it contributes to the complexity. The maximum angle deviation is 90 degrees. However, we can also see that it is not just the accumulated phase deviation that matters. If we have a mode where all phase angles are 90 degrees, then that mode is not complex at all. Therefore, we can start to define a parameter as a measure of modal complexity and call it the modal complexity factor. This factor quantifies a normalized phase deviation of a complex mode shape by calculating the mean value of the inter-element phase differences. The percentage ratio between this value and the maximum deviation of 90 degrees indicates the degree of modal complexity.

Literature

1. Asher, G.W. 1958: A Method of Normal Mode Excitation Utilising Admittance Measurements. *Proceedings of National Specialists' Meeting in Dynamics and Aeroelasticity, Institute of the Aeronautical Sciences*, 69–76.
2. Brock, J.E. 1968: Optimal Matrices Describing Linear Systems. *AIAA Journal*, **6**(7), 1292–1296.
3. Chopra, A.K. 1996: Modal analysis of linear dynamic systems: physical interpretation. *Journal of Structural Engineering*, **122**, 517–527.
4. Ewins, D.J. 1979: Whys and Wherefores of Modal Testing. *SEE Journal*, 1–13.
5. Imregun, M. and Ewins, D.J. 1995: Complex modes – origins and limits. In: *Proceedings of the 13th International Modal Analysis Conference*, Nashville, USA, 496–506.
6. Lewis, R.C. and Wrisley, D.L. 1950: A System for the Excitation of Pure Natural Modes of Complex Structures. *Journal of the Aeronautical Sciences*, **17**(11), 705–722.
7. Lord Rayleigh 1945: *Theory of Sound*, Dover Publications, 2nd edition, New York.
8. Maia, N.M.M., Silva, J. and Ribeiro, A.M.R. 1994: A new concept in modal analysis: the characteristic response function (CRF). *International Journal of Analytical and Experimental Modal Analysis*, **9**(3), 191–202.
9. Mitchell, L.D. 1990: Complex Modes: a Review. *Proceedings of the 8th International Modal Analysis Conference*, Orlando, FL 891–899.
10. Natke, H.G. and Rotert, D. 1985: Determination of Normal Modes from Identified Complex Modes. *Z. Flugwiss. Weltraumforsch*, **9**(2), 82–88.
11. Silva, J.M.M. and Maia, N.M.M. (eds) 1998: *Modal Analysis and Testing*. NATO Science Series E: Applied Sciences, Vol. 363.
12. Zaveri, K. 1985: *Modal Analysis of Large Structures*. Bruel & Kjaer.

7

Frequency response function measurement

7.1 Introduction

The measurement for experimental modal analysis is to acquire frequency response function data from a test structure. For some time-domain analysis it is to obtain either the free decaying impulse response or the response due to ambient excitations. While there are a variety of methods available to carry out measurement, only a simple method is explained in this chapter. This method is to excite a structure with a known input force and measure both the force and responses on the structure. As a result, we obtain the data for a group of FRFs that can be used later for modal analysis to derive the modal model of the structure.

Experimental modal analysis is a system identification endeavour. The structure is a 'black box' that needs to be deciphered. The traditional approach is to provide the 'black box' with a known input, measure the output and proceed with the identification. For our measurement, we use force input so that the FRF can be derived directly from the force and response information. The excitation force can be random, sinusoidal, periodic or impact ones. Theoretically, the type of force does not matter as the FRF is defined as the ratio between the response and force. In practice, whenever practical we want to use a force that has sufficient energy and frequency components to excite all vibration modes of interest and to allow minimum errors in signal processing, leading to the formation of accurate FRF data. We are also limited by the capacities of the hardware available for measurement.

The assumption that the test structure behaves linearly is essential to attaining accurate FRF measurement. This is not difficult to verify in an experiment. If the excitation force level is controllable, then we should be able to double or multiple the force input level and observe the repeatability of the FRF data. For a structure which does not follow linearity assumption exactly, we know that normal modal analysis from the measured FRF data may only represent a linearized mathematical model for the structure. The other essential assumption is the reciprocity property of the structure. Again this assumption is generally not difficult to verify. Sometimes we have to take care that these assumptions also involve the instrument part of the measurement set-up, too. So, ideally, we should like to see the whole measurement set-up (including the structure) follow linearity and reciprocity. In addition, FRF measurement cannot succeed if the dynamic properties of a structure vary during the measurement. For this reason, we require the structure to be time-invariant. This condition is usually met in FRF measurement.

Rapid improvement in measurement hardware and computing power in the last couple of decades has enabled us to make FRF measurement with multiple force inputs and multiple response outputs simultaneously. With multiple force inputs, it becomes possible to make the structure vibrate with reasonably uniform amplitudes rather than having great disparity of amplitudes across it under a single input. This type of measurement, if used properly, can result in more accurate FRF data and subsequently modal data. Time saving is accomplished. The demand on greater resources to conduct multiple input tests confines it to laboratories of sizable institutions.

FRF data are not the only type of data acquired for modal analysis. As we have known, for a special category of modal analysis that utilizes the responses in time history, either free vibration response or impulse response function data are needed. There are also practical cases where laboratory set-up for force excitations is not feasible. Instead, vibration response due to ambient excitations can be readily measured.

7.2 A general measurement set-up

A typical measurement set-up in a laboratory environment should have three constituent parts. Take a simple single input and single output case as an example. The first part is responsible for generating the excitation force and applying it to the test structure; the second part is to measure and acquire the response data; and the third part provides signal processing capacity to derive FRF data from the measured force and response data.

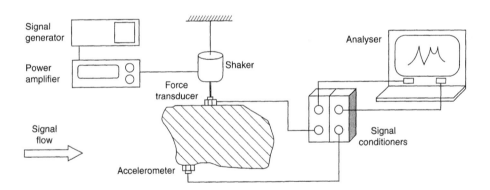

Figure 7.1 A measurement set-up with shaker excitation

7.2.1 Excitation mechanism

The first part of the measurement set-up is an excitation mechanism that applies a force of sufficient amplitude and frequency contents to the structure. There are different types of excitation equipment that are able to excite a structure. The two most common ones are shaker and hammer.

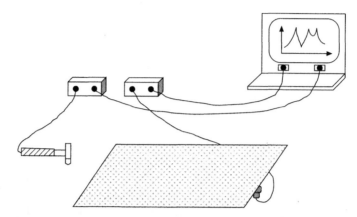

Figure 7.2 A measurement set-up with hammer excitation

A hammer is a device that produces an excitation force pulse to the test structure. It consists of hammer tip, force transducer, balancing mass and handle. The hammer tip can be changed to alter the hardness. Typical materials for the tip are rubber, plastic and steel. The hardness of the tip, together with that of the structure surface to be tested, is directly related to the frequency range of the input pulse force. For a hard tip striking on a hard surface, we can expect the force pulse to distribute energy to a wide range of spectrum. This is the only mechanism to control the frequency components of excitation in a hammer test.

An electromagnetic shaker, also known as an electrodynamic shaker, is the most common type of shaker used in modal testing. It consists of a magnet, a moving block and a coil in the magnet. When an electric current from a signal generator passes through the coil inside the shaker, a force proportional to the current and the magnetic flux density is generated which drives the moving block. An electromagnetic shaker has a wide frequency, amplitude and dynamic range. For low frequency and large amplitude excitation, an electrohydraulic exciter can be used.

7.2.2 Accelerometer

An accelerometer is the most common sensor for modal testing. It measures acceleration of a test structure and outputs the signal in the form of voltage. This signal will be transformed by a signal conditioner before it is processed by an analyser, other hardware or software. The accelerometer does not assume the properties of the measured structure such as linearity. An accurate accelerometer only records faithfully the acceleration at the measurement location.

There are two aspects in the acceleration measurement that a sensor needs to be capable of dealing with. One is the frequency and the other is the amplitude. Both are reflected in the input–output relationship of an accelerometer. An ideal accelerometer should have a linear input and output relationship in order to ensure that the amplitude content of the acceleration signal at different frequencies is truthfully recorded. The FRF of the accelerometer should be uniformly flat so that the amplitude of no frequency is distorted. The accelerometer should also impose zero phase shift to the signal measured.

Most accelerometers come with amplitude-frequency and phase-frequency charts to show their characteristics. Figure 7.3 shows a typical frequency response curve of an accelerometer.

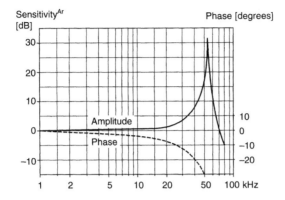

Figure 7.3 A typical chart for an accelerometer

The characteristics of an accelerometer are its potential. They can be fully realized if the sensor is connected rigidly on the structure. In reality, this is not to be the case. An accelerometer has to be mounted non-rigidly on a structure for measurement. If considered as a rigid mass block, the accelerometer and its mount can be modelled as an SDoF system as shown in Figure 7.4.

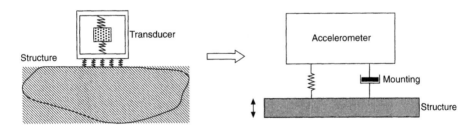

Figure 7.4 An SDoF model of an accelerometer and its mounting

The accuracy of the acceleration measurement depends largely on the mounting which is modelled by a spring and a damper. The accelerometer is of course more than just a mass block. As Figure 7.3 shows, it has its own natural frequency. This frequency is usually much higher than the frequency of the SDoF system in Figure 7.4. The best accuracy would arise if the mounting were rigid. The flexibility of the mounting means that the characteristics of the accelerometer are compromised somewhat. Because of it, acceleration from the structure may be different from that experienced by the accelerometer. However, if the natural frequency of this SDoF system is five times or more of the frequency of the acceleration signal from the measured structure, then there is effectively no magnitude and phase distortion. The acceleration measured is seen as identical to that from the structure.

A most common type of accelerometer is the piezoelectric one as illustrated in Figure 7.5. Piezoelectricity means the phenomenon of strain inducing a change in the shape of a crystal thus leading to the change of electric charge. At the low frequency end, piezoelectric accelerometers do not respond to DC signal. At the high frequency end, the accuracy of measurement is degraded by the natural frequency of the sensor. When selecting an accelerometer for modal testing, a number of factors need to be thought through. The main parameters affecting the performance of a piezoelectric accelerometer are: the frequency response property; the sensitivity and its stability under temperature change; cross-axial sensitivity and base strain.

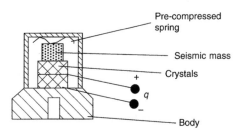

Figure 7.5 Diagram of a piezoelectric accelerometer

The frequency response property determines the linearity of the sensor. The sensitivity of an accelerometer dictates the signal to noise ratio. Large and stable sensitivity means accurate measurement. Cross-axial sensitivity causes inaccuracy in measurement. Base strain is caused by the flexure of the accelerometer base interacting non-rigid structure surface. Usually, a more sensitive accelerometer is more bulky. The accelerometer mass has the potential to change the characteristics of the test structure. This is particularly so if the accelerometer is located at or close to an anti-nodal point of a vibration mode where a minute mass change can cause significant natural frequency shift. Thus, when selecting a better sensitivity, we need to be mass conscious.

7.2.3 Force transducer

A force transducer is another type of sensor used in modal testing. Like an accelerometer, a piezoelectric force transducer generates an output charge or voltage that is proportional to the force applied to the transducer (Figure 7.6). Unlike an accelerometer though, a force transducer does not have an inertial mass attached to the transducing element. It has to be physically compressed or stretched so that the transducing part can generate output. For a shaker test, a force transducer has to be connected between the structure surface and the shaker. For a hammer test, the transducer is located at the hammer tip and is compressed when impact is applied to.

The ways the characteristics of a force transducer affect measurement accuracy

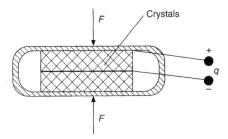

Figure 7.6 A diagram of a piezoelectric force transducer

are similar to that for an accelerometer. They include frequency response characteristics, sensitivity and cross-axial sensitivity. The mass of the force transducer can also potentially affect the measurement outcome. For both force and acceleration sensors, this problem is most acute when the structure resonates.

The main consideration in selecting a force transducer is to understand how it interacts with the excitation device to which it connects. For example, when a force transducer is used on an impact hammer, variation of the hammer tip and the mass of the hammer handle can cause a different force transducer calibration. When the force transducer is used with an excitation shaker, the presence of its mass may cause significant distortion to the force signal measured at structural resonance. The extent of distortion is dependent on the mass difference between the transducer and the structure. The mass of the transducer may also be responsible for the sensitivity the transducer has to bending moments.

7.3 Preparation of the test structure

A real structure is often connected to its surroundings. Therefore, its dynamic characteristics *in situ* are determined by the boundary conditions as well as by itself. When the FRF measurement is to be carried out, the question to be answered is: 'Under what conditions do we want to test the structure, stand alone or *in situ*?' The answer to this question hinges on two considerations when preparing the structure for test: (1) do we need the modal model of the structure in its working condition or in the laboratory environment; and (2) is it realistic to test the structure *in situ* or in laboratory?

For consideration (1), the answer is that it depends on what we need the derived modal model for. For some structures, we need to know its dynamic behaviour exactly in its working conditions or with its surroundings. For example, to study the dynamic behaviour of the tool carriage of a lathe and its relation to the cutting accuracy, it is necessary that the tool carriage be actually on the lathe. At other times, the modal model derived from the test structure may be used as a constituent part for further prediction analysis such as structural coupling analysis. Then, the model should be derived when the structure is isolated in a laboratory environment.

For consideration (2), we need to assess the feasibility for a structure to be tested under the desired conditions. Some structures cannot be tested *in situ* because of deficiency in current measurement technology or inaccessibility. We need to use the modal data obtained from a laboratory measurement to analyse and simulate the dynamic behaviour *in situ*. Other structures cannot fit into a laboratory environment for FRF measurement. They have to be tested *in situ*. In both cases, it is vital to ensure that the test conditions are stable and repeatable so that the measured FRF data are reliable and representative.

If an FRF measurement in the laboratory is desired and feasible, a structure is often prepared to simulate free or grounded boundary conditions. The subsequent use of the modal model to be derived from the measurement determines which boundary conditions to use. It is impossible to imitate perfect free or grounded conditions. These conditions can only be approximated in the laboratory with reasonable accuracy.

The free boundary condition is simulated by supporting the structure with soft materials such as springs or elastic bands. Such an arrangement creates one or more rigid body modes from the stiffness of these supporting materials and the total mass of the structure. If the natural frequencies of these rigid body modes are far from the first natural frequency of the structure, the measured FRF data should not be affected by this boundary condition. Figure 7.7 shows a simple plate supported by four soft springs. For the same principle, a heavy structure can be put on top of an inflated car tyre tube or a few layers of thick porous packaging material.

The grounded boundary condition is more difficult to simulate in the laboratory. Theoretically, a grounded condition means that all the six degrees of freedom at the boundary are rigidly fixed. This can almost never be achieved in reality. In modal testing, the arrangement is often to 'fix' the structure to a much more rigid and heavier object such as a concrete floor.

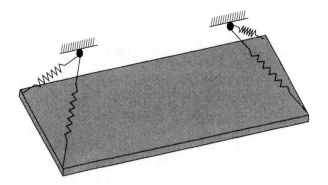

Figure 7.7 A simple plate supported by four soft springs

It is only for some special cases that a true grounded condition can be simulated. For example, accurate measurement of the cantilever in Figure 7.8 needs a rigid boundary condition at the built-in end which may be difficult to realize. The alternative is to measure the beam on the right which is freely supported with a doubled length. Any odd number of modes from this beam will be the equivalent modes for the cantilever. This is due to the fact that for these modes, the middle cross-section of the free–free beam is truly 'fixed'.

For the simulation of the free boundary condition, we often know the limitation

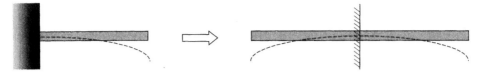

Figure 7.8 A cantilever and a free–free beam with a doubled length

from the natural frequencies of the rigid body modes. For the simulation of the grounded condition, however, we often do not know how rigid (or non-rigid) the boundary condition is and therefore the limitation. It is only when we are convinced that the boundary condition is rigid enough not to affect the frequency range of interest, can we be confident with the FRF data measured from this structural support.

7.4 Selection of excitation forces

The excitation method is important to conducting accurate modal testing. Although theoretically the FRF data should not depend on the excitations (and responses), in practice the accuracy and quality of FRF data do depend, among many factors, on the choice of excitation. The decision on excitation is based first on the optimum method for accurate measurement data that suit test objectives and the structure. It is therefore imperative for modal analysts to appreciate all available excitation methods so as to be able to select the most suitable one for testing. It is unrealistic and unnecessary to expect to have a perfect measurement set-up.

Time constraint is often a determining factor paralleled to hardware. Different test objectives require different amounts of time to conduct. In the following, only brief comments on some of the most common excitation methods are given. For a detailed description of these excitation methods, refer to the published books and papers on vibration and modal testing and signal processing listed at the end of this chapter.

7.4.1 Sinusoidal excitation

Sinusoidal excitation way the most traditional method for modal testing. Today, it is still a most popular one. The force contains one single frequency at a time and the excitation sweeps from one frequency to another with a given step, allowing the structure to engage in one harmonic vibration at a time. This excitation is effective for exciting a structure with high vibration level, for characterizing nonlinearity of a structure, and for exciting normal vibration modes of a damped structure.

Single input sinusoidal excitation could be very time-consuming, despite its superior effectiveness. The advent of multi-input multi-output (MIMO) testing partially rectified this problem. With the ability to perform real-time FRF estimation on many channels, sinusoidal excitation becomes a fast as well as reliable excitation method for modal testing.

The dynamics of structures are physically decomposed by frequency and position. When tuned to near a natural frequency, the structural response is dominated by that vibration mode. This natural segregation provides an avenue for direct parameter identification with usually satisfactory signal to noise ratio. This is a possibility other excitation methods do not offer to modal analysts.

7.4.2 Random excitation

The force signal for random excitation is an ergodic, stationary random signal with Gaussian distribution. It contains all frequencies within the frequency range. For a structure that behaves nonlinearly, random excitation has the tendency to linearize the behaviour from the measurement data. The frequency response function derived from random excitation measurement will then be the linearized FRF. This FRF, though failing to provide more information on the nonlinearity, is actually a very useful function as it is perceived as the 'best' FRF estimate for the structure. It correctly models the amount of energy dissipation of the structure during vibration. However, this linear model is best only for a particular force spectral density. Therefore, we may have a series of linearized FRFs for varying excitation force levels.

The fact that for random excitation neither the force signal nor the response are periodic with infinite time history gives rise to an error called leakage. Some signal processing measures have been available for minimizing this error.

7.4.3 Pseudo random excitation

The force signal in this excitation is an ergodic, stationary random signal that consists of discrete frequencies formed by integer multiples of the frequency resolution used by the Fourier transform. It is a periodic signal with random amplitude and phase distribution. This excitation has eliminated the leakage problem a random excitation commonly encounters. It usually results in more accurate FRF data for modal analysis. Nevertheless, pseudo random excitation requires a special device to generate such a unique force signal and it requires more time than random excitation to implement.

An improved version of pseudo random excitation is called burst random excitation. A burst random signal is created by gating on and off a true random source such that the measurement begins with the source and continues after the source is switched off. The spectrum of a burst random has random amplitude and phase and contains energy throughout the frequency range. By choosing the time for truncation of the random source, the measurement resembles pseudo random excitation but without the need to wait for the decay of the transient portion of the response.

7.4.4 Impact excitation

The time domain force signal for impact excitation is a pulse with uncontrollable frequency contents. In terms of hardware involved, impact excitation is a relatively simple excitation technique compared with shaker excitation. It is convenient to use and very portable for field and laboratory tests. Because of no physical connection between the excitation and the structure, impact test avoids the problem of interaction between them. This means it is possible to measure accurate damping quantities if the signal processing does not add windows or if the additional damping from windowing can be removed.

The main disadvantages of impact excitation are as notable as its advantages. It is difficult to control either the force level or the frequency range of the impact. This could affect the signal to noise ratio in the measurement, thus resulting in poor quality data. The impact cannot usually be repeated unless a special device such as a swing for the impact hammer is used. In addition, some structures are too delicate to be hammered upon.

7.5 Different estimates of an FRF and effects of noise

For any measurement, noise exists at both the input and the output. For frequencies close to resonance, the vibration response is significant so noise at the output can be ignored. For frequencies close to anti-resonance, the excitation signal is significant so noise at the input can be ignored. For other frequencies, we can expect noise from both ends to affect the FRF estimation. This prompted the effort to find different FRF estimators to derive most accurate FRF data with the presence of noise. However, for an FRF with prominent resonances and anti-resonances, it is impossible to find a single FRF estimator which produces most accurate FRF data through the frequency range.

For the idealized measurement situation shown in Figure 7.9 where no noise and measurement errors are present, the FRF can be defined simply as the ratio of two

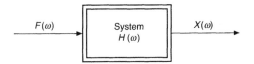

Figure 7.9 A system with single input and output

Fourier transforms. The first is that of the response $x(t)$ and denoted as $X(\omega)$. The second is that of the input force $f(t)$ and denoted as $F(\omega)$:

$$H(\omega) = \frac{X(\omega)}{F(\omega)} \tag{7.1}$$

From dual channel spectral analysis an FRF is defined as the cross-spectrum of excitation and response divided by the auto-spectrum of the excitation. This leads to the following noise-free FRF estimator:

$$H_1(\omega) = \frac{S_{FX}(\omega)}{S_{FF}(\omega)} \tag{7.2}$$

Here, $S_{FX}(\omega)$ is the cross-spectrum of excitation and response and $S_{FF}(\omega)$ is the auto-spectrum of the excitation. It has been found in spectral analysis that the same FRF can be estimated from the ratio of the auto-spectrum of the excitation $S_{XX}(\omega)$ and the cross-spectrum of excitation and response $S_{FX}(\omega)$:

$$H_2(\omega) = \frac{S_{XX}(\omega)}{S_{XF}(\omega)} \tag{7.3}$$

When the test structure satisfies all assumptions and when noise and measurement errors do not exist, these two different FRF estimators should be equal to the correct FRF:

$$H_1(\omega) = H_2(\omega) = H(\omega) \tag{7.4}$$

In reality, FRF measurement cannot be noise free. Figure 7.10 shows that the measured force $\hat{F}(\omega)$ is a combined signal of the genuine force $F(\omega)$ and the noise from the input $M(\omega)$. These two are normally inseparable in the time domain but they are not correlated to each other. Therefore, we have $S_{MF}(\omega) = 0$. This property is useful in deriving different FRF estimators to combat noise. The same happens at the output end. The measured response $\hat{X}(\omega)$ encompasses both the true response $X(\omega)$ and the noise from the output $N(\omega)$. They are not correlated so that $S_{NX}(\omega) = 0$.

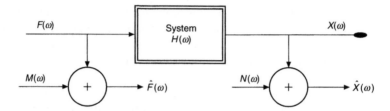

Figure 7.10 FRF estimation with input and output noise

The FRF using $\hat{X}(\omega)$ and $\hat{F}(\omega)$ can be estimated as:

$$\hat{H}_1(\omega) = \frac{\hat{S}_{XF}(\omega)}{\hat{S}_{FF}(\omega)} = H(\omega)\left(1 + \frac{S_{MM}(\omega)}{S_{FF}(\omega)}\right)^{-1} \tag{7.5}$$

or

$$\hat{H}_2(\omega) = \frac{\hat{S}_{XX}(\omega)}{\hat{S}_{XF}(\omega)} = H(\omega)\left(1 + \frac{S_{NN}(\omega)}{S_{XX}(\omega)}\right) \tag{7.6}$$

Neither estimator accounts for the accurate FRF. In terms of amplitude, $\hat{H}_1(\omega)$ is an underestimator and $\hat{H}_2(\omega)$ is an overestimator. More insights can be made if we examine $\hat{H}_1(\omega)$ and $\hat{H}_2(\omega)$ at vicinities of resonances and anti-resonances where most dramatic changes of signal to noise ratio occur either at the input or the output end.

At a resonance, the interaction between the excitation shaker and the structure usually results in a notch in the force signal. This allows measurement noise to dominate the input end while high level of response ensures a large signal to noise ratio at the output end, resulting in a significant $S_{MM}(\omega)$ compared with $S_{FF}(\omega)$. Consequently, the FRF estimator $\hat{H}_1(\omega)$ underestimates the true FRF $H(\omega)$ (as shown in Figure 7.11) and the estimator $\hat{H}_2(\omega)$ is accurate. As many modal analysis methods utilize the measured FRF data at the vicinity of resonances, estimator $\hat{H}_2(\omega)$ should be the choice over $\hat{H}_1(\omega)$.

$$\hat{H}_1(\omega)\big|_{\text{resonance}} = H(\omega)\left(1 + \frac{S_{MM}(\omega)}{S_{FF}(\omega)}\right)^{-1} < H(\omega) \tag{7.7}$$

$$\hat{H}_2(\omega)\big|_{\text{resonance}} = H(\omega)\left(1 + \frac{S_{NN}(\omega)}{S_{XX}(\omega)}\right) \approx H(\omega) \tag{7.8}$$

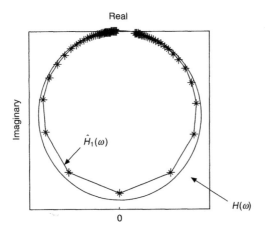

Figure 7.11 The $H(\omega)$ and $\hat{H}_1(\omega)$ plot of an SDoF receptance FRF

At an anti-resonance, the response from the structure usually is insignificant compared with the force input. This allows measurement noise to dominate the output while a high level of force produces a large signal to noise ratio at the input end. This results in a significant $S_{NN}(\omega)$ compared with $S_{XX}(\omega)$ and a small $S_{MM}(\omega)$. Consequently, we have the following estimates for the true FRF $H(\omega)$:

$$\hat{H}_1(\omega)\big|_{\text{anti-resonance}} = H(\omega)\left(1 + \frac{S_{MM}(\omega)}{S_{FF}(\omega)}\right)^{-1} \approx H(\omega) \tag{7.9}$$

$$\hat{H}_2(\omega)\big|_{\text{anti-resonance}} = H(\omega)\left(1 + \frac{S_{NN}(\omega)}{S_{XX}(\omega)}\right) > H(\omega) \tag{7.10}$$

The two FRF estimators $\hat{H}_1(\omega)$ and $\hat{H}_2(\omega)$ are related through the coherence function $\gamma^2(\omega)$. For the input $f(t)$ and output $x(t)$, coherence is defined as:

$$\gamma^2_{FX}(\omega) = \frac{|S_{FX}(\omega)|^2}{S_{FF}(\omega)S_{XX}(\omega)} \tag{7.11}$$

Physically, coherence reflects the causal and linear relationship between the output $x(t)$ and input $f(t)$. Therefore, $\gamma^2(\omega)$ is equal to 0 when the output is not due to the input equal to 1 when the output is caused solely by the input. Poor coherence is indicative of poor signal to noise ratio, measurement errors, nonlinear or time-variant behaviour of the structure, or a combination of them.

From the definitions of two FRF estimators and coherence, we can find that:

$$\gamma^2_{FX}(\omega) = \frac{\hat{H}_1(\omega)}{\hat{H}_2(\omega)} \qquad (7.12)$$

Therefore, if a measurement set-up only provides $\hat{H}_1(\omega)$ data, we can convert it into $\hat{H}_2(\omega)$ using coherence data for modal analysis.

Another source of errors existing in measured FRF data is caused not by noise, as we have discussed, but by leakage in the Fourier transform. Leakage occurs when the free vibration of a resonance has to be truncated prematurely. As a result, both the output spectrum $S_{XX}(\omega)$ and cross-spectrum $S_{XF}(\omega)$ will contain errors. Only the input spectrum $S_{FF}(\omega)$ can be unaffected. When using the $\hat{H}_1(\omega)$ estimator, the ratio of inaccurate $S_{XF}(\omega)$ and accurate $S_{FF}(\omega)$ will inherit leakage errors in the FRF data. When using the $\hat{H}_2(\omega)$ estimator, however, the ratio of inaccurate $S_{XF}(\omega)$ and inaccurate $S_{XX}(\omega)$ cancel out the errors, leaving more accurate FRF data.

7.6 Two incompletenesses of measured data

Theoretically, a real structure has an infinite number of DoFs and, therefore, vibration modes. No measurement is able to cover all. From this sense, the measured FRF data from a structure are always incomplete. When we deal with a structure which has been discretized by an analytical model such as a finite element model, the number of DoFs employed is usually much greater than the number of locations measurement can realistically afford to take. In this case, the measured FRF data are again incomplete.

First, the measured data can only cover a limited frequency range encompassing only a part of the vibration modes. Therefore, from the measured data we can derive an incomplete number of vibration modes. The frequency range covered in measurement is determined by the interest, by the measurement hardware limitation such as the accelerometer mounting, or by the range within which the structure's dynamic behaviour is modal. This can be called the *frequency incompleteness*.

Second, the number of coordinates used in FRF measurement is usually far less than the number of DoFs used in an analytical model. This means that the spatial description of vibration modes is not complete. This is the *spatial incompleteness*. The number of coordinates used in measurement is determined by the cost and time involved in conducting measurement, the accessibility of coordinates on the structure or the ability to measure at all desired coordinates. For vibration modes at low frequencies, a modest number of coordinates are sufficient to describe the mode shapes. High frequency modes have to be outlined using more coordinates. An insufficient number of coordinates may fail to describe the vibration mode shapes adequately or cause spatial aliasing where a high frequency mode is mistaken as a low frequency one.

The number of coordinates used in modal testing and the number of vibration modes covered in the measurement are rarely the same. If the former is greater than the latter, we can assume sufficient spatial resolution exists to distinguish between the measured vibration modes. If the reverse is true, then more coordinates are needed in order to better describe the measured vibration modes.

Assume N coordinates are needed to describe the vibration modes of a structure adequately. Then the FRF matrix of the ideal mathematical model will be dimensioned $N \times N$. By moving all measured coordinates to the upper rows in the matrix, we can illustrate the two incompletenesses in the matrix as shown in Figure 7.12.

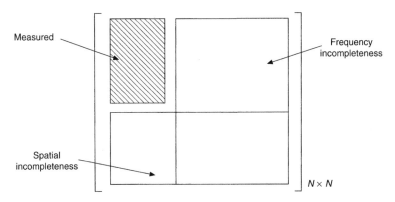

Figure 7.12 Two incompletenesses in measured FRF data shown in the FRF matrix

7.7 Initial assessment of measured FRF data

The quality of modal analysis relies critically on the quality of the measured FRF data or impact response data. Although modal analysis methods can try to minimize the effect of inaccuracy carried in the measured data, no method is able to rectify fundamental errors or measurement mistakes through the analysis. The modal properties derived from erroneous FRF data are susceptible to unacceptable errors. To compound the situation, often we have few effective means of identifying the errors in the derived modal model. This means that assessment on the quality of measured FRF data becomes fundamentally essential for experimental modal analysis.

The assessment of measured FRF data is basically to ascertain two things: (1) the structure satisfies the assumptions modal analysis requires; and (2) human and system errors are minimized or eliminated. Basically, the structure needs to comply with reciprocity, time invariance and linearity so that consistent modal properties exist in the measured FRF data which can be revealed by the subsequent analysis. If, however, these assumptions are not verified, then confidence in the derived modal properties will be eroded. The assessment is also to ensure that human errors such as incorrect bookkeeping are not made and errors caused by signal processing are minimized.

Although there are methods to assist in identifying potential errors in measured FRF data, some of them cannot be identified. For example, if the measurement system is not properly calibrated, then it would be impossible to find it just from the measured FRF data.

7.7.1 Repeatability check of the measured FRF data

The simplest, but not the least useful, assessment of FRF measurement is to ascertain the repeatability of the measurement. This is mainly to ensure that the structure's dynamic behaviour and the whole measurement set-up system are time-invariant. For selected force input and response locations, a linear structure should yield identical FRF curves for every measurement. By selection of a pair of input and output locations and testing conditions, a number of measurements can be carried out with time intervals. Typically, selected FRF measurements can be repeated before and after the whole structure is tested. This seemingly trivial process can be quite helpful to ensure that not only the structure's behaviour is constant, but also that the testing conditions are kept unaltered throughout the measurement.

7.7.2 Reciprocity check of the measured FRF data

A linear and time-invariant structure honours reciprocity property. For a single input, this means that the FRF data from a measurement should be identical if we exchange the locations of force and response. Theoretically, this property can be traced back to the symmetry of mass, stiffness and damping matrices. Because of the symmetry, the FRF matrix, which is mathematically the inverse of the dynamic stiffness matrix, will also be symmetric. The reciprocity property of the FRF can be used to assess the reliability and accuracy of the measured FRF data.

In measurement, this reciprocity may not be applicable to some part of a structure. For example, if we conduct a shaker test, certain coordinates may be inaccessible to the shaker because of physical constraints. This precludes the reciprocity check of some FRFs. To duplicate exact test conditions may also be an issue in the reciprocity check, such as the force condition and precision of two locations. It is not unusual to find that the anti-resonances of two FRFs from the same pair of coordinates do not agree well.

7.7.3 Linearity check of the measured FRF data

Perhaps the most important assumption of modal analysis is that the structure measured for FRF data behaves linearly. Without this assumption, modal analysis results will not be meaningful.

The ultimate check of linearity is to ensure that the FRF data are independent of excitation amplitudes. This can be achieved either qualitatively or quantitatively. For the former, FRF data from the same locations can be measured repeatedly with different but uncontrolled changes of excitation amplitudes. The measured FRF data can be overlaid to verify the uniformity of the curves. For the latter, controlled measurement is used to understand the nonlinearity existing in the structure. For example, an FRF measurement with constant response amplitude has the capacity to linearize nonlinearity.

7.7.4 Special characteristics of an FRF

From the theory of modal analysis, we can derive some characteristics of an FRF and use them to assess the FRF data measured in modal testing. This assessment has the potential to detect measurement errors or mistakes made during the testing.

The first characteristic is that for a point FRF measurement we expect to see an anti-resonance between two adjacent resonances. The rationale of this characteristic has been explained in Chapter 5. Therefore, if this characteristic is not fully observed on a point measurement, it is likely that the force and response transducers are not actually at the same coordinate, as they should be. Any minor offset of the two could degenerate some of the anti-resonances.

For a grounded structure, at very low frequency range, the predominant characteristic of the structure is its static stiffness. Therefore, at the beginning of the FRF, we should see a stiffness line before the first resonance appears. On the other hand, for a freely supported structure, the prevailing characteristic at very low frequency is the mass and inertia. This means we should see a mass line at the beginning of the FRF. Figures 7.13 and 7.14 illustrate these two cases.

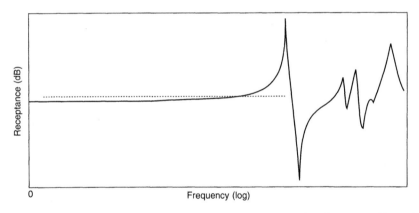

Figure 7.13 A point FRF of a grounded structure with the dotted stiffness line

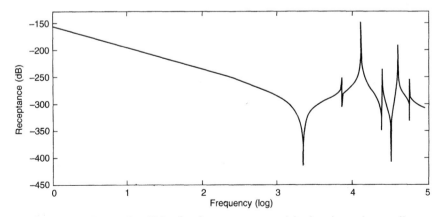

Figure 7.14 A point FRF of a free structure with the dotted mass line

Literature

1. Allemang, H.J., Host, H.W. and Brown, D.L. 1983: Multiple Input Estimation of Frequency Response Functions: Excitation Considerations. *ASME Paper*, No. 83-DET -73.
2. Asher, G.W. 1958: A Method of Normal Mode Excitation Utilising Admittance Measurements. *Proceedings of National Specialists' Meeting in Dynamics and Aeroelasticity*, 69–76.
3. Avitabile, P. and Haselton, D. 1997: A new procedure for selecting modal testing reference locations. *Sound and Vibration*, **31**, 24–29.
4. Beliveau, J.G., Vigneron, Y.S. and Draisey, D. 1986: Adaptation of modal testing procedure to shaker tables. *Proceedings of the 3rd Conference on Dynamic Response of Structures*, 513–521.
5. Bishop, R.E.D. and Gladwell, G.M.L. 1963: An Investigation into the Theory of Resonance Testing. *Philosophical Trans. of the Royal Society of London*, **255A**(1055), 241–280.
6. Bono, R.W. *et al.* 1999: New Developments in Multi-channel Test Systems. *Proceedings of the 17th International Modal Analysis Conference*, Kissimmee, FL, USA, 518.
7. Broch, J.T. 1980: *Mechanical Vibration and Shock Measurements*. Bruel & Kjaer.
8. Cawley, P. 1986: Rapid Measurement of Modal Properties Using FFT Analysers with Random Excitation. *Trans. ASME, Journal of Vibration, Acoustics, Stress and Reliability in Design*, **108**, 394–398.
9. Cogger, N.D. and Webb, R.V. 1983: Frequency Response Analysis. *Technical Report 010/83*, Solartron.
10. Comstock, T.R. 1999: Improving Exciter Performance in Modal Testing. *Proceedings of the 17th International Modal Analysis Conference*, Kissimmee, FL, USA, 1770–1775.
11. Cooley, J.W. and Tukey, J.W. 1965: An Algorithm for the Machine Calculation of Complex Fourier Series. *Mathematics of Computation*, **19**, 297–301.
12. Comstock, T. 1999: Improving exciter performance in modal testing. *Proceedings of the 17th International Modal Analysis Conference*, Kissimmee, FL, USA, 1770–1776.
13. Corelli, D. and Brown, D.L. 1984: Impact Testing Considerations. *Proceedings of the 3rd International Modal Analysis Conference*, Orlando, FL, 735–742.
14. Craig Jr, R.R. and So, Y.W.T. 1974: On Multiple-Shaker Resonance Testing. *AIAA Journal*, **12**(7), 924–931.
15. Der Auweraer, H.V. *et al.* 1991: Multiple excitation sine testing: an integrated approach. *DE-Vol. 38, Modal Analysis, Modelling, Diagnostics, and Control, SAME*, 31–37.
16. Ewins, D.J. 1984: *Modal Testing – Theory and Practice*. Research Studies Press, UK.
17. Ewins, D.J. and Griffin, J. 1981: A State-of-the-Art Assessment of Mobility Measurement Techniques – Results for the Mid-Range Structures. *Journal of Sound and Vibration*, **78**(2), 197–222.
18. Ewins, D.J. and Imregun, M. 1986: State-of-the-Art Assessment of Structural Dynamic Response Analysis Methods (DYNAS). *Shock and Vibration Bulletin*, **56**, Part I.
19. Farrar, C.R. *et al.* 1999: A review of methods for developing accelerated testing criteria. *Proceedings of the 17th International Modal Analysis Conference*, Kissimmee, FL, USA, 608.
20. Farrar, C.R. *et al.* 1999: Excitation methods for bridge structures. *Proceedings of the 17th International Modal Analysis Conference*, Kissimmee, FL, USA, 1063–1069.
21. Formenti, D. 2000: Excitation direction in modal testing – is it important? *Sound and Vibration*, **34**(3), 14–16.
22. French, M. 1998: Contact problems in modal analysis. *Sound and Vibration*, January, 26–30.
23. Friswell, M.I. and Penny, J.E.T. 1990: Stepped multisine modal testing using phased components. *Mechanical Systems and Signal Processing*, **4**(2), 145–156.
24. Good, M. 1984: Summary of Excitation Signals for Structural Testing. *Proceedings of the 3rd International Modal Analysis Conference*, Orlando, FL 566–571.
25. Hallauer Jr, W.L. and Stafford, J.F. 1978: On the Distribution of Shaker Forces in Multiple-Shaker Modal Testing. *The Shock & Vibration Bulletin*.

26. Halvorsen, W.G. and Brown, D.L. 1997: Impulse Technique for Structural Frequency Response Testing. *Journal of Sound and Vibration*, November, 8–21.

27. Harris, C.M. 1995: *Shock and Vibration Handbook*. McGraw-Hill, ISBN 0-07-026920-3.

28. Hunt, D.L. *et al.* 1984: Optimal Selection of Excitation Methods for Enhanced Modal Testing. *The AIAA Dynamics Conference*, Palm Springs, CA, AIAA Paper No. 84-1068, 549–553.

29. Hutin, C. 2000: Modal analysis using appropriated excitation techniques. *Sound and Vibration*, **34**(10), 18–25.

30. Jenkins, G. and Watts, D. 1968: *Spectral Analysis and its Application*. Holden-Day, San Francisco.

31. Johnson, J.L. 1996: Demystifying frequency response. *Hydraulics and Pneumatics*, **49**, 16.

32. Kabe, A.M. 1984: Multi-Shaker Mode Testing. *AIAA Journal of Guidance*, **7**(6), 740–746.

33. Kaiser, A. and Goovaerts, F. 1990: Automating large modal tests. *Sound and Vibration*, August, 20–24.

34. Landgraf, M.W. 1997: Vibration testing of heavy machinery structures. *Sound and Vibration*, **31**, 16–18.

35. Lewis, R.D. and Wrisley, D.L. 1950: A System for the Excitation of Pure Nature Modes of Complex Structures. *Journal of Aeronautical Sciences*, **17**(11), 705–722.

36. McConnell, K. 1995: *Vibration Testing, Theory and Practice*. John Wiley and Sons Inc., ISBN 0-471-30435-2.

37. McConnell, K. 1990: Errors in Using Force Transducers. *Proceedings of the 10th International Modal Analysis Conference*, San Diego, CA, 884–890.

38. Mitchell, L.D. 1982: Improved Method for the Fast Fourier Transform (FFT) Calculation of the Frequency Response Function. *Transactions of ASME, Journal of Mechanical Design*, **104**, 277–279.

39. Mitchell, L.D. *et al.* 1988: An Unbiased Frequency Response Function Estimator. *Journal of Modal Analysis*, **3**(1), 12–19.

40. Mitchell, L.D. and Elliott, K.B. 1984: A method for designing stingers for use in mobility testing. *Proceedings of the 2nd International Modal Analysis Conference*, Orlando, FL, 872–876.

41. Newland, D.E. 1984: *An Introduction to Random Vibrations and Spectral Analysis*. Longman, 2nd edition.

42. Olsen, N. 1984: Excitation Functions for Structural Frequency Response Measurements. *Proceedings of the 3rd International Modal Analysis Conference*, Orlando, FL, 894–902.

43. Pavic, A., Pimentel, R.L. and Waldron, P. 1998: Instrumented Sledge Hammer Impact Excitation: Worked Examples. *Proceedings of the 16th International Modal Analysis Conference*, Santa Barbara, CA, 929–935.

44. Peterson, E.L. and Mouch, T.A. 1990: Modal Excitation: Force Drop-off at Resonances. *Proceedings of the 8th International Modal Analysis Conference*, Kissimmee, FL, 1226–1231.

45. Phillips, A.W., Allemang, R.J. and Zucker, A.T. 1998: A New Excitation Method: Combining Burst Random Excitation with Cyclic Averaging. *Proceedings of the 16th International Modal Analysis Conference*, 891–899.

46. Phillips, A.W., Zucker, A.T. and Allemang, R.J. 1999: A Comparison of MIMO-FRF Excitation/Averaging Techniques on Heavily and Lightly Damped Structures. *Proceedings of the 17th International Modal Analysis Conference*, Kissimmee, FL, USA, 1161–1168.

47. Randall, R. 1977: *Frequency Analysis*. Bruel & Kjaer.

48. Rao, D.K. 1987: Electrodynamic Interaction between a Resonating Structure. *Proceedings of the 5th International Modal Analysis Conference*, London, UK, 1142–1145.

49. Reynolds, P. and Pavic, A. 2000: Impulse hammer versus shaker excitation for the modal testing of building floors. *Experimental Techniques*, **24**(3), 39–44.

50. Rogers, J.D. 1990: An introduction to shaker shock simulations. *Proceedings of the 10th International Modal Analysis Conference*, San Diego, CA, 905–911.

51. Remmers, G. and Belsheim, R.O. 1964: Effects of Technique on Reliability of Mechanical Impedance Measurement. *Shock and Vibration Bulletin*, **34**.

52. Shelley, S. *et al.* 1988: Investigation and Modelling of Shaker Dynamic Effects. *Proceedings of the 13th International Seminar on Modal Analysis*. Katholieke Universiteit, Leuven.
53. Wang, M.L. and Subia, S.R. 1991: Displacement time histories by direct numerical integration of acceleration data. *DE-Vol. 37, Vibration Analysis, SAME*, 29–35.
54. To, W.M. and Ewins, D.J. 1990: The characteristics of frequency response function estimators using random excitations. *Proceedings of the 10th International Modal Analysis Conference*, San Diego, CA, 1101–1107.
55. Tomlinson, G.R. 1979: Force Distortion in Resonance Testing of Structures with Electro-Dynamic Vibration Exciters. *Journal of Sound and Vibration*, **163**(3), 337–350.
56. Trail-Nash, R.W. 1958: On the Excitation of Pure Natural Modes in Aircraft Resonance Testing. *Journal of Aerospace Sciences*, **01**(25), 775–778.
57. Walter, P.L. 1999: Accelerometer Selection for and Application to Modal Analysis. *Proceedings of the 17th International Modal Analysis Conference*, Kissimmee, FL, USA, 566.
58. White, R.G. 1971: Evaluation of the dynamic characteristics of structures by transient testing. *Journal of Sound and Vibration*, **15**(2), 147–161.
59. White, R.G. 1972: Specially shaped transient forcing functions for frequency response measurement. *Journal of Sound and Vibration*, **23**(3), 307–318.
60. Wicks, A.L. and Vold, H. 1986: The Hs Frequency Response Function Estimation. *Proceedings of the 4th International Modal Analysis Conference*, Los Angeles, CA.
61. Zaveri, K. 1984: *Modal Analysis of Large Structures – Multiple Exciter Systems*. Bruel & Kjaer Publications.

8

Modal analysis methods – frequency domain

8.1 Introduction

Modal analysis is a process of extracting modal parameters (natural frequencies, damping loss factors and modal constants) from measured vibration data. Since the measured data can be in the form of either frequency response functions or of impulse responses, there are frequency domain modal analysis and time domain modal analysis.

The fundamental of modal analysis using measured frequency response function data is about curving fitting the data using a predefined mathematical model of the measured structure. This model assumes the number of DoFs of the structure, its damping type and possibly the number of vibration modes within the measured frequency range. These assumptions should dictate the mathematical expression of each FRF curve from measurement. As a result, the subsequent work will be a curve-fitting process trying to derive all modal parameters in a mathematical formula of an FRF using measurement data.

The accuracy of modal analysis is not a simple question of how a measured FRF curve is best fitted in a pure mathematical sense. Obviously, the more accurate the measured FRF data are, the better chance we have to get more accurate curve fitting. In mathematics, the accuracy or successfulness of a curve-fitting endeavour can usually be appraised by defining an error function and aiming to minimize it. This approach is only valid if the correct mathematical formula is used in the curve fitting. If, however, an incorrect mathematical model is used, the curve-fitting outcome is doomed to be a bad one if not a failure, even if the error function is actually minimized numerically.

For frequency domain modal analysis, the mathematical model is the analytical expression of a frequency response function that is truly representative of the FRF data from measurement. This raises a few questions. First, a structure has an infinite number of DoFs. As a result, the measured FRF data contain the modal information of the modes beyond the frequency range of measurement as well as those within. Assume a structure can be discretized as an 'N' DoF structural system. An analytical expression of an FRF of the system is:

$$\alpha_{ij}(\omega) = \sum_{r=1}^{N} \frac{{}_r A_{ij}}{\omega_r^2 - \omega^2 + \omega_r^2 \eta_r j} \tag{8.1}$$

If only the first '*m*' modes fall within the frequency range of a baseband measurement (frequency commences from zero), then the expression of

$$\alpha_{ij}(\omega) \approx \sum_{r=1}^{m} \frac{{}_r A_{ij}}{\omega_r^2 - \omega^2 + \omega_r^2 \eta_r j} \tag{8.2}$$

precludes the impact of modes beyond the measurement range. Since those modes are not measured, there are no data with which they can be analysed. One solution is to elect that those invisible modes be represented by a high frequency residual term $R_{ij}(\omega)$ so that:

$$\alpha_{ij}(\omega) = \sum_{r=1}^{m} \frac{{}_r A_{ij}}{\omega_r^2 - \omega^2 + \omega_r^2 \eta_r j} + R_{ij}(\omega) \tag{8.3}$$

This residual term is usually made to be a linear function so that modal analysis can be carried out without much additional problems caused by invisible modes. However, this highlights the fact that the mathematical model derived by modal analysis from measured FRF data does have inherent inadequacy.

Like any curve fitting in numerical analysis, the error function is usually defined as the difference between the FRF estimate from the identified vibration modes and the measured FRF data:

$$e_{ij}(\omega) = \tilde{\alpha}_{ij}(\omega) - \alpha_{ij}(\omega) \tag{8.4}$$

where $\tilde{\alpha}_{ij}(\omega)$ is the measured FRF data. Many modal analysis methods are based on the minimization of this error function. As discussed above, the validity and accuracy of the outcome depends not only on the accuracy of the measured data, but also on the validity of the analytical expression of the FRF $\alpha_{ij}(\omega)$.

8.2 Detection of vibration modes from measured FRF data

No matter what modal analysis methods we will be using, a very essential question in the analysis is how many vibration modes within a selected frequency range are we dealing with, and, if a particular peak of the FRF represents a genuine vibration mode. There is an intrinsic dilemma here. The exact number of modes may become clear if proper analysis has been completed but a proper analysis is often impossible without first knowing the exact number of modes at present. Another relevant question is weather we really need to know the exact number within the selected frequency range to obtain meaningful and useful analysis results.

With the theory from previous chapters (especially from Chapters 5 and 6), we know that an FRF can be presented in a number of different forms. Some of them are more useful for identifying a vibration mode. It is equally helpful to understand what forms of FRF are not conducive for identifying vibration modes so as to avoid using them in analysis.

The modulus vs frequency plot of an FRF is sometimes perceived to be the crucial plot for identifying vibration modes as it manifests resonance peaks. This fallacy does not accord with two scenarios: (1) not every mode will appear on every measured

FRF; and (2) not every peak of an FRF is an authentic mode. For example, the measured FRF shown in Figure 8.1 is from a structure consisting of beams. It contains a few peaks which are due to noise, and it also has some modes invisible in the frequency range.

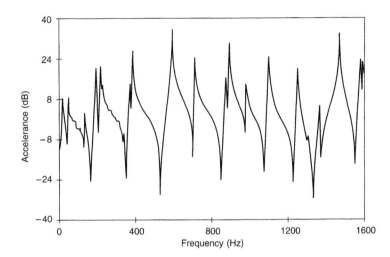

Figure 8.1 Modulus of a point FRF measured from a beam structure

The identification of vibration modes prior to analysis needs to be based on several FRF curves. A convenient solution is to add all the FRF curves in dB scales together. This will ensure that vibration modes become more prominent in the total FRF plot. This summation bears little physical meaning. It is a pure numerical treatment. It does not question if all the prominent peaks are modes although unauthentic peaks do not prevail in the simulation.

The phase plot of an FRF may provide additional help. Theoretically, we accept that a vibration mode would cause a 180 degree phase change. With the presence of damping, the change of phase is less than clear cut. For a real measurement, the phase plot is very sensitive to noise. This means sometimes there will be many phase changes due to reasons other than a resonance. In this case, the phase plot is not a useful source of information for identifying vibration modes.

There are other methods such as the SVD-based. All possible methods could fail in case of close modes. If two modes are really close, then it will be very difficult to tell them from one mode. This situation happens more often than usually thought because many real structures show some form of geographical symmetry. The presence of this symmetry will theoretically lead to pairs of identical modes. However, due to structural imperfection or measurement errors, these pairs of identical modes become pairs of close modes. In a measured FRF curve, each pair appears to be a single mode.

The Nyquist plot can lend a big hand in identifying modes. A genuine mode should show the symptom of a circle in the Nyquist plot. A false resonance peak will simply trace out a straight line on the Nyquist plane.

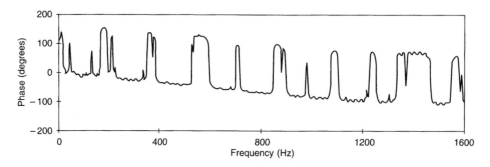

Figure 8.2 Phase of the point FRF in Figure 8.1

8.3 Derivation of modal data from FRF data – SDoF methods

All SDoF modal analysis methods are based on the SDoF assumption: at the vicinity of a resonance, the FRF is dominated by the contribution of that vibration mode and the contributions of other vibration modes are negligible. If this assumption holds, then the FRF from an MDoF system or a real structure can be treated as the FRF from an SDoF system momentarily. The simplicity of the mathematical model for an SDoF system can then be used in the curve fitting to derive the modal parameters mode by mode.

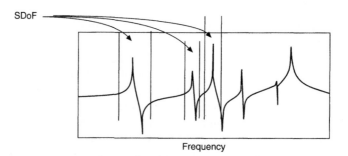

Figure 8.3 SDoF assumption on an MDoF FRF

This SDoF assumption can actually be extended. Instead of assuming the negligence of the contributions made by all other modes at the vicinity, these contributions can be taken as a complex constant. This is reasonably true if modes are well separated. If this extended SDoF assumption holds, then some modal analysis methods can be successfully applied. In Figure 8.5, we will see how this SDoF assumption works for the circle fit method.

8.3.1 Peak-picking method

The 'peak-picking method' is perhaps the simplest SDoF method for modal analysis. It is also called the 'half-power method'. It relies on the strict compliance of the SDoF assumption. The method treats the FRF data at the vicinity of a resonance as the data from an SDoF system. The procedure of using the peak-picking method is:

(a) Estimating the natural frequency
    The natural frequency of the rth mode selected for analysis is identified from the peak value of the FRF $|\alpha_r(\omega)|_{max}$ as $\omega_r = \omega_{peak}$.

(b) Estimating damping
    For estimating damping, the half power points at ω_a and ω_b are located first from each side of the identified peak with amplitude $\dfrac{\alpha_{max}}{\sqrt{2}}$. The damping loss factor or the damping ratio can then be estimated from the width of the resonance peak as:

$$\eta_r = \frac{\omega_b^2 - \omega_a^2}{2\omega_r^2} \approx \frac{\omega_b - \omega_a}{\omega_r} \tag{8.5}$$

or

$$\zeta_r = \frac{\omega_b^2 - \omega_a^2}{4\omega_r^2} \approx \frac{\omega_b - \omega_a}{2\omega_r} \tag{8.6}$$

(c)   Estimating the modal constant
    From the SDoF model, the FRF at the peak is known to be $\alpha_{max} = \dfrac{A_r}{\eta_r \omega_r^2}$. The modal constant A_r can be estimated from $A_r = \alpha_{max}\eta_r\omega_r^2$. For viscous damping model, this becomes $A_r = 2\alpha_{max}\zeta_r\omega_r^2$.

Due to its remarkable simplicity, the peak-picking method (Figure 8.4) can derive quick analysis results. However, it is not capable of producing accurate modal data. This method relies on the peak FRF value, which is very difficult to measure accurately, to estimate the natural frequency and modal constant. Damping is estimated from half power points only. No other FRF data points are used. The half power points have to be interpolated, as it is unlikely that they are two of the measured data points.

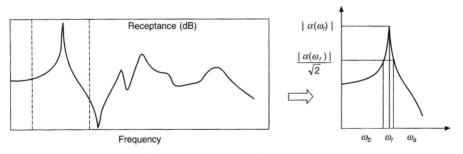

Figure 8.4 Peak-picking method

It is also evident that there is no mechanism for this method to deal with noise in the measured FRF data.

The peak-picking method is suitable only for lightly damped FRF data with well-separated modes and good frequency resolution. It can perform a survey study leading to a more sophisticated analysis.

8.3.2 Circle fit method

The circle fit method is the most used SDoF modal analysis method. It is based on the circularity of the Nyquist plot of an SDoF FRF discussed in Chapter 4. With structural damping, the receptance FRF $\alpha(\omega)$ traces a perfect circle on the Nyquist plane.

$$(\text{Re}(\alpha))^2 + \left(\text{Im}(\alpha) + \frac{1}{2h} \right)^2 = \left(\frac{1}{2h} \right)^2 \tag{4.27}$$

Such a unique property gives rise to a mathematically convenient and accurate curve-fitting model for modal analysis.

When this circularity property of a SDoF system is used to analyse an FRF from an MDoF system, the SDoF assumption should apply first. Of course we can assume that at the vicinity of a mode, the contributions of all other modes are negligible. This assumption is difficult to meet in reality. However, the circle fit method can rely on a more relaxed assumption. The receptance FRF of an 'N' DoF system in modal form is given in Chapter 6 as:

$$\alpha_{jk}(\omega) = \sum_{r=1}^{N} \frac{_r A_{jk}}{\lambda_r^2 - \omega^2} \tag{8.7}$$

If the rth mode is to be analysed, we can single it out from the summation so that:

$$\alpha_{jk}(\omega) = \frac{_r A_{jk}}{\lambda_r^2 - \omega^2} + \sum_{\substack{s=1 \\ s \neq r}}^{N} \frac{_s A_{jk}}{\lambda_s^2 - \omega^2} \tag{8.8}$$

If the summation term in this equation can be approximated by a complex constant such that:

$$\alpha_{jk}(\omega) \approx \frac{_r A_{jk}}{\omega_r^2 - \omega^2 + j\eta_r \omega_r^2} + B_{jk} \tag{8.9}$$

then the circularity of the Nyquist plot shall not change except that the circle is shifted a distance away from the origin of the complex plane by the complex constant B_{jk}, as shown in Figure 8.5.

If the modal data are to be derived on the circle, then the relative position of the circle on the complex plane becomes irrelevant to the analysis. The SDoF assumption is now extended. It means that when we discuss the analysis method, we can simply use the mathematical model of the SDoF FRF without needing to consider the complex constant B_{jk}.

The derivation of the modal data from circle fitting follows the same sequential procedure as the peak-picking method, i.e. from natural frequency to damping loss factor and to modal constant. However, the accuracy of the derived modal data is

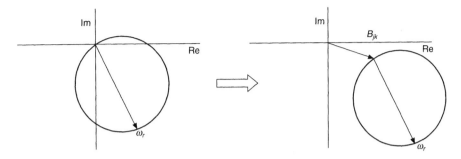

Figure 8.5 The shift of a Nyquist circle by a complex constant

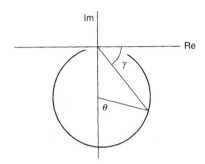

Figure 8.6 Derivation of natural frequency

improved significantly. In order to estimate the natural frequency, the first step is to fit a circle from the selected FRF data points. These points are selected at the vicinity of the resonance peak.

Mathematically, it can be found that the natural frequency ω_r is at the location where maximum arc change occurs on the Nyquist circle. From Figure 8.6 we have:

$$\tan \frac{\theta}{2} = \tan(90° - \gamma) = \frac{\text{Re}(\alpha)}{\text{Im}(\alpha)} = \frac{\omega_r^2 - \omega^2}{\omega_r^2 \eta_r} \tag{8.10}$$

or

$$\omega^2 = \omega_r^2 \left(1 - \eta_r \tan \frac{\theta}{2}\right) \tag{8.11}$$

It can be found that:

$$\frac{d\theta}{d\omega^2} = \frac{-2}{\omega_r^2 \eta_r} \frac{\eta_r^2 \omega_r^4}{\eta_r^2 \omega_r^4 + (\omega_r^2 - \omega^2)^2} \tag{8.12}$$

For the maximum derivative of $\dfrac{d\theta}{d\omega^2}$, it can be derived that the frequency:

$$\omega^2 = \frac{1 + \sqrt{1 + 3(1 + \eta_r^2)}}{3} \omega_r^2 \tag{8.13}$$

When the modal damping loss factor η_r is small, the following approximation is sufficient:

$$\omega_r = \omega \bigg|_{\frac{d\theta}{d\omega^2} = \max} \qquad (8.14)$$

This means that the rate at which the locus sweeps around the circular arc takes the maximum value at resonance ω_r. Since the frequency resolution of the FRF data points may not be sufficiently fine, interpolation can be used to enhance frequency resolution so that the natural frequency can be located accurately.

Once the natural frequency ω_r is accurately located on the Nyquist circle, the damping loss factor can be estimated from any FRF data point using:

$$\eta_r = \frac{\omega_r^2 - \omega_a^2}{\omega_r^2} \frac{1}{\tan \dfrac{\theta_a}{2}} \qquad (8.15)$$

Theoretically, we should get an identical damping loss factor using any FRF data point. However, due to measurement noise, nonlinearity or errors, the damping loss factor estimated from different FRF points varies. Thus, the estimates of η_r become a useful indicator for the accuracy of the analysis. For example, as Figure 8.7 shows, if the natural frequency ω_r is not located accurately, then the estimate of η_r will exhibit a systematic change from the FRF data points before or after ω_r.

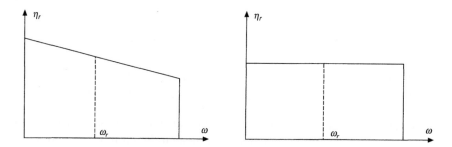

Figure 8.7 Damping plot with inaccurate and accurate ω_r estimate

It is also possible to use one FRF point before the estimated natural frequency and one after that to estimate the damping loss factor.

If the data points correspond to frequencies ω_b and ω_a, as shown in Figure 8.8, then from equation (8.10) we have:

$$\tan \frac{\theta_a}{2} = \frac{\omega_r^2 - \omega_a^2}{\omega_r^2 \eta_r} \qquad (8.16)$$

and

$$\tan \frac{\theta_b}{2} = \frac{\omega_r^2 - \omega_b^2}{\omega_r^2 \eta_r} \qquad (8.17)$$

Combining these formulae leads to an estimate of the damping loss factor:

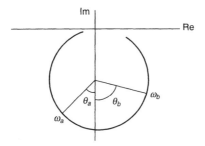

Figure 8.8 Damping estimation from the Nyquist circle

$$\eta_r = \frac{\omega_a^2 - \omega_b^2}{\omega_r^2} \frac{1}{\tan\dfrac{\theta_a}{2} + \tan\dfrac{\theta_b}{2}} \tag{8.18}$$

By selecting different pairs of FRF data points, one from before and one from after the natural frequency, we have an array of damping estimates. These values, when plotted against the number of data points, form a three-dimensional damping plot. For an ideal case, this 3-D plot is a horizontal flat plate. However, the plot is sensitive to errors in the natural frequency estimation, noise and nonlinearity in the FRF data. It is a useful indicator for assessing the quality of the derived modal data. Figure 8.9 shows the three-dimensional damping plot with inaccurate and accurate ω_r estimate.

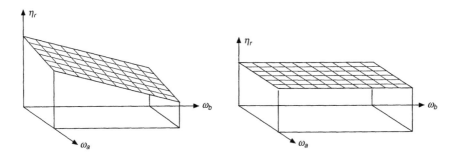

Figure 8.9 A 3-D damping plot with inaccurate and accurate ω_r estimate

The last to be estimated is the modal constant. This is done from the diameter of the Nyquist circle. The diameter, if denoted as $_rD_{jk}$ for the rth mode of the FRF $\alpha_{jk}(\omega)$, is conveniently quantified at the location of natural frequency where $\omega = \omega_r$. Therefore, the

$$_rD_{jk} = \frac{|_rA_{jk}|}{\omega_r^2 \eta_r} \tag{8.19}$$

After the circle fit, the diameter is known. Therefore, the modal constant can be estimated as:

$$|_r A_{jk}| = {}_r D_{jk} \omega_r^2 \eta \tag{8.20}$$

The phase angle of the modal constant $_r A_{jk}$ is given by the location of the natural frequency.

To summarize, the procedure of the SDoF circle fit method follows the succeeding steps:

(1) select a vibration mode from the resonance peak of the receptance FRF;
(2) select data points of the receptance FRF on the Nyquist plane;
(3) fit a circle using selected points;
(4) locate the natural frequency of the mode from the maximum arc rate change and estimate the phase angle for the modal constant;
(5) estimate the damping loss factor;
(6) determine the modulus of the modal constant from the diameter of the circle and its phase from the location of the natural frequency.

8.3.3 Inverse FRF method

An alternative to the SDoF circle fit method is a method to make use of the linearity property of the inverse FRF data. It was discussed in Chapter 4 that the inverse receptance FRF of an SDoF system shows a remarkable simplicity. Its real and imaginary parts can be linear functions of either ω or ω^2. This property gives rise to curve-fitting a straight line rather than a circle. However, when dealing with the FRF of an MDoF system, that simplicity does not hold any more, as mode complexity and contributions of other modes compromise it. It is only when these factors become insignificant that the simplicity of straight lines can be relied upon for modal analysis.

If we can assume that at the vicinity of the rth mode, contributions of other modes are negligible, then the receptance FRF $\alpha_{jk}(\omega)$ can be approximated as:

$$\alpha_{jk}(\omega) \approx \frac{{}_r A_{jk}}{\omega_r^2 - \omega^2 + j\eta_r} \tag{8.21}$$

Furthermore, if this vibration mode is nearly real, then the modal constant $_r A_{jk}$ is a real constant. The inverse FRF becomes very simple mathematical function:

$$\frac{1}{\alpha_{jk}(\omega)} \approx \frac{1}{{}_r A_{jk}} (\omega_r^2 - \omega^2) + \frac{\eta_r}{{}_r A_{jk}} j \tag{8.22}$$

Both the real and imaginary parts can be plotted as straight lines shown in Figure 8.10.

These two plots are both straight lines. The intercept of the real plot indicates the natural frequency of the rth mode and its slope gives the inverse of the modal constant. The intercept of the imaginary plot indicates the amount of structural damping.

If the vibration mode is a well isolated but complex mode, then the inverse FRF method needs to be reformulated. Since the modal constant is complex, we can express it in a polar form as:

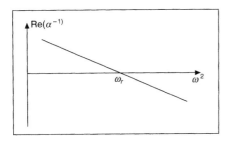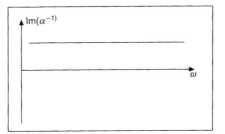

Figure 8.10 Plots of the real and imaginary parts of the inverse FRF

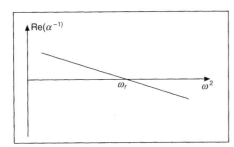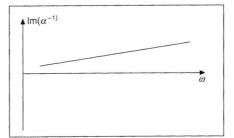

Figure 8.11 Real and imaginary parts of the inverse FRF

$$_rA_{jk} = |_rA_{jk}|e^{j\theta} \tag{8.23}$$

The inverse receptance can be formulated as:

$$\alpha_{jk}^{-1}(\omega) \approx \frac{\omega_r^2 - \omega^2 + j\eta_r}{|_rA_{jk}|} e^{j\theta} = (\text{Re}(\alpha^{-1}) + j\,\text{Im}(\alpha^{-1}))e^{j\theta} \tag{8.24}$$

It can be shown that

$$\alpha_{jk}^{-1}(\omega) \approx (\text{Re}(\alpha^{-1})\cos\theta - \text{Im}(\alpha^{-1})\sin\theta) + (\text{Re}(\alpha^{-1})\sin\theta - \text{Im}(\alpha^{-1}\quad \theta)j$$

$$= X + Yj \tag{8.25}$$

and
$$\frac{dX}{d\omega^2} = \frac{1}{|_rA_{jk}|}\cos\theta \tag{8.26}$$

$$\frac{dY}{d\omega^2} = \frac{-1}{|_rA_{jk}|}\sin\theta \tag{8.27}$$

$$\frac{dY}{dX} = \tan\theta \tag{8.28}$$

Therefore, the inverse FRF method can be used to derive the modal data of the rth mode from the FRF data using the following procedure:

(1) Plot the real and imaginary parts of the inverse FRF separately against frequency square.

(2) Curve fit both plots with straight lines.
(3) From the intercept of the real plot determine the natural frequency.
(4) From the ratio of two slopes (tan θ) estimate the phase of the modal constant.
(5) From either slope estimate the modulus of the modal constant.
(6) From either the real or imaginary part estimate the damping loss factor.

8.3.4 Least-squares method

This method aims to find the best estimates of modal data of a selected mode that minimizes a defined error function. This function converts the curve fitting of a nonlinear mathematical model into a weighted linear model. Such a conversion makes the curve fitting simpler and the solution easier to derive.

We commence from the modal expression of a mobility FRF:

$$Y_{jk}(\omega) = \sum_{r=1}^{N} \frac{{}_r A_{jk}}{\lambda_r^2 - \omega^2} j\omega \tag{8.29}$$

To simplify the notation during this derivation, we omit the subscript jk so that the FRF is written as:

$$Y(\omega) = \sum_{r=1}^{N} \frac{A_r}{\lambda_r^2 - \omega^2} j\omega \tag{8.30}$$

To differ, assume the mobility FRF data measured is denoted as $H(\omega)$. If the rth mode is to be analysed, then for an FRF data point at $\overline{\omega}_i$, an error function can be defined as:

$$E_i = \frac{A_r}{\lambda_r^2 - \overline{\omega}_i^2} j\overline{\omega}_i + \sum_{\substack{s=1 \\ s \neq r}} \frac{A_r}{\lambda_s^2 - \overline{\omega}_i^2} j\overline{\omega}_i - H(\overline{\omega}_i) = \frac{A_r}{\lambda_r^2 - \overline{\omega}_i^2} j\overline{\omega}_i + R(\overline{\omega}_i) \tag{8.31}$$

This is a nonlinear error function for the complex modal data A_r and λ_r. The expansion of the function into a Taylor series would bring about poor numerical outcomes for the estimation. However, the function can be transformed as:

$$E_i = \frac{1}{\lambda_r^2 - \overline{\omega}_i^2} (j\overline{\omega}_i A_r + (\lambda_s^2 - \overline{\omega}_i^2) R(\overline{\omega}_i)) \tag{8.32}$$

If the complex natural frequency λ_r is known approximately, then the term $\frac{1}{\lambda_r^2 - \omega^2}$ in equation (8.32) can be treated as a weighting for the linear function of parameters A_r and λ_r^2. It is hereafter denoted as w_i. The modulus squared error is given as:

$$E^2 = E_i E_i^* = |w_i|^2 (j\overline{\omega}_i A_r + (\lambda_s^2 - \overline{\omega}_i^2) R(\overline{\omega}_i))(-j\overline{\omega}_i A_r^* + (\lambda_s^2 - \overline{\omega}_i^2) R^*(\overline{\omega}_i)) \tag{8.33}$$

Here, the bar on the top indicates the complex conjugate. When using a number of the FRF data points in the least-squares estimate, the summation of the errors is given by:

$$E^2 = \sum_i |w_i|^2 (j\overline{\omega}_i A_r + (\lambda_s^2 - \overline{\omega}_i^2) R(\overline{\omega}_i))(-j\overline{\omega}A_r^* + (\lambda_s^2 - \overline{\omega}_i^2) R^*(\overline{\omega}_i)) \tag{8.34}$$

This error function can be minimized in order to derive the best estimates for the complex parameters A_r and λ_r^2. By taking the derivatives of equation (8.34) with respect to the unknown parameters A_r and λ_r^2, we can obtain the following two complex linear simultaneous equations:

$$\begin{cases} A_s \sum |w_i|^2 \overline{\omega}_i^2 - j\lambda_s^2 \sum |w_i|^2 \overline{\omega}_i R(\overline{\omega}_i) + j \sum |w_i|^2 j\overline{\omega}_i^2 R(\omega_i) = 0 \\ jA_s \sum |w_i|^2 \overline{\omega}_i R(\overline{\omega}_i) + \lambda_s^2 \sum |w_i|^2 |R(\overline{\omega}_i)|^2 - \sum |w_i|^2 \overline{\omega}_i^2 |R(\overline{\omega}_i)|^2 = 0 \end{cases} \tag{8.35}$$

The complex natural frequency λ_r is linked to the natural frequency and damping loss factor ($\lambda_r^2 = \omega_r^2 (1 + \eta_r j)$). From these equations, the modal constant A_r, natural frequency ω_r and damping loss factor η_r of the rth mode are obtained. This estimation can be repeated for all the measured modes of an FRF.

This method relies on initial approximation of natural frequencies in order to incorporate the weighting function used in equation (8.33). These approximations can be determined from the simple peak-picking analysis. Accurate modal data shall be determined from equation (8.35).

8.3.5 Dobson's method

The inverse FRF method can be developed into a more sophisticated method. This development is based on the extended SDoF assumption used by the circle fit method, i.e. the collective contribution of other modes is a complex constant. Like the inverse FRF method, it is based on curve-fitting straight lines. If we analyse the rth mode of an FRF, then the function can be written as:

$$\alpha_{jk}(\omega) = \frac{A_r}{\lambda_r^2 - \omega^2} + R \tag{8.36}$$

Here, R is a complex constant. If an FRF data point at the vicinity of the rth mode is $\alpha(\omega)$, then the difference between this data point and others nearby is defined as:

$$\Delta(\Omega) = \frac{A_r}{\lambda_r^2 - \omega^2} - \frac{A_r}{\lambda_r^2 - \Omega^2} \tag{8.37}$$

This difference does not assist much in estimating the modal data. However, if we define a new function from $\Delta(\Omega) = \dfrac{A_r}{\lambda_r^2 - \omega^2} - \dfrac{A_r}{\lambda_r^2 - \Omega^2}$ as:

$$\Theta(\omega^2) = \frac{\omega^2 - \Omega^2}{\alpha(\omega) - \alpha(\Omega)} \tag{8.38}$$

we find from equation (8.37) that

$$\Theta(\omega^2) = \frac{\omega^2 - \Omega^2}{\alpha(\omega) - \alpha(\Omega)} = \frac{1}{A_r}(\lambda_r^2 - \omega^2)(\lambda_r^2 - \Omega^2) \tag{8.39}$$

What follows is a clear algebra to convert this seemingly complicated complex function into two linear equations of variable ω^2. Since modal constant A_r can be written in its real and imaginary parts as $A_r = P + Qj$ and the complex natural frequency square as $\lambda_r^2 = \omega_r^2(1 + \eta_r j)$, equation (8.39) can be detached into its real and imaginary parts:

$$\text{Re}(\Theta) = c_R + t_R \omega^2 \tag{8.40}$$

$$\text{Im}(\Theta) = c_I + t_I \omega^2 \tag{8.41}$$

Here the slope of these two linear functions of variable ω^2 are given as:

$$t_R = -\frac{P\omega_r^2 + Q\eta_r\omega_r^2}{P^2 + Q^2} + \frac{P}{P^2 + Q^2}\Omega^2 \tag{8.42}$$

$$t_I = -\frac{P\eta_r\omega_r^2 - Q\omega_r^2}{P^2 + Q^2} - \frac{Q}{P^2 + Q^2}\Omega^2 \tag{8.43}$$

The intercepts are given by:

$$c_R = \frac{P(\omega_r^2(\omega_r^2 - \Omega^2) - \eta_r^2\omega_r^4) + Q\eta_r\omega_r^2(2\omega_r^2 - \Omega^2)}{P^2 + Q^2} \tag{8.44}$$

$$c_I = \frac{P\eta_r\omega_r^2(2\omega_r^2 - \Omega^2) - Q(\omega_r^2(\omega_r^2 - \Omega^2) - \eta_r^2\omega_r^4)}{P^2 + Q^2} \tag{8.45}$$

By selecting frequency Ω around the natural frequency ω_r, we can obtain a series of straight lines for both the real and imaginary parts of function $\Theta(\omega^2)$. Equations (8.42) to (8.45) are not easy to use to derive modal parameters. However, we note that both slopes in equations (8.42) and (8.43) are linear functions of Ω^2. Therefore, by selecting a different frequency Ω, we can derive a series of slopes in these equations. These slopes can be graphically presented as two slope plots against Ω^2. Fitting straight lines from these two plots, the modal parameters can be identified from the slopes and intercepts of the two straight lines given by equations (8.42 and 8.43).

Figure 8.12 shows a receptance FRF of an MDoF system. To analyse its first mode, we select the data points shown in the Nyquist plot on the right and use equations (8.40) and (8.41) to calculate a family of straight lines shown in Figure 8.13.

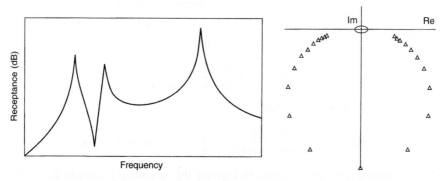

Figure 8.12 An FRF and the first mode of an MDoF system

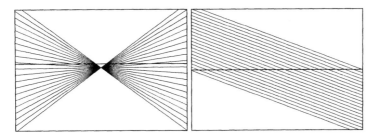

Figure 8.13 Real and imaginary parts of function $\Theta(\omega^2)$ with different frequency Ω

The slopes of the lines in Figure 8.13 (both real and imaginary parts) can be estimated and plotted against Ω^2. From equations (8.42) and (8.43) we know that these plots are again straight lines, as shown in Figure 8.14.

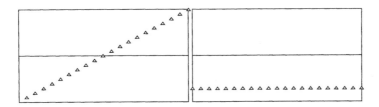

Figure 8.14 Slopes of the lines in Figure 8.13 plotted against Ω^2

The two straight lines in Figure 8.14 can be curve fitted. Using equations (8.42) and (8.43) we can derive the modal parameters from the slopes of the lines in Figure 8.14. The FRF can then be constructed using the derived modal parameter. Figure 8.15 shows the Nyquist plot of the FRF data (the thin line) and the regenerated FRF data points.

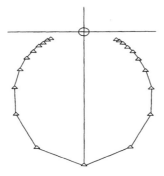

Figure 8.15 Overlay Nyquist plots of the original and regenerated FRF data

8.4 Derivation of modal data from FRF data – MDoF methods

8.4.1 Rational fraction polynomials

Rational fraction polynomial method is an MDoF modal analysis method using measured FRF data. The idea of the rational fraction polynomial method is to express an FRF in terms of rational fraction polynomials, and through numerical manipulations, the coefficients of these polynomials can be identified. The links between these coefficients and the modal parameters of the FRF can be established. This leads to the identification of these parameters.

The description of the rational fraction polynomial method commences with the transfer function of an MDoF system. Such a transfer function can be expressed as a function of the Laplace variable 's' so that:

$$\alpha_{ij}(s) = \frac{N(s)}{D(s)} = \frac{a_0 + a_1 s + a_2 s^2 + \ldots + a_m s^m}{b_0 + b_1 s + b_2 s^2 + \ldots + s^n} \tag{8.46}$$

Here, the order of the denominator 'n' is higher than that of the numerator 'm' by 2. For the sake of simplicity, the subscript of the transfer function is omitted and the following notations are adopted:

$$p_0(s) = 1, \ p_1(s) = s, \ p_2(s) = s^2, \ldots, \ p_m(s) = s^m$$
$$q_0(s) = 1, \ q_1(s) = s, \ q_2(s) = s^2, \ldots, \ q_n(s) = s^n$$

Then, we have:

$$\alpha(s) = \frac{\sum\limits_{k=0}^{m} a_k p_k(s)}{\sum\limits_{k=0}^{n} b_k q_k(s)} \quad \text{where} \quad b_n = 1 \tag{8.47}$$

When dealing with a frequency response function, this equation becomes:

$$\alpha(j\omega) = \frac{\sum\limits_{k=0}^{m} a_k p_k(j\omega)}{\sum\limits_{k=0}^{n} b_k q_k(j\omega)} \quad \text{where} \quad b_n = 1 \tag{8.48}$$

For the subsequent analysis, assume there are 'p' measurement data points corresponding to frequencies at $\omega_1, \omega_2, \ldots, \omega_p$. It is also assumed that there are 'p' negative frequency points $\omega_p, \omega_{p-1}, \ldots, \omega_{-1}$. The FRF at the negative frequencies can be shown as:

$$\alpha(j\omega_{-i}) = \alpha(-j\omega_i) = \alpha^*(j\omega_i) \quad i = 1, 2, \ldots, p \tag{8.49}$$

The purpose of using these negative frequencies will become clear in the subsequent analysis. To commence the analysis of the rational fraction polynomial method, we begin with the total error defined in equation (8.50):

$$e_{ij}(\omega) = \alpha_{ij}(\omega) - \tilde{\alpha}_{ij}(\omega) \tag{8.50}$$

Omitting the subscript and substituting equation (8.48) into this equation yields:

$$e(j\omega) = \frac{\sum\limits_{k=0}^{m} a_k p_k(j\omega)}{\sum\limits_{k=0}^{n} b_k q_k(j\omega)} - \tilde{\alpha}(j\omega) \tag{8.51}$$

This error is not a linear function of coefficients a_k and b_k. For the ease of analysis and more accurate results, a new error function can be defined as:

$$\hat{e}(j\omega) = e(j\omega) \sum\limits_{k=0}^{n} b_k q_k(j\omega) = \sum\limits_{k=0}^{m} a_k p_k(j\omega) - \tilde{\alpha}(j\omega) \left(\sum\limits_{k=0}^{n-1} b_k q_k(j\omega) + q_n(j\omega) \right) \tag{8.52}$$

The total error with respect to the whole column of FRF becomes:

$$\{E\} = \{\hat{e}_{ij}(\omega_{-p}), \ldots, \hat{e}_{ij}(\omega_{-1}), \hat{e}_{ij}(\omega_1), \ldots, \hat{e}_{ij}(\omega_p)\}^T \tag{8.53}$$

or,
$$\{E\} = [U]\{A\} - [V]\{B\} - \{W\} \tag{8.54}$$

where

$$[U] = \begin{bmatrix} p_0(j\omega_{-p}) & p_1(j\omega_{-p}) & \cdots & p_m(j\omega_{-p}) \\ \cdots & \cdots & \cdots & \cdots \\ p_0(j\omega_{-1}) & p_1(j\omega_{-1}) & \cdots & p_m(j\omega_{-1}) \\ p_0(j\omega_1) & p_1(j\omega_1) & \cdots & p_m(j\omega_1) \\ \cdots & \cdots & \cdots & \cdots \\ p_0(j\omega_p) & p_1(j\omega_p) & \cdots & p_m(j\omega_p) \end{bmatrix} \tag{8.55}$$

$$[V] = \begin{bmatrix} \tilde{\alpha}(j\omega_{-p})q_0(j\omega_{-p}) & \tilde{\alpha}(j\omega_{-p})q_1(j\omega_{-p}) & \cdots & \tilde{\alpha}(j\omega_{-p})q_{n-1}(j\omega_{-p}) \\ \cdots & \cdots & \cdots & \cdots \\ \tilde{\alpha}(j\omega_{-1})q_0(j\omega_{-1}) & \tilde{\alpha}(j\omega_{-1})q_1(j\omega_{-1}) & \cdots & \tilde{\alpha}(j\omega_{-1})q_{n-1}(j\omega_{-1}) \\ \tilde{\alpha}(j\omega_1)q_0(j\omega_1) & \tilde{\alpha}(j\omega_1)q_1(j\omega_1) & \cdots & \tilde{\alpha}(j\omega_1)q_{n-1}(j\omega_1) \\ \cdots & \cdots & \cdots & \cdots \\ \tilde{\alpha}(j\omega_p)q_0(j\omega_p) & \tilde{\alpha}(j\omega_p)q_1(j\omega_p) & \cdots & \tilde{\alpha}(j\omega_p)q_{n-1}(j\omega_p) \end{bmatrix} \tag{8.56}$$

$$\{A\} = \{a_0, a_1, \ldots, a_m\}^T \tag{8.57}$$

and
$$\{B\} = \{b_0, b_1, \ldots, a_{n-1}\}^T \tag{8.58}$$

$$[W] = \{\tilde{\alpha}(j\omega_{-p})q_n(j\omega_{-p}) \ldots \tilde{\alpha}(j\omega_{-1})q_n(j\omega_{-p})\tilde{\alpha}(j\omega_1)q_n(j\omega_{-p}) \ldots$$
$$\tilde{\alpha}(j\omega_p)q_n(j\omega_{-p})\}^T \tag{8.59}$$

The total deviation of the error is defined as:

$$J = \{E\}^H\{E\} \tag{8.60}$$

To determine the coefficient vectors $\{A\}$ and $\{B\}$ that will minimize the total deviation, the following partial derivatives should be equated to zeros:

$$\frac{\partial J}{\partial \{A\}} = \frac{\partial J}{\partial \{B\}} = 0 \tag{8.61}$$

This will lead to the following equation for estimation of coefficient vectors

$$\begin{bmatrix} \frac{1}{2}([P]^H[P] + [P]^T[P]^*) & -\mathrm{Re}([P]^H[Q]) \\ (-\mathrm{Re}([P]^H[Q]))^T & \frac{1}{2}([Q]^H[Q] + [Q]^T[Q]^*) \end{bmatrix} \begin{Bmatrix} \{A\} \\ \{B\} \end{Bmatrix}$$

$$= \begin{Bmatrix} \mathrm{Re}([P]^H\{W\}) \\ \mathrm{Re}([Q]^H\{W\}) \end{Bmatrix} \tag{8.62}$$

Coefficients $\{A\}$ and $\{B\}$ are related to the modal data of the FRF.

8.4.2 Lightly damped structures

When a structure is very lightly damped, it becomes difficult to obtain accurate FRF data near resonances. The Nyquist plot of an FRF is not very useful since all FRF data points are amassed along the real axis. Modal analysis methods that rely on the FRF data around resonances are not effective any more. The alternative is a method to use the FRF data away from resonances. The mathematical derivation of the method is straightforward. Using the structural damping model, we know the FRF can be written as:

$$\alpha_{jk}(\omega) = \sum_{r=1}^{N} \frac{{}_rA_{jk}}{\omega_r^2 - \omega^2 + j\eta_r\omega_r^2} \tag{8.63}$$

Because of the insignificant presence of damping, this expression can be approximated into:

$$\alpha_{jk}(\omega) \approx \sum_{r=1}^{N} \frac{{}_rA_{jk}}{\omega_r^2 - \omega^2} \tag{8.64}$$

This approximation leads to a convenient way to estimate modal constants of the FRF. At a measured frequency Ω_1, the FRF can be written as:

$$\alpha_{jk}(\Omega_1) = \left\{ \frac{1}{\omega_1^2 - \Omega_1^2} \quad \frac{1}{\omega_2^2 - \Omega_1^2} \quad \cdots \quad \frac{1}{\omega_n^2 - \Omega_1^2} \right\} \begin{Bmatrix} {}_1A_{jk} \\ {}_2A_{jk} \\ \vdots \\ {}_NA_{jk} \end{Bmatrix} \tag{8.65}$$

By using the FRF data at different frequencies, we can form the following matrix equation:

$$\left\{ \begin{array}{c} \alpha_{jk}(\Omega_1) \\ \alpha_{jk}(\Omega_2) \\ \vdots \end{array} \right\} = \left[\begin{array}{cccc} \dfrac{1}{\omega_1^2 - \Omega_1^2} & \dfrac{1}{\omega_2^2 - \Omega_1^2} & \cdots & \\ \dfrac{1}{\omega_1^2 - \Omega_2^2} & \cdots & & \cdots \\ \vdots & \vdots & \vdots & \vdots \end{array} \right] \left\{ \begin{array}{c} _1 A_{jk} \\ _2 A_{jk} \\ \vdots \end{array} \right\} \tag{8.66}$$

or simply:

$$\{\alpha_{jk}(\Omega)\} = [R(\Omega)]\{A_{jk}\} \tag{8.67}$$

From here, the modal constants can be estimated as:

$$\{A_{jk}\} = [R(\Omega)]^{-1}\{\alpha_{jk}(\Omega)\} \tag{8.68}$$

It can be seen that to select 'N' points from all the measured data points needs very careful consideration in order to obtain an accurate modal constant array $\{A_{jk}\}$.

The procedure for this method is as follows:

(1) measure FRF over the frequency range of interest;
(2) locate resonances graphically;
(3) select 'N' points to analyse 'N' modes (for 'N' modal constants);
(4) compute 'N' modal constants;
(5) regenerate FRF to check the accuracy of estimated modal data.

Literature

1. Adcock, J. and Potter, R. 1985: A Frequency Domain Curve Fitting Algorithm with Improved Accuracy. *Proceedings of the 3rd International Model Analysis Conference,* Orlando, Florida, 28–31.
2. Barone, P. and Ramponi, A. 2000: A new estimation method in modal analysis. *IEEE Transactions on Signal Processing,* **48**(4), 1002–1014.
3. Bishop, R.E.D. and Gladwell, G.M.L. 1963: An Investigation into the Theory of Resonance Testing. *Philosophical Trans. of the Royal Society of London,* **255A**(1055), 241–280.
4. Brittingham, J.N., Miller, E.K. and Willows, J.L. 1980: Pole Extraction from Real Frequency Information. *Proceedings of IEEE,* **68**(2), 263–273.
5. Brown, D.L. *et al.* 1979: Parameter Estimation Methods for Modal Analysis. *SAE Paper* 790221.
6. Chipwadia, K.S., Zimmerman, D.C. and James III, G.H. 1999: Evolving autonomous modal parameter estimation. *Proceedings of the 17th International Modal Analysis Conference,* Kissimmee, FL, USA, 819–826.
7. Cooley, J.W. and Tukey, J.W. 1965: An algorithm for the machine calculation of complex Fourier series. *Mathematics and Computations,* **19**(90), 297–301.
8. Coppolino, R.N. 1981: A Simultaneous Frequency Domain Technique for Estimation of Modal Parameters from Measured Data. *SAE Technical Paper Series* No. 811046.
9. Coppolino, R.N. and Stroud, R.C. 1986: A Global Technique for Estimation of Modal Parameters from Measured Data. *Proceedings of the 4th International Modal Analysis Conference,* Los Angeles, California, 674–681.
10. Dobson, B.J. 1984: Modal Analysis Using Dynamic Stiffness Data. *Royal Naval Engineering College (RNEC),* TR-84015.

11. Dobson, B.J. 1987: A straight-line technique for extracting modal properties from frequency response data. *Mechanical Systems and Signal Processing*, **1**(1) 29–40.

12. Elliott, K.B. and Mitchell, L.D. 1984: The Improved Frequency Response Function and its Effect on Modal Circle Fits. *Transactions of ASME, Journal of Applied Mechanics*, **51**, 657–663.

13. Ensminger, R. and Turner, M.J. 1979: Structural Parameter Identification from Measured Vibration Data. *AIAA/ASME/ASCE/AHS 20th Structural Dynamics Materials Conference*, St Louis, USA, AIAA Paper 79-0829, 410–416.

14. Ewins, D.J. and Gleeson, P.T. 1982: A method for modal identification of lightly damped structures, *Journal of Sound and Vibration*, **84**(1), 57–79.

15. Ewins, D.J. and Sidhu, J. 1982: Modal Testing and the Linearity of Structures. *Mecanique, Materiaux, Electricite*, No. 389-390-391.

16. Fillekrug, U. 1987: Survey of Parameter Estimation Methods in Experimental Modal Analysis. *Proceedings of the 5th International Modal Analysis Conference*, London, UK, 460–467.

17. Fillod, R. *et al.* 1985: Global Method of Modal Identification. *Proceedings of the 3rd International Modal Analysis Conference*, Orlando, Florida, 1145–1151.

18. Forsythe, G.E. 1957: Generation and Use of Orthogonal Polynomials for Data-Fitting with a Digital Computer. *Journal of the Society for Industrial and Applied Mathematics*, **5**.

19. Fullekrug, U. 1987: Survey of Parameter Estimation Methods in Experimental Modal Analysis. *Proceedings of the 5th International Modal Analysis Conference*, London, UK, 460–467.

20. Gaukroger, D.R., Skingle, C.W. and Heron, K.H. 1973: Numerical Analysis of Vector Response Loci. *Journal of Sound and Vibration*, **29**(3), 341–353.

21. Goyder, H.G. 1980: Methods and application of structural modelling from measured structural frequency response data. *Journal of Sound and Vibration*, **68**(2), 209–230.

22. Ibrahim, S.R. 1983: Computation of Normal Modes from Identified Complex Modes. *AIAA Journal*, **21**(3), 446–451.

23. Ibrahim, S.R. and Mikulcik, E.C. 1977: A Method for the Direct Identification of Vibration Parameters from Free Responses. *Shock and Vibration Bulletin*, **147**(4), 183–198.

24. Kennedy, C.C. and Pancu, C.D.P. 1947: Use of Vectors in Vibration Measurement and Analysis. *Journal of the Aeronautical Sciences*, **14**(11), 603–625.

25. Kirshenboim, J. and Ewins, D.J. 1984: A Method for the Derivation of Optimal Modal Parameters from Several Single-Point Excitation Tests. *Proceedings of the 2nd International Modal Analysis Conference*, Orlando, Florida, 991–997.

26. Leuridan, J.M. and Kundrat, Jo A. 1982: Advanced Matrix Methods for Experimental Modal Analysis – A Multi-Matrix Method for Direct Parameter Excitation. *Proceedings of the 1st International Modal Analysis Conference*, Orlando, Florida, 192–200.

27. Lewis, R.C. and Wrisley, D.L. 1950: A System for the Excitation of Pure Natural Modes of Complex Structures. *Journal of the Aeronautical Sciences*, **17**(11), 705–722.

28. Lieven, N.A.J. and Ewins, D.J. 1988: Spatial Correlation of Mode Shapes, the Coordinate Modal Assurance Criterion (COMAC). *Proceedings of the 6th International Modal Analysis Conference*, Kissimmee, FL, 690–695.

29. Link, M. 1986: Identification of Physical System Matrices Using Incomplete Vibration Test Data. *Proceedings of the 4th International Modal Analysis Conference*, Los Angeles, California.

30. Maia, N.M.M. 1999: Modal Identification Methods in the Frequency Domain. *NATO ASI Series E: Applied Sciences*, **363**, 251–264.

31. Maia, N.M.M. and Ewins, D.J. 1987: Modal Analysis of Double Modes: A First Approach to Intelligent Curve-Fitting. *Proceedings of the 5th International Modal Analysis Conference*, London, UK, 1302–1308.

32. Maia, N.M.M. and Ewins, D.J. 1989: A new approach for the modal identification of lightly damped structures. *Mechanical Systems and Signal Processing*, **3**(2), 173–193.

33. Marples, V. 1973: The Derivation of Modal Damping Ratios from Complex-Plane Response Plots. *Journal of Sound and Vibration*, **31**(1), 105–117.

34. Mitchell, L.D. 1988: A Perspective View of Modal Analysis. Keynote Address. *Proceedings of the 6th International Modal Analysis Conference,* Kissimmee, Florida, xvii–xxi.

35. Niedbal, N. 1984: Analytical Determination of Real Normal Modes from Measured Complex Responses. *Proceedings of the 25th Structures, Structural Dynamics and Materials Conference,* Palm Springs, California, 292–295.

36. Pendered, J.W. and Bishop, R.E.D. 1963: A Critical Introduction to Some Industrial Resonance Testing Techniques. *Journal of Mechanical Engineering Science,* **5**(4), 345–367.

37. Pendered, J.W. 1965: Theoretical Investigation into the Effects of Close Natural Frequencies in Resonance Testing. *Journal of Mechanical Engineering Science,* **7**(4), 372–379.

38. Petrick, L. 1984: Obtaining Global Frequency and Damping Estimates Using Single Degree-of-Freedom Real Mode Methods. *Proceedings of the 2nd International Modal Analysis Conference,* Orlando, Florida, 425–431.

39. Rades, M. 1985: Frequency Domain Experimental Modal Analysis Techniques. *The Shock and Vibration Digest,* **17**(6), 3–15.

40. Richardson, M.H. and Formenti, D.L. 1982: Parameter estimation from frequency response measurements using rational fraction polynomials. *Proceedings of the 1st International Modal Analysis Conference,* Orlando, Florida, 176–181.

41. Richardson, M.H. and Formenti, D.L. 1985: Global curve fitting of frequency response measurements using the rational fraction polynomial method. *Proceedings of the 3rd International Modal Analysis Conference,* Orlando, Florida, January, 390–397.

42. Richardson, M.H. 1986: Global Frequency and Damping Estimates from Frequency Response Measurements. *Proceedings of the 4th International Modal Analysis Conference,* Los Angeles, California.

43. Schmerr, L.W. 1982: A New Complex Exponential Frequency Domain Technique for Analysing Dynamic Response Data. *Proceedings of the 1st International Modal Analysis Conference,* Orlando, Florida, 8–10.

44. Shih, C.Y. *et al.* 1988: A Frequency Domain Global Parameter Estimation Model for Multiple Reference Frequency Response Measurements. *Proceedings of the 6th International Modal Analysis Conference,* Kissimmee, Florida, 389–396.

45. Silva, J.M.M. and Maia, N.M.M. 1988: Single Mode Identification Techniques for Use with Small Micro-Computers. *Journal of Sound and Vibration,* **124**(1), 13–26.

46. Smiley, R.G., Wey, Y.S. and Hogg, K.D. 1984: A Simplified Frequency Domain MDoF Curve-Fitting Process. *Proceedings of the 2nd International Modal Analysis Conference,* Orlando, Florida.

47. Stroud, R.C. 1985: Excitation, Measurement and Analysis Methods for Modal Testing: Combined Experimental Analytical Modelling of Dynamic Structural Systems. *The Joint ASCE/ASME Mechanics Conference,* Albuquerque, New Mexico, AMD Vol. 67, 49–78.

48. Traill-Nash, R.W., Long, G. and Bailey, C.M. 1967: Experimental Determination of the Complete Dynamical Properties of a Two-Degree-of-Freedom Model Having Nearly Coincident Natural Frequencies. *Journal of Mechanical Engineering Science,* **9**(5), 402–413.

49. Vold, H. and Leuridan, J. 1982: A Generalised Frequency Domain Matrix Estimation Method for Structural Parameter Identification. *Proceedings of the 7th International Seminar on Modal Analysis,* K.U. Leuven, Belgium.

50. Woodcock, D.L. 1963: On the Interpretation of the Vector Plots of Forced Vibrations of a Linear System with Viscous Damping. *The Aeronautical Quarterly,* 45–62.

51. Zhang, L. *et al.* 1985: A Polyreference Frequency Domain Method for Modal Parameter Identification. *ASME Paper 85-DET-106,* 1–6.

9

Modal analysis methods – time domain

Time domain methods belong to another category of modal analysis methods. Frequency domain modal analysis relies on measured frequency response function (FRF) data. Time domain modal analysis uses time response data. The methods were developed on the advances of modern control theory and computer technology.

FRF data are not the 'raw' measurement data. They are the outcomes of a transformation from measured response time history to data in frequency domain. The data directly measured by response transducers are the real time acceleration and force signals. We know from spectral analysis theory that the FRF and the impulse response function (IRF) of a structure are a Fourier transform pair. Therefore, if we can deduce the impulse response information from measured time responses or the FRF data, then it is possible to extract the modal parameters from there.

Time domain modal analysis methods use response data either from known excitation sources or from ambient excitation. With known excitation, the measurement conditions are similar to the frequency domain approach. The data of known inputs enable correlation analysis and usually assure better accuracy of the analysis. The data from ambient excitation, however, grant a special appeal to the time domain approach. This is because it is impractical to excite some structures in a controlled laboratory environment. The ambient excitation, which is often unmeasurable, becomes the only excitation source in measurement. This scenario often disqualifies many frequency domain methods. For example, the vibration of a bridge can be measured using the ambient traffic as the excitation. Sea wave impact can be the excitation of a ship. Wind excitation can be the input source for a tall building. These excitations cannot be quantified. The independence of some time domain methods on the excitation data enables them to be used for special applications.

Time domain modal analysis has several advantages over its frequency domain counterpart. It has been mentioned that it does not depend on excitation and equipment needed to obtain designed excitation forces. Generally, it needs fewer response data. It is possible to use this technique for modal analysis based on line damage detection and machine health diagnosis. Because the analysis is not based on the FRF data (although some methods can utilize the IRF data from the inverse Fourier transform of the FRF data), it can be an alternative for analysing close vibration modes. However, time domain modal analysis requires a very high signal and noise ratio from measured data. Unlike FRF data where vibration modes are almost visible, the time response does not manifest vibration modes. This demands more stringent analysis skills. The data inaccuracy and presence of noise may lead to computational modes in the calculation. They are not genuine system vibration modes but nevertheless are 'identified'.

In the following, we discuss several time domain methods which have been used by practising engineers for modal analysis. Practical application cases can be found from the literature.

9.1 Least-squares time domain method

This method is based on the model of free decay vibration response for an MDoF system. The measured vibration response data are used for a curve-fitting effort against a defined mathematical model in order to derive the modal parameters. For an MDoF dynamic system, its free vibration response is known to be in the form:

$$x(t) = \sum_{i=1}^{N} e^{-n_i t}(a_i \sin \omega_{di} t + b_i \cos \omega_{di} t) \tag{9.1}$$

Here, a_i and b_i are the components of the vibration amplitude. ω_{di} is the damped natural frequency and n_i is related to damping ratio. All are the parameters to be identified. If we sample the response with a time series $t_j (j = 1, 2, \ldots, P)$ and the time resolution Δt, then the measured data points can be denoted as $\tilde{X}(t_j)$ or $\tilde{X}_j$ for short. The corresponding data from the mathematical model given in equation (9.1) can be denoted as $X(t_j)$ or X_j.

The response model in equation (9.1) is a nonlinear and transcendental function for the parameters we need to identify. To simplify the problem, let us give the initial values for parameters ω_{di} and n_i as $\omega_{di}^{(0)}$ and $n_i^{(0)}$, and try to estimate parameters a_i and b_i arranged vertically in a vector $\{\theta\} = \begin{Bmatrix} a \\ b \end{Bmatrix}_{2N \times 1}$. Since the data $X(t)$ are linear functions of parameters a_i and b_i, we can form a set of linear simultaneous equations to solve for vector $\{\theta\}$:

$$\{X_j\} = [S]_{P \times 2N} \{\theta\}_{2N \times 1} \tag{9.2}$$

where

$$[S] = \begin{bmatrix} e^{-n_1 t_1} \sin \omega_{d1} t_1, \; e^{-n_1 t_1} \cos \omega_{d1} t_1, \ldots, e^{-n_N t_1} \sin\omega_{dN} t_1, \; e^{-n_N t_1} \cos\omega_{dN} t_1 \\ \ldots \\ e^{-n_1 t_P} \sin \omega_{d1} t_P, \; e^{-n_1 t_P} \cos \omega_{d1} t_P, \ldots, e^{-n_N t_P} \sin\omega_{dN} t_P, \; e^{-n_N t_P} \cos\omega_{dN} t_P \end{bmatrix} \tag{9.3}$$

and

$$\{X_j\} = \{X_1, X_2, \ldots, X_P\}^T \tag{9.4}$$

Usually, we have more data points than twice the number of modes (i.e. $P > 2N$) so the least squares solution for vector $\{\theta\}$ can be derived as:

$$\{\theta\} = ([S]^T [S])^{-1} [S]^T \{\tilde{X}_j\} \tag{9.5}$$

Equation (9.2) does not have to use all P data points. Nor does it have to contain all N modes at once. Once vector $\{\theta\}$ is estimated, we can use them to estimate parameters ω_{di} and n_i. From Taylor series expansion and ignoring the higher order terms we have:

$$X(t_j, \omega_i, n_i) \approx X(t_j, \omega_{di}^{(0)}, n_i^{(0)}) + \sum_{i=1}^{N} \left(\frac{\partial X_j}{\partial \omega_{di}} \bigg|_0 \Delta \omega_{di} + \frac{\partial X_j}{\partial n_{di}} \bigg|_0 \Delta n_i \right) \quad (9.6)$$

Using this equation with data at different time sampling points, we can establish a set of linear simultaneous equations for parameters $\Delta \omega_{di}$ and Δn_i. The solution, when combined with the initial value $\omega_{di}^{(0)}$ and $n_i^{(0)}$, will yield the estimates of parameters ω_{di} and n_i.

The whole process described above can be iterated for more accurate modal parameter estimation. This method relies heavily on the initial values $\omega_{di}^{(0)}$ and $n_i^{(0)}$ and the number of modes 'N' used.

9.2 Ibrahim time domain (ITD) method

The Ibrahim time domain method, known as the ITD method, uses IRF data to identify modal parameters. It constructs an eigenvalue problem from the IRF data and solves the problem to derive the natural frequencies, damping loss factors and modal constants. The method, when originally proposed, constructs the eigenvalue problem using displacement, velocity and acceleration IRF data. As a result, numerical integration is needed to obtain displacement and velocity time histories from acceleration measurement at each response point. This approach was later improved by using displacement, velocity or acceleration free response data only. The deficiency of the amount of data needed is remedied by sampling the displacement data with a selected time delay. The free response of a structure means either the free decays measured from random execution of the structure or the impulse response obtained from the inverse of an FRF.

We begin the discussion from the equation of motion of an 'N' DoF system:

$$[M]\{\ddot{x}\} + [C]\{\dot{x}\} + [K]\{x\} = \{0\} \quad (9.7)$$

As in Chapter 6, we can define a new variable vector to transform the equation into a standard eigenvalue problem. The new vector is:

$$\{y\} = \begin{Bmatrix} x \\ \dot{x} \end{Bmatrix}_{2N \times 1} \quad (9.8)$$

Equation (9.7) can be transformed into a new eigenvalue problem:

$$\{\dot{y}\} = \{A\}\{y\} \quad (9.9a)$$

where
$$[A] = \begin{bmatrix} [0] & [I] \\ -[M]^{-1}[K] & -[M]^{-1}[C] \end{bmatrix} \quad (9.9b)$$

Matrix $[A]$ is not known but it is possible to derive it from time domain response data. Assume we have the free vibration response data in the form of displacement $x(t)$, velocity $\dot{x}(t)$ and acceleration $\ddot{x}(t)$. Otherwise, the displacement and velocity can be derived from integrating the acceleration data. Sampling these data, we can form two matrices:

$$[Y] = \begin{bmatrix} x(t_1) & x(t_2) & \ldots & x(t_{2N}) \\ \dot{x}(t_1) & \dot{x}(t_2) & \ldots & \dot{x}(t_{2N}) \end{bmatrix} \tag{9.10}$$

and

$$[\dot{Y}] = \begin{bmatrix} \dot{x}(t_1) & \dot{x}(t_2) & \ldots & \dot{x}(t_{2N}) \\ \ddot{x}(t_1) & \ddot{x}(t_2) & \ldots & \ddot{x}(t_{2N}) \end{bmatrix} \tag{9.11}$$

It is evident that every column of $[Y]$ and its corresponding column in $[\dot{Y}]$ follows equation (9.9). Therefore, we arrive at the following equation:

$$[\dot{Y}] = [A][Y] \tag{9.12}$$

If the free vibration response $x(t)$ contains information for 'N' vibration modes, then $[Y]$ is non-singular. As a result, we can derive the matrix $[A]$ from the response data as:

$$[A] = [\dot{Y}][Y]^{-1} \tag{9.13}$$

Once matrix $[A]$ is determined, we can use equation (9.9) to solve the eigenvalue problem and derive the modal parameters of the system. This approach relies on sampling the measurement data in displacement, velocity and acceleration forms.

A different approach of the ITD method uses a delayed sampling to obtain sufficient amounts of data. These sampled data are then used to construct an eigenvalue problem to derive modal data. To introduce this approach, let us refresh the modal analysis theory for an MDoF system governed by equation (9.7). It has $2N$ complex conjugate eigenvalues and corresponding complex eigenvectors. The eigenvalues, denoted as

$$\lambda_r = -n_r + j\omega_{dr}, \tag{9.14}$$

are the rth complex natural frequency, and the eigenvectors, denoted as $\{\varphi\}_r$, are the rth complex mode shapes.

If the test structure behaves modally, then it is reasonable to assume that there is a theoretical model of 'N' degrees of freedom that describes the dynamic characteristics of the structure and the MDoF modal analysis theory applies. The response from any given measurement point on the structure should then have a combined contribution of all vibration modes. For the sake of simplicity, assume first that the number of measurement points match with the number of DoFs. For the ith measurement point, the time response can be sampled to obtain '$2N$' data points: $\{x(t_1), x(t_2), \ldots, x(t_{2N})\}^T$. If the same sampling is carried out on data from all 'N' measurement points, then a matrix of sampled free vibration response can be formed as:

$$[X] = \begin{bmatrix} x_1(t_1) & x_1(t_2) & \ldots & x_1(t_{2N}) \\ x_2(t_1) & x_2(t_2) & \ldots & x_2(t_{2N}) \\ \ldots & \ldots & \ldots & \ldots \\ x_N(t_1) & x_N(t_2) & \ldots & x_N(t_{2N}) \end{bmatrix}_{N \times 2N} \tag{9.15}$$

Each data point should follow the theory as:

$$x_i(t) = \sum_{r=1}^{2N} \varphi_{ir} e^{\lambda_r t} \tag{9.16}$$

Each column of $[X]$ can be written in a matrix form:

$$\{X_j\} = [\{\varphi\}_1, \{\varphi\}_2, \ldots, \{\varphi\}_{2N}] \begin{Bmatrix} e^{\lambda_1 t_j} \\ e^{\lambda_2 t_j} \\ \ldots \\ e^{\lambda_{2N} t_j} \end{Bmatrix} \tag{9.17}$$

Equation (9.15) can now be recast into:

$$\begin{bmatrix} x_1(t_1) & \ldots & x_1(t_{2N}) \\ \ldots & \ldots & \ldots \\ x_N(t_1) & \ldots & x_N(t_{2N}) \end{bmatrix} = [\{\varphi\}_1 \ldots \{\varphi\}_{2N}] \begin{bmatrix} e^{\lambda_1 t_1} & \ldots & e^{\lambda_1 t_{2N}} \\ \ldots & \ldots & \ldots \\ e^{\lambda_{2N} t_1} & \ldots & e^{\lambda_{2N} t_{2N}} \end{bmatrix} \tag{9.18}$$

or simply:

$$[X] = [\Psi][\Lambda] \tag{9.19}$$

This equation establishes direct links between the complex natural frequencies, mode shapes and the time response of the system. It alone, however, is insufficient to solve for modal data. The approach is to sample the same response data but with a duration τ shift. For the ith measurement, this means:

$$x_i(t + \tau) = \sum_{r=1}^{2N} \varphi_{ir} e\lambda_r^{(t+\tau)} \tag{9.20}$$

For all the response points, a vector form of equation (9.20) can be written as:

$$\{Y_j\} = \{X(t_j + \tau)\} = [\{\varphi\}_1 e^{\lambda_1 \tau}, \{\varphi\}_2 e^{\lambda_2 \tau}, \ldots, \{\varphi\}_{2N} e^{\lambda_{2N} \tau}] \begin{Bmatrix} e^{\lambda_1 t_j} \\ e^{\lambda_2 t_j} \\ \ldots \\ e^{\lambda_{2N} t_j} \end{Bmatrix} \tag{9.21}$$

Again, taking the same sampling time $(t_1, t_2, \ldots, t_{2N})$ but with shift τ, equation (9.17) will take its matrix form:

$$\begin{bmatrix} x_1(t_1 + \tau) & \ldots & x_1(t_{2N} + \tau) \\ \ldots & \ldots & \ldots \\ x_N(t_1 + \tau) & \ldots & x_N(t_{2N} + \tau) \end{bmatrix}$$

$$= [\{\varphi\}_1 e^{\lambda_1 \tau} \ldots \{\varphi\}_{2N} e^{\lambda_{2N} \tau}] \begin{bmatrix} e^{\lambda_1 t_1} & \ldots & e^{\lambda_1 t_{2N}} \\ \ldots & \ldots & \ldots \\ e^{\lambda_{2N} t_1} & \ldots & e^{\lambda_{2N} t_{2N}} \end{bmatrix} \tag{9.22}$$

or $$[Y] = [P][\Lambda] \qquad (9.23)$$

If the system has unique natural frequencies, then vectors $\{\varphi_{i1}, \varphi_{i2}, \ldots, \varphi_{i2N}\}^T$ and $\{\varphi_{i1}e^{\lambda_1\tau}, \varphi_{i2}e^{\lambda_2\tau}, \ldots, \varphi_{i2N}e^{\lambda_{2N}\tau}\}^T$ are linearly independent of each other. This means that by shifting the time interval τ and sampling a set of '$2N$' values, we have effectively 'created' more measurement data points.

Likewise, we can increase the time shift to 2τ and sample the response data again. For all the response points, a vector can be written as:

$$\{Z_j\} = \{X(t_j + 2\tau)\} = [\{\varphi\}_1 e^{2\lambda_1\tau}, \{\varphi\}_2 e^{2\lambda_2\tau}, \ldots, \{\varphi\}_{2N} e^{2\lambda_{2N}\tau}] \begin{Bmatrix} e^{\lambda_1 t_j} \\ e^{\lambda_2 t_j} \\ \ldots \\ e^{\lambda_{2N} t_j} \end{Bmatrix} \qquad (9.24)$$

Again, taking the same sampling time $(t_1, t_2, \ldots, t_{2N})$ but with shift 2τ, equation (9.24) will take its matrix form:

$$\begin{bmatrix} x_1(t_1 + 2\tau) & \ldots & x_1(t_{2N} + 2\tau) \\ \ldots & \ldots & \ldots \\ x_N(t_1 + 2\tau) & \ldots & x_N(t_{2N} + 2\tau) \end{bmatrix}$$

$$= [\{\varphi\}_1 e^{2\lambda_1\tau} \ldots \{\varphi\}_{2N} e^{2\lambda_{2N}\tau}] \begin{bmatrix} e^{\lambda_1 t_1} & \ldots & e^{\lambda_1 t_{2N}} \\ \ldots & \ldots & \ldots \\ e^{\lambda_{2N} t_1} & \ldots & e^{\lambda_{2N} t_{2N}} \end{bmatrix} \qquad (9.25)$$

or $$[Z] = [Q][\Lambda] \qquad (9.26)$$

From this analysis, we have constructed three sampled data matrices $[X]$, $[Y]$, $[Z]$ and three modal vector matrices $[\varphi]$, $[P]$, $[Q]$. The next task is to assemble an eigenvalue problem from these matrices that shall derive the complex natural frequencies and mode shapes. To do that, we need to regroup these matrices and define the following:

$$[V] = \begin{bmatrix} [X] \\ [Y] \end{bmatrix} \qquad (9.27)$$

$$[W] = \begin{bmatrix} [Y] \\ [Z] \end{bmatrix} \qquad (9.28)$$

$$[\mu] = \begin{bmatrix} [\varphi] \\ [P] \end{bmatrix} \qquad (9.29)$$

$$[v] = \begin{bmatrix} [P] \\ [Q] \end{bmatrix} \qquad (9.30)$$

Also, we know that the following equalities exist:

$$[P] = [\varphi][\,^\backprime\Delta_.], \qquad (9.31)$$

$$[Q] = [P][\,\dot{\Delta}.], \tag{9.32}$$

$$[v] = [\mu][\,\dot{\Delta}.] \tag{9.33}$$

where
$$[\,\dot{\Delta}.] = \begin{bmatrix} e^{\lambda_1\tau} & & & 0 \\ & e^{\lambda_2\tau} & & \\ & & \cdots & \\ 0 & & & e^{\lambda_{2N}\tau} \end{bmatrix} \tag{9.34}$$

and
$$[V] = [\mu][\Lambda] \tag{9.35}$$

$$[W] = [v][\Lambda] \tag{9.36}$$

From equations (8.31) to (8.36) we can derive:

$$[W][V]^{-1}[\mu] = [\mu][\,\dot{\Delta}.] \tag{9.37}$$

This equation constitutes a complex eigenvalue problem if matrix $[V]$ is non-singular. From the beginning, we assumed that the system we are studying has no repeated eigenvalues. This means that columns in matrix $[\mu]$, which consist of the complex eigenvectors, are independent of each other and therefore are non-singular. Matrix $[\Lambda]$ is also non-singular because of no repeated eigenvalues. As a result, matrix $[V]$ is indeed non-singular.

The upper half of $[\mu]$ is matrix $[\varphi]$, i.e. the complex mode shape matrix. Matrix $[\,\dot{\Delta}.]$ contains the complex eigenvalues. These eigenvalues can be converted to the natural frequencies and damping ratio. Assume the rth eigenvalue is in the form:

$$\Delta_r = e^{\lambda_r\tau} = \alpha_r + j\beta_r \tag{9.38}$$

Combining this equation with equation (9.14) will lead to the solutions:

$$n_r = -\frac{1}{2\tau}\ln(\alpha_r^2 + \beta_r^2) \tag{9.39}$$

$$\omega_{dr} = \frac{1}{\tau}\tan^{-1}\frac{\beta_r}{\alpha_r} \tag{9.40}$$

From these solutions, we can derive the undamped natural frequency and damping ratio:

$$\omega_r = \sqrt{\omega_{dr}^2 + n_r^2} \tag{9.41}$$

$$\zeta_r = \frac{n_r}{\omega_r} \tag{9.42}$$

The solution approach of the ITD method can be summarized as:

(1) record or generate free vibration responses in the form of displacement, velocity or acceleration;
(2) sample the response data;
(3) use the sampled free response data to construct matrices $[V]$ and $[W]$;
(4) use these matrices to form a complex eigenvalue problem; and
(5) solve the eigenvalue problem to identify modal data.

The ITD method involves intensive numerical analysis. Many methods have been proposed in the literature in order to reduce the effects of measurement noise and improve the accuracy of the identified modal data. One of the methods is to make better use of the existing measured data by introducing more sampling points. If we sample the measured data by '*m*' points ($m > 2N$), then more system information will be used. In this case, both matrices $[V]$ and $[W]$ will be non-square. The inverse $[V]^{-1}$ in equation (9.37) can be replaced by the pseudo inverse. An alternative is to obtain the inverse using singular value decomposition.

One of the uncertainties in modal analysis of both frequency and time domain is the number of vibration modes involved. For frequency domain modal analysis, we may select a frequency range of interest at a time and decide the number of modes to analyse. There are a number of methods to help identify mode numbers. For time domain analysis, usually less confidence exists on assuming the number of modes. In fact, the free vibration response of a structure should theoretically contain information of an infinite number of vibration modes. In order to identify sufficient number of modes, pseudo measurement data can be created from the measured free response data. By selecting a time delay τ' different to the shift τ, we can obtain the following set of data:

$$\{X_j'\} = \{X(t_j + \tau')\} = [\{\varphi\}_1 e^{\lambda_1 \tau'}, \{\varphi\}_2 e^{\lambda_2 \tau'}, \ldots, \{\varphi\}_{2N} e^{\lambda_{2N} \tau'}] \begin{Bmatrix} e^{\lambda_1 t_j} \\ e^{\lambda_2 t_j} \\ \ldots \\ e^{\lambda_{2N} t_j} \end{Bmatrix} \quad (9.43)$$

This set of data is a pseudo measurement. It can be augmented to the real measurement to form a new set of 'measurement' data:

$$\{\tilde{X}_j\} = \begin{Bmatrix} \{X_j\} \\ \{X_j'\} \end{Bmatrix} \quad (9.44)$$

Using data $\{\tilde{X}_j\}$ to replace $\{X_j\}$ in the analysis described from equations (9.15) to (9.42) will lead to the identification of many numerical vibration modes as well as genuine modes. To distinguish the latter from the former, a modal confidence factor (MCF) can be defined. For a selected measurement point, if the modal constant of the *i*th mode identified is φ_i and for the same point it is identified as $\overline{\varphi}_i$ with a time delay Δt, theoretically we have:

$$\overline{\varphi}_i = \varphi_i e^{\lambda_i \Delta t} \quad (9.45)$$

When the identified mode is not a genuine one, this relationship does not hold. Therefore, the modal confidence factor can be defined as:

$$MCF = \left| \frac{\overline{\varphi}_i}{\varphi_i e^{\lambda_i \Delta t}} \right| \quad (9.46)$$

If the mode is a genuine one, *MCF* will be close to unity. Otherwise, it is likely the mode is a numerical one.

9.3 Random decrement method

The random decrement technique (Figure 9.1) was initially developed to extract free-decay response data from the response of a dynamic system subjected to random or ambient excitation. The random decrement method for modal analysis relies on the free-decay response data extracted to continue with modal data identification. This marks a main difference between this method and other time domain methods.

Figure 9.1 A flow chart of data process for the random decrement technique

Dynamic response of a linear structure subjected to random or ambient excitation contains free and forced responses. At time $t + t_0$, the response should consist of three parts: (1) the step response from the initial displacements at time t_0; (2) the IRF from the initial velocity at time t_0; and (3) the random response from the excitation applied to the structure in the period between t_0 and $t + t_0$. These three parts of the responses can be denoted in the following equation:

$$x(t + t_0) = x(t + t_0)|_{x(t_0)} + x(t + t_0)|_{\dot{x}(t_0)} + x(t + t_0)|_{f(t)} \qquad (9.47)$$

If we pick a constant time segment from the response history every time the response amplitude reaches a selected value X_0 (these segments are allowed to overlap), we obtain a number of response history segments of equal lengths and they all begin with the response amplitude X_0. Figure 9.2 shows some of the segments taken from a response history. The first segment begins at time t_1 and ends at $(t_1 + \tau)$. The whole segment can be denoted as $X_0(t_1 + \tau)$.

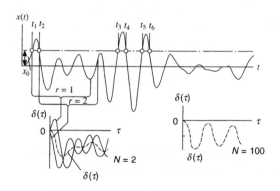

Figure 9.2 Segments taken from a response history

When these segments are averaged, an interesting outcome emerges. The average of N segments is denoted as $\delta_N(\tau)$ and is given as:

$$\delta_N(\tau) = \frac{1}{N} \sum_{r=1}^{N} X_0 (t_r + \tau) \qquad (9.48)$$

As the number N increases, the part of the responses in each segment that corresponds to the initial velocity (impulse response) will be averaged out since the sign of the initial velocity appears alternatively. The random parts of each segment are also averaged out because of the random nature of the response, as shown in Figure 9.3. This is to assume that the average of the random response is zero – an assumption often being valid.

$$\left. \sum_{r=1}^{N} X_0 (t_r + \tau) \right|_{\dot{x}(t_0)} \to 0 \qquad \left. \sum_{r=1}^{N} X_0 (t_r + \tau) \right|_{f(t)} \to 0$$

What is left after the average is the response due to the initial displacement step input. This output can be denoted $\delta(\tau)$. It is known as the *randomdec signature*. The physical meaning of the randomdec signature can be explained. If we apply to a structure a constant load at coordinate 'j', there will be a static local displacement. When the load is suddenly withdrawn, the response at the coordinate is that due to the initial displacement step input.

If we select the value X_0 to be zero, then a sufficient average of $\delta_N(\tau)$ will yield the impulse response. This is the essential data for time domain modal analysis. This technique of deriving the impulse response from the response measured under ambient vibration excitation is the random decrement technique (Figure 9.3).

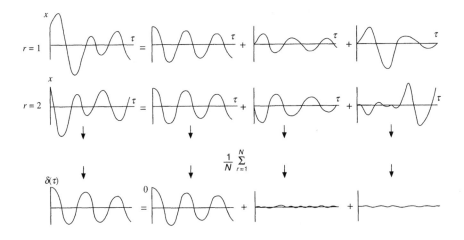

Figure 9.3 Average of segments taken from a response history

The above discussion is based on a single random response of a structure. To extract the free response of all measured coordinates, the random decrement technique needs to be applied to all the random responses. Take two random responses $x_1(t)$ and $x_2(t)$ as an example. Both are due to ambient random excitations to the measured structure and measurements were synchronized. If the randomdec signatures of these responses are $\delta^{(1)}(\tau)$ and $\delta^{(2)}(\tau)$, then for a sufficiently large N,

$$\delta^{(1)}(\tau) = \frac{1}{N} \sum_{r=1}^{N} X_1(t_r + \tau) \tag{9.49}$$

and

$$\delta^{(2)}(\tau) = \frac{1}{N} \sum_{r=1}^{N} X_2(t_r + \tau) \tag{9.50}$$

Assume we select the initial value of the response 1 to be X_1 in order to derive response segments from random response $x_1(t)$. Then the randomdec signature of response $x_1(t)$ is the free response at coordinate 1 due to the initial displacement input X_1. Since measurements $x_1(t)$ and $x_2(t)$ are synchronized, for each segment of random response $x_1(t)$, there is a corresponding response segment (in the same time period) of random response $x_2(t)$. The average of these segments is the free response at coordinate 2 due to the initial displacement input X_1 at coordinate 1. This process can be repeated on all measured coordinates. For a structure with m measurement coordinates, we can select an initial displacement X_j at coordinate j, and sample the response history to derive the randomdec signature for the coordinate. At the same time, corresponding segments can be obtained from the response histories of all other coordinates. At the end, $\delta^{(r)}(\tau)$ ($r = 1, 2, \ldots, m$) are the free decay responses of the structure at all coordinates due to the displacement input X_j at coordinate j.

Now we will show analytically that the randomdec signatures of an MDoF system can be derived. For an N DoF system with random excitations, the equation of motion is given as:

$$[M]\{\ddot{x}\} + [C]\{\dot{x}\} + [K]\{x\} = \{f\} \tag{9.51}$$

If we define an operator $\breve{p}_{ij}$ as:

$$\breve{p}_{ij}(x) = m_{ij}\frac{dx_j}{dt^2} + c_{ij}\frac{dx_j}{dt} + k_{ij}x_j \tag{9.52}$$

then the matrix equation of motion can be recast as:

$$\begin{bmatrix} \breve{p}_{11} & \breve{p}_{12} & \cdots & \breve{p}_{1N} \\ \breve{p}_{21} & \breve{p}_{22} & \cdots & \breve{p}_{2N} \\ \cdots & \cdots & \cdots & \cdots \\ \breve{p}_{N1} & \breve{p}_{N2} & \cdots & \breve{p}_{NN} \end{bmatrix} \begin{Bmatrix} x_1 \\ x_2 \\ \vdots \\ x_N \end{Bmatrix} = \begin{Bmatrix} f_1 \\ f_2 \\ \vdots \\ f_N \end{Bmatrix} \tag{9.53}$$

If we select an initial displacement of response x_j, we can take L segments of the response as $x_{jr}(\tau)$ ($r = 1, 2, \ldots, L$). We also have corresponding response segments from all other responses and all the inputs. For each segment, equation (9.53) holds. Therefore, we can average all L segments to form:

$$\frac{1}{L}\sum_{r=1}^{N} \begin{bmatrix} \breve{p}_{11} & \breve{p}_{12} & \cdots & \breve{p}_{1N} \\ \breve{p}_{21} & \breve{p}_{22} & \cdots & \breve{p}_{2N} \\ \cdots & \cdots & \cdots & \cdots \\ \breve{p}_{N1} & \breve{p}_{N2} & \cdots & \breve{p}_{NN} \end{bmatrix} \begin{Bmatrix} x_{1r}(\tau) \\ x_{2r}(\tau) \\ \vdots \\ x_{Nr}(\tau) \end{Bmatrix} = \frac{1}{L}\sum_{r=1}^{L} \begin{Bmatrix} f_{1r}(\tau) \\ f_{2r}(\tau) \\ \vdots \\ f_{Nr}(\tau) \end{Bmatrix} \tag{9.54}$$

Since the excitations of the system are random signals with zero means, the right-hand side of the equation approaches zero with a sufficient number of averages. For the left-hand side, the operations of summation and differentiation can swap so that:

$$\left(\frac{1}{L}\sum_{r=1}^{L}\breve{p}_{ij}(x_{jr}(\tau))\right) = \breve{p}_{ij}\left(\frac{1}{L}\sum_{r=1}^{L}x_{jr}(\tau)\right) \tag{9.55}$$

Therefore, equation (9.54) becomes:

$$\begin{bmatrix} \breve{p}_{11} & \breve{p}_{12} & \cdots & \breve{p}_{1N} \\ \breve{p}_{21} & \breve{p}_{22} & \cdots & \breve{p}_{2N} \\ \cdots & \cdots & \cdots & \cdots \\ \breve{p}_{N1} & \breve{p}_{N2} & \cdots & \breve{p}_{NN} \end{bmatrix} \begin{Bmatrix} \delta_1(\tau) \\ \delta_2(\tau) \\ \vdots \\ \delta_N(\tau) \end{Bmatrix} = \begin{Bmatrix} 0 \\ 0 \\ \vdots \\ 0 \end{Bmatrix} \tag{9.56}$$

This equation is identical to the equation of motion of the system for free vibration. Thus, variables $\delta_r(\tau)$ ($r = 1, 2, \ldots, N$) are the free vibration responses due to a single initial displacement applied at coordinate j.

Once the free decay or impulse responses of the structure are extracted from the random response measurements using the random decrement technique, time domain modal analysis methods can be used to derive the modal data.

9.4 ARMA time series method

The ARMA time series method is a system identification method. It can be used for modal analysis of a structure from measured random responses. The basic idea of time series analysis is to identify a system and predict its present and future response from the information of its past inputs and outputs. The ARMA model introduced in Chapter 2 suggests that a stationary random signal with zero means can be sampled at time ($t = 1, 2, \ldots$) for samples ($x_1, x_2, \ldots$). Here the subscripts of x and a denote the time of the sample. And these samples can be related by a finite difference equation as:

$$x_t - \sum_{r=1}^{n}\phi_r x_{t-r} = a_t - \sum_{s=1}^{m}\theta_s a_{t-s} \quad n > m \tag{9.57}$$

Here, a_{t-s} ($s = 1, 2, \ldots, m$) is the random input series. n is the order of auto-regressive model and m is the order of the moving average model. ϕ_r are auto-regressive coefficients and θ_s are moving average coefficients. Understanding the physical meaning of the ARMA model is necessary in order to appreciate its application in modal analysis. The term $\phi_r x_{t-r}$ in equation (9.57) signifies the weighted contribution of the historical sample x_{t-r} to the present response x_t. The term $\theta_s a_{t-s}$ represents the weighted contribution of the historical input a_{t-s} to the present response x_t. Therefore, the ARMA model in equation (9.57) describes the input–output relationship of a measured structure. This is an alternative to other forms of input–output relationship functions such as the FRF and IRF.

The ARMA model in equation (9.57) can be derived from the FRF of a structure. From Chapter 6 we know that the transfer function of an N DoF dynamic system can be in the following form:

$$\alpha_{ij}(s) = \frac{x_i(s)}{a_j(s)} = \frac{b_0 + b_1 s + \ldots + b_{2N-2} s^{2N-2}}{a_0 + a_1 s + \ldots + a_{2N} s^{2N}} \tag{9.58}$$

Applying the z-transform to the transfer function will reveal the z-function:

$$\alpha_{ij}(z) = \frac{x(z)}{a(z)} = \frac{\theta_0 + \theta_1 z^{-1} + \ldots + \theta_{2N-1} z^{-2N+1}}{1 + \phi_1 z^{-1} + \ldots + \phi_{2N} z^{-2N}} \tag{9.59}$$

or $(1 + \phi_1 z^{-1} + \ldots + \phi_{2N} z^{-2N})x(z) = (\theta_0 + \theta_1 z^{-1} + \ldots + \theta_{2N-1} z^{-2N+1})a(z)$

$$\tag{9.60}$$

Upon using the inverse z-transform, equation (9.60) can be converted to the time domain:

$$x_t + \phi_1 x_{t-1} + \ldots + \phi_{2N} x_{t-2N} = a_t + \theta_1 a_{t-1} + \ldots + \theta_{2N-1} a_{t-2N+1} \tag{9.61}$$

Taking the first $n + 1$ terms from the left-hand side and $m + 1$ terms from the right-hand side, and embedding negative signs into coefficients ϕ_r $(r = 1, 2, \ldots, n)$ and θ_r $(r = 1, 2, \ldots, m)$, equation (9.61) becomes identical to equation (9.57) for the ARMA model. However, for an N DoF system, the ARMA model in equation (9.57) should be written as:

$$x_t - \phi_1 x_{t-1} - \ldots - \phi_{2N} x_{t-2N} = a_t - \theta_1 a_{t-1} - \ldots - \theta_{2N-1} a_{t-2N+1} \tag{9.62}$$

Modal analysis using the ARMA model involves the following steps: (1) identifying the auto-regressive coefficients ϕ_r; (2) identifying complex roots of a characteristic equation; (3) identifying modal parameters from the roots. For an N DoF system subjected to random inputs of zero means (so $a_1 = 0$), we know from the ARMA model that the response at time t is given as:

$$x_t = \{-x_{t-1} \ldots -x_{t-2N} \; a_{t-1} \ldots a_{t-2N+1}\} \begin{Bmatrix} \phi_1 \\ \vdots \\ \phi_{2N} \\ \theta_1 \\ \vdots \\ \theta_{2N-1} \end{Bmatrix} \tag{9.63}$$

Obviously, we do not have samples $x_h (h < 0)$. If we sample the response $x(t)$ and input $a(t)$ from $t = 2N + 1$ sequentially to $t = 2N + L$ for a sufficiently large number L, then equation (9.63) can be used to construct a set of linear equations from where auto-regressive coefficients and moving average coefficients can be estimated. Knowing the moving average coefficients, the residues of the transfer function in equation (9.58) can be identified.

The characteristic equation of the system is given as:

$$1 + \phi_1 z^{-1} + \ldots + \phi_{2N} z^{-2N} = 0 \tag{9.64}$$

With the estimated auto-regressive coefficients, equation (9.64) can be used to identify roots z which come in complex conjugate pairs. These roots are related to the natural frequencies and damping ratios of the system. From the relationship between the Laplace transform and the z-transform we know that:

$$z_r = e^{s_r \Delta} \tag{9.65}$$

Here Δ is sampling time resolution. Also the complex roots s_r follow:

$$s_r = -\zeta_r \omega_r + j\sqrt{1 - \zeta_r^2}\,\omega_r \tag{9.66}$$

$$s_r^* = -\zeta_r \omega_r - j\sqrt{1 - \zeta_r^2}\,\omega_r \tag{9.67}$$

Therefore, the natural frequency and damping ratio of the rth mode can be estimated from:

$$\omega_r = \frac{1}{\Delta}\sqrt{\ln z_r \, \ln z_r^*} \tag{9.68}$$

and

$$\zeta_r = \frac{-\ln(z_r z_r^*)}{2\omega_r \Delta} \tag{9.69}$$

Selecting the optimal order for the ARMA model is crucial to the success of using the ARMA model for modal analysis.

9.5 Least-squares complex exponential (LSCE) method

Least-squares complex exponential method is another time domain modal analysis method. It explores the relationship between the IRF of an MDoF system and its complex poles and residues through a complex exponential. By establishing the analytical links between the two, we can construct an AR model. The solution of this model leads to the establishment of a polynomial whose roots are the complex roots of the system. Having estimated the roots (thus the natural frequencies and damping ratios), the residues can be derived from the AR model for mode shapes. The IRF can be derived from the inverse Fourier transform of an FRF or from random decrement process.

The LSCE method begins with the transfer function of an MDoF system given in Chapter 6:

$$\alpha_{ij}(s) = \sum_{r=1}^{N} \left(\frac{{}_rA_{ij}}{s - s_r} + \frac{{}_rA_{ij}^*}{s - s_r^*} \right) \tag{9.70}$$

or

$$\alpha_{ij}(s) = \sum_{r=1}^{2N} \frac{{}_rA_{ij}}{s - s_r} \quad \text{for } r > N: \; {}_rA_{ij} = {}_rA_{ij}^*; \, s_r = s_r^* \tag{9.71}$$

The inverse Laplace transform of this transfer function is the IRF given as:

$$h_{ij}(t) = \sum_{r=1}^{2N} {}_rA_{ij} e^{s_r t} \tag{9.72}$$

For simplicity, the subscript of the transfer function ij is omitted from the following analysis. If this IRF is sampled at a series of equally spaced time intervals $k\Delta$ ($k = 0, 1, \ldots, 2N$), then we will have a series of sampled IRF data:

$$h(k\Delta) = \sum_{r=1}^{2N} {}_rA_{ij} e^{s_r k\Delta} \quad (k = 0, 1, \ldots, 2N) \tag{9.73}$$

or
$$h_k = \sum_{r=1}^{2N} {}_rA_{ij}\, z_r^k \quad (k = 0, 1, \ldots, 2N) \quad z_r^k = e^{s_r k\Delta} \tag{9.74}$$

All these samples are real-valued data, although the residues ${}_rA_{ij}$ and the roots s_r are complex quantities. It is easy to show that all imaginary parts will cancel each other because of the complex conjugates for both ${}_rA_{ij}$ and s_r. The next step is to estimate the roots and residues from the sampled data. This solution is aided by the conjugacy of the roots s_r, therefore z_r. Mathematically, this means that z_r are roots of a polynomial with only real coefficients:

$$\beta_0 + \beta_1 z_r + \beta_2 z_r^2 + \ldots + \beta_{2N-1} z_r^{2N-1} + \beta_{2N} z_r^{2N} = 0 \tag{9.75}$$

This equation is known as the Prony equation. The coefficients can be estimated from the samples of the IRF data. Since there are $2N + 1$ equalities in equation (9.74), we can multiply each equality with a corresponding coefficient β and add all equalities together to form the following equation:

$$\sum_{k=0}^{2N} \beta_k h_k = \sum_{k=0}^{2N} \beta_k \sum_{r=1}^{2N} {}_rA_{ij} z_r^k \tag{9.76}$$

or
$$\sum_{k=0}^{2N} \beta_k h_k = \sum_{r=1}^{2N} {}_rA_{ij} \sum_{k=0}^{2N} \beta_k z_r^k \tag{9.77}$$

From equation (9.75) we know that the right-hand side of equation (9.77) becomes zero when Z_r is a root of polynomial in equation (9.75). This will lead us to a simple relationship between the coefficients β and the IRF samples, namely:

$$\sum_{k=0}^{2N} \beta_k h_k = 0 \tag{9.78}$$

This equation offers a numerical way of estimating the β coefficients. In equation (9.75) we can assign β_{2N} to be one. Taking a set of $2N$ samples of IRF, one linear equation for β is formed in equation (9.78). Taking $2N$ sets of $2N$ samples of IRF, we have $2N$ linear equations. The coefficients can be estimated from the linear simultaneous equations:

$$\begin{bmatrix} h_0 & h_1 & h_2 & \cdots & h_{2N-1} \\ h_1 & h_2 & h_3 & \cdots & h_{2N} \\ \vdots & \vdots & \vdots & \vdots & \vdots \\ h_{2N-1} & h_{2N} & h_{2N+1} & \cdots & h_{4N-2} \end{bmatrix} \begin{Bmatrix} \beta_0 \\ \beta_1 \\ \vdots \\ \beta_{2N-1} \end{Bmatrix} = \begin{Bmatrix} h_{2N} \\ h_{2N+1} \\ \vdots \\ h_{4N-1} \end{Bmatrix} \tag{9.79}$$

The selection of IRF data samples can vary provided that the h elements in each row are evenly spaced in sampling and sequentially arranged. No two rows have identical h elements. The number of rows in equation (9.79) can exceed the number of β coefficients for the least-square solutions.

With known β coefficients, equation (9.75) can be solved to yield the z_r roots. These roots are related to the system complex natural frequencies s_r, as given in equation (9.74). Since the complex natural frequencies s_r are determined by the undamped natural frequencies ω_r and damping ratios ζ_r, as shown below:

$$s_r = -\zeta_r \omega_r + j\sqrt{1 - \zeta_r^2}\,\omega_r \tag{9.80}$$

$$s_r^* = -\zeta_r \omega_r - j\sqrt{1 - \zeta_r^2}\,\omega_r, \tag{9.81}$$

we can derive the natural frequency and damping ratio of the rth mode as:

$$\omega_r = \frac{1}{\Delta}\sqrt{\ln z_r \, \ln z_r^*} \tag{9.82}$$

and
$$\zeta_r = \frac{-\ln(z_r\, z_r^*)}{2\omega_r \Delta} \tag{9.83}$$

To determine the mode shapes of the system from the IRF data, we can simply rewrite equation (9.74) in a new form as:

$$\begin{bmatrix} 1 & 1 & \cdots & 1 \\ z_1 & z_2 & \cdots & z_{2N} \\ \vdots & \vdots & \vdots & \vdots \\ z_1^{2N-1} & z_2^{2N-1} & \cdots & z_{2N}^{2N-1} \end{bmatrix} \begin{Bmatrix} _1 A_{ij} \\ _2 A_{ij} \\ \vdots \\ _{2N} A_{ij} \end{Bmatrix} = \begin{Bmatrix} h_0 \\ h_1 \\ \vdots \\ h_{2N-1} \end{Bmatrix} \tag{9.84}$$

The solution to this set of linear equations will yield the residues. Alternatively, we can first denote the complex natural frequency (already known by now) and the residues (unknown) by their real and imaginary parts:

$$s_r = \alpha_r + i\beta_r \qquad s_r^* = \alpha_r - i\beta_r \tag{9.85, 86}$$

$$_r A_{ij} = U_r + iV_r \qquad _r A_{ij}^* = U_r - iV_r \tag{9.87, 88}$$

The sampled IRF data can now be expressed as:

$$h_k = \left(2\sum_{r=1}^{N} e^{\alpha_r k\Delta} \cos(\beta_r k\Delta)\right) U_r - \left(2\sum_{r=1}^{N} e^{\alpha_r k\Delta} \sin(\beta_r k\Delta)\right) V_r \tag{9.89}$$

This is a linear equation for parameters U_r and V_r. By taking N sets of N samples of IRF, we can form N linear equations from where these parameters are determined and so are the mode shapes. The procedure of the LSCE method can be illustrated in Figure 9.4.

The above analysis describes the main thrust of the LSCE method and its execution. The method can be improved by using more IRF data samples in equations (9.79), (9.84) and (9.89) with more sophisticated numerical techniques for solving linear simultaneous equations.

9.6 Summary of time domain modal analysis

The development of time domain modal analysis methods relies upon the progress of modern control theory. Theoretical and mathematical aspects of these methods are similar to that in system identification developed for modern control engineering. When applying modern control engineering to structural dynamics and modal analysis,

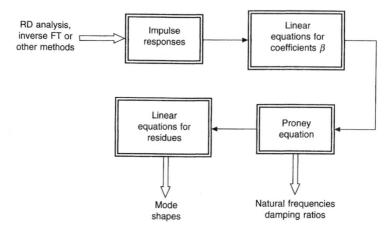

Figure 9.4 Procedure of the LSCE method

additional requirements are placed on vibration testing, data acquisition and signal processing, especially when dealing with large structures.

Traditional frequency domain modal analysis is based on frequency analysis theory. The physical interpretation of this modal analysis is usually evident. For example, an FRF exhibits resonances and anti-resonances of a test structure and the amount of damping for each vibration mode is evidenced by the sharpness of resonance peaks. There are developed techniques to combat measurement noise and other errors. For example, the average of records and application of windows are effective means of reducing random and systematic errors in FRF estimation. To obtain the response measurement for the estimation of FRF data, a structure normally needs to undergo excitations which are accurately measured. This type of measurement may require costly measurement equipment and a laboratory environment. It is not suitable for online measurement and analysis.

Time domain modal analysis is able to make use of vibration responses due to ambient excitations. This is a clear advantage over frequency domain modal analysis. The time responses obviously do not exhibit clear modal information such as resonances as FRF data do. Noise experienced in measurement is more a problem in time domain than in frequency domain because we do not have the luxury to utilize FRF data around resonances for better signal to noise ratio. Modes with less vibration energy at a measurement location can be inundated by noise and therefore disappear from analysis radar screen. This poses great demands on modal analysis methods to improve their algorithms in order to deliver accurate modal analysis results.

Ideally, time domain modal analysis requires data measured from all measurement points on a structure simultaneously. This avoids unnecessary phase difference among responses and ensures that all responses come from the same initial forcing conditions. In reality, it may not be practical to measure many channels of responses at the same time due to hardware limitations. One solution is to divide measurement points into groups. Each group can be analysed separately using the time domain modal analysis methods. By having one or more overlapping points between groups, it is possible later to scale all the mode shape data consistently.

Literature

1. Batill, S.M. and Hollkamp, J.J. 1989: Parameter identification of discerte time series models for structural response prediction. *AIAA Journal*, **27**(11), 1636–1649.

2. Brincker, R., De Steffano, A. and Piombo, B. 1996: Ambient data to analyse the dynamic behaviour of bridges: A first comparison between different techniques. *Proceedings of the 14th International Modal Analysis Conference*, Dearborn, USA, 477–482.

3. Budwantord, B. and Jezequel, L. 1990: Comparison of time domain modal identification methods. *Proceedings of the 8th International Modal Analysis Conference*, Kissimmee, FL, 540–546.

4. Cole, H.A. 1968: On-the-line analysis of random vibrations. *AIAA Paper* No. 68–288.

5. Hollkamp, J.J. and Batill, S.M. 1991: Automated parameter identification and order reduction for discrete time series models. *AIAA Journal*, **29**(1), 96–103.

6. Ibrahim, S.R. 2000: Fundamentals of Time Domain Modal Identification. *NATO ASI Series E: Applied Sciences*, 241–250.

7. Ibrahim, S.R. and Mikulcik, E.C. 1973 A time domain modal vibration test technique. *The Shock and Vibration Bulletin*, Bulletin 43, Part 4, June.

8. Ibrahim, S.R. and Mikulcik, E.C. 1976: The Experimental Determination of Vibration Parameters from Time Responses. *The Shock and Vibration Bulletin*, **46**(5), 187–196.

9. Ibrahim, S.R. 1978: Modal Confidence Factor in Vibration Testing. *The Shock and Vibration Bulletin*, **48**(1), 65–75.

10. Koh, C.G., Hong, B. and Liaw, C.-Y. 2000: Parameter identification of large structural systems in time domain. *Journal of Structural Engineering*, **126**(8), 957–963.

11. Mickleborough, N.C. and Pi, Y.L. 1989: Modal parameter identification using z-transforms. *International Journal for Numerical Methods in Engineering*, **28**, 2307–2321.

12. Pappa, R.S. and Ibrahim, S.R. 1981: A Parametric Study of the Ibrahim Time Domain Identification Algorithm. *The Shock and Vibration Bulletin*, **51**(3), 43–72.

13. Silva, J.M.M and Maia, N.M.M. (eds) 1998: Modal Analysis and Testing. *NATO Science Series E: Applied Sciences*, Vol. 363.

14. Smith, W.R. 1981: Least squares time-domain methods for simultaneous identification of vibration parameters from multiple free-response records. *AIAA/ASME/ASCE/AHS 22nd SDM Conference*, April, 194–201.

15. Spitznogle, F.R. and Quazi, A.H. 1970: Representation and analysis of time-limited signals using a complex exponential algorithm. *Journal of the Acoustical Society of America*, Vol. 47, No. 5, 1150–1155.

16. Zaghlool, S.A. 1980: Single-Station Time-Domain Vibration Testing Technique: Theory and Application. *Journal of Sound and Vibration*, **72**(2), 205–234.

10

Multi-input multi-output modal analysis methods

10.1 Introduction

In the last two chapters, we have studied several frequency domain and time domain modal analysis methods using test data acquired with single input and multi-output configuration. These methods are more applicable for small to medium structures. However, for many large structures such as aeroplanes, ships, missiles, offshore platforms, mining machines and space vehicles, single excitation input often fails to provide sufficient vibration energy in order to produce meaningful data. The response collected far from the excitation location vanishes. In general measured data have a less than satisfactory signal to noise ratio. Simple enlargement of the excitation strength could relegate vibration to a nonlinear range. When excitation is situated at a nodal point, a vibration mode becomes invisible. In addition, the FRF data used in modal analysis are just one column of the FRF matrix. This also limits the accuracy of modal analysis and hampers the ability to identify closely positioned vibration modes. Against this background, the multi-input multi-output (also known as MIMO) approach for modal analysis was developed.

MIMO modal analysis can be conducted both in frequency domain and in time domain. Frequency domain methods use measured FRF data for modal parameter identification while time domain methods use impulse response function data or vibration response data to establish the time domain model before identification proceeds.

In the frequency domain MIMO category, the poly-reference frequency domain method is one example. In the time domain MIMO category, there have been several methods available. Typically, there are the poly-reference complex exponential method, the direct parameter identification method and the eigensystem realization method. All MIMO methods are based on simultaneous and global identification. This makes it possible to obtain more consistent modal parameters and reduce human decision making during the analysis. Whether this automatically leads to more accurate modal data is application dependent.

For frequency domain MIMO methods, obtaining accurate FRF data is of paramount importance, as the accuracy of FRF data dictates the quality of the modal analysis outcome. The following section introduces several methods for estimating FRFs.

10.2 Estimation of FRFs for MIMO testing

For an MDoF structural system with P input forces and L output responses, the input and output spectra can be denoted respectively as $F_1(\omega)$, $F_2(\omega)$, ..., $F_P(\omega)$ and $X_1(\omega)$, $X_2(\omega)$, ..., $X_L(\omega)$. These spectra contain noise components. The noise spectra from the input sides are denoted as $M_1(\omega)$, $M_2(\omega)$, ..., $M_P(\omega)$ and that from the output sides as $N_1(\omega)$, $N_2(\omega)$, ..., $N_L(\omega)$. The noise free input and output spectra can be denoted as $U_1(\omega)$, $U_2(\omega)$, ..., $U_P(\omega)$ and $V_1(\omega)$, $V_2(\omega)$, ..., $V_L(\omega)$ respectively. These quantities can be grouped together using vector notations:

$$\{F\} = \{F_1(\omega), F_2(\omega), \ldots, F_P(\omega)\}^T \tag{10.1}$$

$$\{X\} = \{X_1(\omega), X_2(\omega), \ldots, X_L(\omega)\}^T \tag{10.2}$$

$$\{M\} = \{M_1(\omega), M_2(\omega), \ldots, M_P(\omega)\}^T \tag{10.3}$$

$$\{N\} = \{N_1(\omega), N_2(\omega), \ldots, M_L(\omega)\}^T \tag{10.4}$$

$$\{U\} = \{U_1(\omega), U_2(\omega), \ldots, U_P(\omega)\}^T \tag{10.5}$$

$$\{V\} = \{V_1(\omega), V_2(\omega), \ldots, V_L(\omega)\}^T \tag{10.6}$$

A MIMO system with all these quantities can be shown by Figure 10.1.

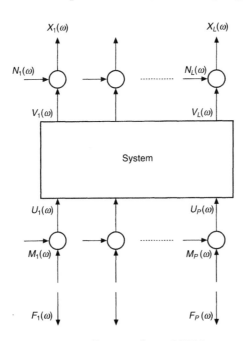

Figure 10.1 A diagram for a MIMO system

When no noise exists, the output of the system at the ith coordinate is:

$$X_i(\omega) = V_i(\omega) = \sum_{j}^{p} H_{ij}(\omega)F_j(\omega) \tag{10.7a}$$

When the system has noise, however, the same response becomes:

$$X_i(\omega) = V_i(\omega) + N_i(\omega) = \sum_j^p H_{ij}(\omega)(F_j(\omega) - M_j(\omega)) + N_i(\omega)$$

$$= \sum_j^p H_{ij}(\omega)F_j(\omega) + E_i(\omega) \tag{10.7b}$$

Here $E_i(\omega)$ is the total measurement error at the ith point of the system which can comprise the input and output measurement errors caused by noise and test disturbance, signal processing errors such as truncation and leakage errors and nonlinearity induced errors. When using equation (10.7) to estimate the FRF, the accuracy is dictated by the assumption of error types and by how errors are quantified. If all possible errors are taken into account and the methods are found to minimize estimation errors, then a most accurate FRF estimate is possible. In reality, all errors are not included. In addition, accuracy and cost oppose each other. As a result, it is necessary to find the optimal FRF estimate to suit different scenarios.

10.2.1 No input noise – $[\hat{H}_1(\omega)]$ model

This estimation of FRF assumes that there is no input noise (so $\{M\} = \{0\}$) and the output noise does not correlate with the system input. The response will then consist of the genuine response due to force inputs and the noise as:

$$\{X(\omega)\}_{L \times 1} = [H(\omega)]_{L \times P} \{F(\omega)\}_{P \times 1} + \{N(\omega)\}_{L \times 1} \tag{10.8}$$

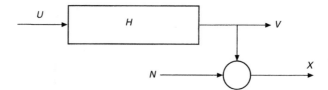

Figure 10.2 A case of output noise only

Post-multiplying both sides of the equation by $\{F(\omega)\}^H$, the Hermitian transpose of the force vector, the expected values of the responses are in the form of power spectra:

$$[G_{XF}(\omega)]_{L \times P} = [H(\omega)]_{L \times P} [G_{FF}(\omega)]_{P \times P} + [G_{NF}(\omega)]_{L \times P} \tag{10.9}$$

Here, the power spectra are respectively:

$[G_{XF}(\omega)]_{L \times P}$ cross power spectrum of the input F and output X;
$[G_{FF}(\omega)]_{P \times P}$ auto power spectrum of the input F;
$[G_{NF}(\omega)]_{L \times P}$ cross power spectrum of the input F and noise N.

Where no correlation exists between the input F and the noise N, the cross power spectrum $[G_{NF}(\omega)]$ becomes zero. This leads to an estimate of the FRF matrix:

$$[\hat{H}(\omega)]_{L\times P} = [G_{XF}(\omega)]_{L\times P} \, [G_{FF}(\omega)]^{-1}_{P\times P} \qquad (10.10)$$

This estimate is known as $[\hat{H}_1(\omega)]$ for the FRF matrix. $[\hat{H}_1(\omega)]$ is the traditional estimate and mathematically it is an underestimate. This can be shown below using a single-input single-output (SISO) case for convenience. As we know that the input and output noise do not correlate with each other and nor do the noise with the genuine signals, we have the following spectra:

$$\begin{cases} G_{FF}(\omega) = G_{UU}(\omega) + G_{MM}(\omega) \\ G_{XX}(\omega) = G_{VV}(\omega) + G_{NN}(\omega) \\ G_{XF}(\omega) = G_{VU}(\omega) \end{cases} \qquad (10.11)$$

If the true FRF without any noise is denoted as $H(\omega)$, then:

$$H(\omega) = \frac{G_{VU}(\omega)}{G_{UU}(\omega)} \qquad (10.12)$$

while:

$$\hat{H}_1(\omega) = \frac{G_{XF}(\omega)}{G_{FF}(\omega)} = \frac{1}{1 + \dfrac{G_{MM}(\omega)}{G_{UU}(\omega)}} H(\omega) \qquad (10.13)$$

Obviously the estimate is less than the true FRF. Because of only including the output noise, $[\hat{H}_1(\omega)]$ is prone to error when the noise does exist. Near a vibration resonance where input forces decline dramatically because of impedance mismatch between the structure and the input sources, the signal to noise ratio drops considerably. As a result, the accuracy of the $[\hat{H}_1(\omega)]$ estimate is poor around resonances.

From equation (10.10) it is evident that the auto-power spectrum matrix of the input forces $[G_{FF}(\omega)]$ needs to be inverted in order to estimate the FRF. However, this matrix is not always invertible. There are several reasons why the matrix can become singular:

(1) When one of the input forces $F_j(\omega)$ is zero, the jth row and column of matrix $[G_{FF}(\omega)]$ are full of zeros. The matrix is singular.
(2) When two or more input forces are completely dependent on each other, the rank of matrix $[G_{FF}(\omega)]$ is deficient.
(3) Numerical analysis may be hindered because of the ill-conditioning of matrix $[G_{FF}(\omega)]$.

Therefore, the success of the FRF matrix estimate $[\hat{H}_1(\omega)]$ depends primarily on the invertibility of the force spectrum matrix. In particular, it is essential to assure that input forces are not interdependent. For a MIMO system, a few types of coherence functions can be of great assistance to gaining confidence on the FRF estimates.

1. Ordinary coherence function

Ordinary coherence function is defined as the measure of the causal relationship between two signals with the presence of other signals. For a MIMO system, the coherence function is identical to that for a SISO system. For the ith output and the jth input, the ordinary coherence function is formulated as:

$$\gamma_{ij}^2 = \frac{|G_{XF}(i,j)|^2}{G_{FF}(j,j)G_{XX}(i,i)} \tag{10.14}$$

When the coherence function is equal to one, two signals are completely related. If the coherence function between two input signals is one, the two inputs are completely related. This would hinder the estimation of the FRF matrix.

2. Partial coherence function

Because of the reaction of the test structure or coupling among excitation sources, an input signal of a MIMO system is affected by other input sources. The partial coherence function is defined as the coherence function between an input and an output, or between two inputs, or between two outputs after the cross-influence among input signals is removed (these 'refined' input signals are referred to as the 'conditioned' signals). Assume the spectra of the inputs are $F_1(\omega)$, $F_2(\omega)$, ..., $F_P(\omega)$. Eliminate from spectrum $F_2(\omega)$ the influence of $F_1(\omega)$:

$$F_2^1(\omega) = F_2(\omega) - H_{21}F_1(\omega) \tag{10.15}$$

Here, $H_{21}(\omega) = \dfrac{G_{F_1F_2}(\omega)}{G_{F_1F_1}(\omega)}$ is the nominal FRF between two forces. $F_2^1(\omega)$ represents the force spectrum $F_2(\omega)$ after the influence of $F_1(\omega)$ is removed. Likewise, we can derive $F_3^1(\omega)$, $F_4^1(\omega)$ till $F_P^1(\omega)$. As the ordinary coherence function between $F_1(\omega)$ and $F_2(\omega)$ is defined as:

$$\gamma_{12}^2 = \frac{|G_{F_1F_2}|^2}{G_{F_1F_1}G_{F_2F_2}}, \tag{10.16}$$

the auto-power spectrum of $F_2^1(\omega)$ can be derived from equation (10.15) as:

$$G_{F_2^1F_2^1} = G_{F_2F_2} - \frac{|G_{F_1F_2}|^2}{G_{F_1F_1}} = G_{F_2F_2}(1 - \gamma_{12}^2) \tag{10.17}$$

If the coherence function γ_{12}^2 is close to zero, then the forces $F_1(\omega)$ and $F_2(\omega)$ are almost uncorrelated. Having refined $F_2(\omega)$ from the influence of $F_1(\omega)$, spectrum $F_2^1(\omega)$ can be used to refine forces $F_3^1(\omega)$ and the subsequent forces.

$$F_3^2(\omega) = F_3^1(\omega) - H_{32}F_2^1(\omega) \tag{10.18}$$

Here, $H_{32}(\omega) = \dfrac{G_{F_2^1F_3^1}(\omega)}{G_{F_2^1F_2^1}(\omega)}$ is the nominal FRF between forces 3 and 2. The partial coherence between them is defined as:

$$(\gamma_{23}^2)^1 = \frac{|G_{F_2^1F_3^1}|^2}{G_{F_2^1F_2^1}G_{F_3^1F_3^1}} \tag{10.19}$$

The auto-power spectrum of $F_3^2(\omega)$ can be derived from:

$$G_{F_2^2 F_3^2} = G_{F_3^1 F_3^1} - \frac{|G_{F_2^1 F_3^1}|^2}{G_{F_2^1 F_2^1}} = G_{F_3^1 F_3^1}(1 - (\gamma_{23}^2)^1) \qquad (10.20)$$

To continue this process, the general formula for the partial coherence function can be derived as:

$$(\gamma_{(k+1)(k+2)}^2)^k = \frac{|G_{F_{(k+1)}^k F_{(k+2)}^k}|^2}{G_{F_{(k+1)}^k F_{(k+1)}^k} \, G_{F_{(k+2)}^k F_{(k+2)}^k}} \qquad k = 1, 2, \ldots, (p-2) \quad (10.21)$$

If $(\gamma_{(k+1)(k+2)}^2)^k$ is equal to one, then the input forces $F_{(k+1)}^k(\omega)$ and $F_{(k+2)}^k(\omega)$ are correlated.

A different method of detecting the correlation between input forces is to use the eigenvalue decomposition of the input spectrum matrix $[G_{FF}(\omega)]$. Assume $F_1'(\omega), F_2'(\omega), \ldots, F_P'(\omega)$ denote the auto-spectra of 'P' independent input forces. The force spectra from measurement can be expressed as a linear combination of them such that:

$$[F_1(\omega) \, F_2(\omega) \ldots F_P(\omega)] = [F_1'(\omega) \, F_2'(\omega) \ldots F_P'(\omega)][Q] \qquad (10.22)$$

where matrix $[Q]$ is a complex matrix with dimension $P \times P$. The eigenvalue decomposition of matrix $[G_{FF}(\omega)]$ becomes:

$$[G_{F'F'}(\omega)] = [Q]^H [G_{FF}(\omega)][Q] \qquad (10.23)$$

Here, $[G_{F'F'}(\omega)]$ is a diagonal matrix. Its diagonal elements are called principal power spectra of the input forces and they follow the orthogonality:

$$G_{F_i' F_j'}(\omega) = \begin{cases} F_i' * F_i' & i = j \\ 0 & i \neq j \end{cases} \qquad (10.24)$$

Matrix $[G_{F'F'}(\omega)]$ is the eigenvalue matrix for $[G_{FF}(\omega)]$ while matrix $[Q]$ is its eigenvector matrix. Since $[G_{FF}(\omega)]$ is a Hermitian matrix, its eigenvectors are independent. As a result, matrix $[Q]$ is non-singular. When some eigenvalues of $[G_{FF}(\omega)]$ are zeros, $[Q]$ becomes singular. This means when the principal power spectra of the input forces are minute, some input forces are correlated. Therefore, the estimates of the principal power spectra of the input forces are indicative of the correlation among forces.

3. Multiple coherence function

The multiple coherence function describes the linear and causal links between one output signal and all known input signals. It can be used to evaluate the influence of unknown inputs to an output. These unknown inputs may be measurement noise, energy leakage or nonlinearity. Every output has a multiple coherence. If it is equal to one, the output is correlated with all inputs. If it is equal to zero, the output is due to unknown inputs such as noise.

Multiple coherence is defined as the ratio between two power spectra. They are the cross power spectra between the output and the known inputs and the auto-power spectrum of the output. Physically, this ratio means the proportion of the output signal due to the known inputs.

$$\gamma^2_{X_iF} = \frac{\sum_{j=1}^{P} H_{ij} G_{X_iF_j}}{G_{X_iX_i}} = 1 - \frac{(G_{N_iN_i})_{min}}{G_{X_iX_i}} \tag{10.25}$$

Equation (10.25) is a least square outcome based on the model of no input noise. With the presence of input and output noises, the equation will assume a different form.

In order to be able to calculate the multiple coherence using input cross power spectra and output auto-power spectra, we can define an augmented matrix of cross power spectra for the ith output as:

$$[G_{X_iFF}] = \begin{bmatrix} G_{X_iX_i} & \vdots & G_{X_iF_1} & G_{X_iF_2} & \cdots & G_{X_iF_P} \\ G_{F_1X_i} & \vdots & G_{F_1F_1} & G_{F_1F_2} & \cdots & G_{F_1F_P} \\ G_{F_2X_i} & \vdots & G_{F_2F_1} & G_{F_2F_2} & \cdots & G_{F_2F_P} \\ \vdots & \vdots & \vdots & \vdots & \vdots & \vdots \\ G_{F_PX_i} & \vdots & G_{F_PF_1} & G_{F_PF_2} & \cdots & G_{F_PF_P} \end{bmatrix} \tag{10.26a}$$

The lower right part of the matrix is the force cross-spectrum matrix for input forces which can be denoted as:

$$[G_{FF}] = \begin{bmatrix} G_{F_1F_1} & G_{F_1F_2} & \cdots & G_{F_1F_P} \\ G_{F_2F_1} & G_{F_2F_2} & \cdots & G_{F_2F_P} \\ \vdots & \vdots & \vdots & \vdots \\ G_{F_PF_1} & G_{F_PF_2} & \cdots & G_{F_PF_P} \end{bmatrix} \tag{10.26b}$$

From equation (10.25) we derive:

$$G_{X_iX_i} = \sum_{j=1}^{P} H_{ij} G_{X_iF_j} + G_{N_iN_i} \tag{10.27}$$

Substituting this into equation (10.26a) and calculating its determinant leads to:

$$|[G_{X_iFF}]| = G_{N_iN_i} |[G_{FF}]| \tag{10.28}$$

The multiple coherence function can now be derived from:

$$\gamma^2_{X_iF} = 1 - \frac{|[G_{X_iFF}]|}{G_{X_iX_i} |[G_{FF}]|} \tag{10.29}$$

10.2.2 No output noise – $[\hat{H}_2(\omega)]$ model

When using this model to estimate the FRF matrix, it is assumed that no output noise exists, as shown in Figure 10.3. The input noise is not correlated with the output. The

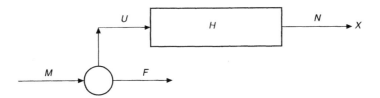

Figure 10.3 A case of input noise only

input and output spectra follows:

$$[X(\omega)] = [H(\omega)]([F(\omega)] - [M(\omega)]) \tag{10.30}$$

Post-multiplying the equation by $[X(\omega)]^H$, the expected response estimate becomes:

$$[G_{XX}(\omega)] = [H(\omega)][G_{FX}(\omega)] \tag{10.31}$$

This yields the new estimate of the FRF matrix as:

$$[\hat{H}_2(\omega)] = [G_{XX}(\omega)][G_{FX}(\omega)]^{-1} \tag{10.32}$$

This solution is correct only when the matrix inverse in the equation exists. When the number of excitation 'P' is less than that of the response points 'L', inverse $[G_{FX}(\omega)]^{-1}$ does not exist. A pseudo inverse has to be used to derive the FRF matrix.

It can also be shown using the SISO case that $[\hat{H}_2(\omega)]$ is an overestimate of the FRF. For a single input f and single output x, the FRF is estimated as:

$$\hat{H}_2(\omega) = \frac{G_{XX}(\omega)}{G_{FX}(\omega)} = \frac{G_{VV}(\omega) + G_{NN}(\omega)}{G_{UV}(\omega)}$$

$$= \frac{G_{VV}(\omega)}{G_{UV}(\omega)} \left(1 + \frac{G_{NN}(\omega)}{G_{VV}(\omega)}\right) = H(\omega)\left(1 + \frac{G_{NN}(\omega)}{G_{VV}(\omega)}\right) \tag{10.33}$$

This shows that $\hat{H}_2(\omega)$ is an overestimate of the correct FRF $H(\omega)$. The same observation applies to the MIMO case.

Estimate $[\hat{H}_2(\omega)]$ is sensitive to the output noise. It is particularly inaccurate when anti-resonances occur with minimum responses. However, it gives an accurate estimate around resonance frequencies. This is important for modal analysis since many methods rely on the FRF data at the vicinity of resonances.

When both the input and the output noise exist, both $[\hat{H}_1(\omega)]$ and $[\hat{H}_2(\omega)]$ are biased estimates. Their accuracy depends on signal to noise ratios of the inputs and outputs. Generally, $[\hat{H}_2(\omega)]$ provides accurate estimates around resonances while $[\hat{H}_1(\omega)]$ are more accurate around anti-resonances and minima. It can be shown that:

$$|\hat{H}_2(\omega)| \geq |H(\omega)| \geq |\hat{H}_1(\omega)| \tag{10.34}$$

10.2.3 Both input and output noise –
$[\hat{H}_3(\omega)]$, $[\hat{H}_4(\omega)]$, $[\hat{H}_v(\omega)]$, $[\hat{H}_s(\omega)]$ models

When both the input and output noise are considered, it is logical to attempt to find the estimate of the FRF $[H(\omega)]$ from an average of the two models $[\hat{H}_1(\omega)]$ and $[\hat{H}_2(\omega)]$. Depending on the type of the average, there will be different estimates. Using the SISO notations again we have the following.

1. Arithmetical average

$$\hat{H}_3(\omega) = \frac{1}{2}\left(\hat{H}_1(\omega) + \hat{H}_2(\omega)\right) = \frac{1}{2}\left(\frac{1}{1 + \dfrac{G_{MM}}{G_{UU}}} + \left(1 + \dfrac{G_{NN}}{G_{VV}}\right)\right) \quad (10.35)$$

2. Geometrical average

$$\hat{H}_4(\omega) = \sqrt{\hat{H}_1(\omega)\,\hat{H}_2(\omega)} = H(\omega)\sqrt{\frac{1 + \dfrac{G_{NN}}{G_{VV}}}{1 + \dfrac{G_{MM}}{G_{UU}}}} \quad (10.36)$$

It is easy to show that:

$$|\hat{H}_2(\omega)| \geq |\hat{H}_3(\omega)| \geq |\hat{H}_4(\omega)| \geq |\hat{H}_1(\omega)| \quad (10.37)$$

3. $\hat{H}_v(\omega)$ estimate for a MIMO system
The FRF matrix of a MIMO system can also be estimated using the least square principle to minimize an error matrix. Like before, assume there are 'P' input forces and 'L' output responses. Then,

$$[X]_{L\times1} - [N]_{L\times1} = [H]_{L\times P}\,([F]_{P\times1} - [M]_{P\times1}) \quad (10.38)$$

Saving the subscripts, the equation can be recast as:

$$[X] - [H][F] = [N] - [H][M] \quad (10.39)$$

Post-multiplying both sides with their Hermitian transpose leads to:

$$([X] - [H][F])([X] - [H][F])^H = ([N] - [H][M])([N] - [H][M])^H \quad (10.40)$$

Using power spectra, the equation can be expressed as:

$$[-[I][H]]\begin{bmatrix}[G_{XX}] & [G_{XF}] \\ [G_{FX}] & [G_{FF}]\end{bmatrix}\begin{bmatrix}-[I] \\ [H]^H\end{bmatrix}$$

$$= [G_{NN}] + [H][G_{MM}][H]^H - [N][H][M]^H - [H][G_{MN}] \quad (10.41)$$

Since the input noise is unrelated to the output noise, the last two terms of the equation will become zero. This leads to:

$$[-[I][H]] \begin{bmatrix} [G_{XX}] & [G_{XF}] \\ [G_{FX}] & [G_{FF}] \end{bmatrix} \begin{bmatrix} -[I] \\ [H]^H \end{bmatrix} = [G_{NN}] + [H][G_{MM}][H]^H \quad (10.42)$$

Assume the input and output noises are of comparable power (i.e. $G_{MM} = G_{NN}$), then their auto-spectra can be seen as equal to each other. This leads to a recast of equation (10.42) into:

$$[-[I][H]]_{L \times (L+P)} \begin{bmatrix} [G_{XX}] & [G_{XF}] \\ [G_{FX}] & [G_{FF}] \end{bmatrix}_{(L+P) \times (L+P)} \begin{bmatrix} -[I] \\ [H]^H \end{bmatrix}_{(L+P) \times L}$$

$$= [G_{NN}]_{L \times L} [-[I][H]]_{L \times (L+P)} \begin{bmatrix} -[I] \\ [H]^H \end{bmatrix}_{(L+P) \times L} \quad (10.43)$$

Namely, we have the following formula for the noise spectrum matrix:

$$[G_{NN}] = [\Delta]^H [G][\Delta]([\Delta]^H[\Delta])^{-1} \quad (10.44)$$

where $\quad [G] = \begin{bmatrix} [G_{XX}] & [G_{XF}] \\ [G_{FX}] & [G_{FF}] \end{bmatrix}_{(L+P) \times (L+P)} \quad$ and $\quad [\Delta] = \begin{bmatrix} -[I] \\ [H]^H \end{bmatrix}_{(L+P) \times L} \quad (10.45, 46)$

Mathematically, the optimal estimate of the FRF matrix should coincide with the minimum trace of the noise spectrum matrix. The trace is the sum of the smallest eigenvalues and matrix $[\Delta]$ consists of the eigenvectors of these smallest eigenvalues. Therefore,

$$\min(Tr[G]) = \sum_{r=P+1}^{P+L} \lambda_r \quad (10.47)$$

From the eigenvalue decomposition of matrix $[G]$, we can obtain $(P + L)$ eigenvalues λ_r and eigenvector Δ_r ($r = 1, 2, \ldots, P + L$). Taking the smallest L eigenvalues and their corresponding eigenvectors to form matrix $[\Delta]$ which has dimension of $(P + L) \times L$. Its upper part is an $L \times L$ identity matrix and the lower part an $P \times L$ $[H]^H$ matrix. The complex conjugate of matrix $[H]^H$ yields an estimate of the FRF matrix known as $\hat{H}_v(\omega)$. This estimate takes into account both the input and output noise and is derived from an overall least square sense. It is expected to be more accurate than the $\hat{H}_1(\omega)$ and $\hat{H}_2(\omega)$ estimates.

For a case study, assume the input and output noise vary within 0% and 20%, equations (10.13), (10.33) and (10.47) can be used to estimate the errors in estimating the FRF. Tables 10.1, 10.2 and 10.3 show the outputs of numerical studies using different FRF estimates with prescribed errors.

These tables show that the output noise has no effect on the $\hat{H}_1(\omega)$ estimate. The input noise has no effect on the $\hat{H}_2(\omega)$ estimate. With the same noise level, $\hat{H}_v(\omega)$ offers the most accurate estimate.

4. $\hat{H}_s(\omega)$ estimate

The $\hat{H}_s(\omega)$ is another model for estimating the FRF of a system. Like the $\hat{H}_v(\omega)$ model, the $\hat{H}_s(\omega)$ estimate also considers both the input and the output noise. Using a SISO case, the signals are as shown in Figure 10.4.

Table 10.1 $\hat{H}_1(\omega)$ estimate with different noise levels

$\hat{H}_1(\omega)$ estimate		Input noise		
		0.0%	10.0%	20.0%
Output noise	0.0%	0.0	− 9.0	− 16.7
	10.0%	0.0	− 9.0	− 16.7
	20.0%	0.0	− 9.0	− 16.7

Table 10.2 $\hat{H}_2(\omega)$ estimate with different noise levels

$\hat{H}_1(\omega)$ estimate		Input noise		
		0.0%	10.0%	20.0%
Output noise	0.0%	0.0	0.0	0.0
	10.0%	10.0	10.0	10.0
	20.0%	20.0	20.0	20.0

Table 10.3 $\hat{H}_v(\omega)$ estimate with different noise levels

$\hat{H}_1(\omega)$ estimate		Input noise		
		0.0%	10.0%	20.0%
Output noise	0.0%	0.0	− 4.6	− 8.7
	10.0%	4.9	0.0	− 4.3
	20.0%	9.5	4.4	0.0

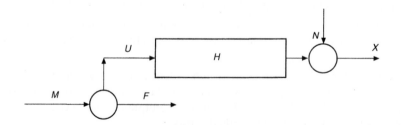

Figure 10.4 Signals for a SISO system

$$X(\omega) - N(\omega) = H(\omega)(F(\omega) - M(\omega)) \tag{10.48}$$

This equation can also be simplified into:

$$HM - N = HF - X \tag{10.49}$$

Post-multiplying each side with its Hermitian transpose and assuming the input and output noise are not correlated will lead to the following error equation:

$$\Sigma NN^H + HH^H \Sigma MM^H = \Sigma(HF - X)(HF - X)^H \qquad (10.50)$$

Usually, $\Sigma NN^H \neq \Sigma MM^H$. Equation (10.50) cannot be used for a least square minimization. However, a scaling factor S between these two error terms can be defined that to facilitate the minimization process.

$$S^2 = \frac{\Sigma NN^H}{\Sigma MM^H} \qquad (10.51)$$

or

$$X - N = SH(F - M) \qquad (10.52)$$

Equation (10.50) can then be written as:

$$S^2 \Sigma MM^H + HH^H S^2 \Sigma MM^H = \Sigma(HSF - X)(HSF - X)^H \qquad (10.53)$$

This leads to the estimation of the error term:

$$\Sigma MM^H = \frac{\Sigma(HSF - X)(HSF - X)^H}{(1 + HH^H)S^2} \qquad (10.53)$$

Minimizing this error term for the elimination of the FRF H will lead to:

$$S \Sigma X^H FH^2 + (S^2 \Sigma FF^H - \Sigma XX^H)H + S\Sigma F^H X = 0 \qquad (10.54)$$

From here we can arrive at the new estimate for FRF:

$$\hat{H}_s = \frac{-(S^2\Sigma FF^H - \Sigma XX^H) + \sqrt{(S^2\Sigma FF^H - \Sigma XX^H)^2 + 4S^2 \Sigma X^H F\Sigma F^H X}}{2S \Sigma X^H F}$$
$$(10.55)$$

10.3 Frequency domain poly-reference modal analysis method

The frequency domain poly-reference method utilizes the FRF data from MIMO testing to identify the modal parameters of the test structure. Since the method processes all the input and output signals simultaneously, the identification can be more accurate and consistency among the derived modal parameters preserved. There are several modal analysis methods that can be categorized as poly-reference methods. In the following, one of them is introduced.

The motion of a general MDoF system subjected to external forces is governed by:

$$[M]_{N\times N} \{\ddot{X}\}_{N\times 1} + [C]_{N\times N} \{\dot{X}\}_{N\times 1} + [K]_{N\times N} \{X\}_{N\times 1} = \{F\}_{N\times 1} \qquad (10.56)$$

For P excitation forces, the impulse response matrix of the system is derived as:

$$[\bar{h}(t)] = [\Psi]e^{[\Lambda]t}[L] \qquad (10.57)$$

Here, $[\Psi]$ is an $N \times 2N$ complex mode shape matrix. $[\Lambda]$ is a $2N \times 2N$ complex eigenvalue matrix. $[L]$ is a $2N \times P$ mode participation matrix while N is the number of DoFs of the system.

The Laplace transform of the left-hand side of equation (10.57) is the FRF matrix $[\alpha(s)]$. This leads to:

$$[\alpha(s)] = [\Psi](s[I] - [\Lambda])^{-1}[L] \tag{10.58}$$

Denote a new matrix:

$$[T(s)] = (s[I] - [\Lambda])^{-1}[L], \tag{10.59}$$

then equation (10.58) becomes:

$$[\alpha(s)] = [\Psi][T(s)] \tag{10.60}$$

The corresponding mobility matrix can be written as:

$$[Y(s)] = s[\alpha(s)] - [\bar{h}(t)]\Big|_{t=0} = s[\alpha(s)] - [\Psi][L]$$
$$= s[\Psi][T(s)] - [\Psi](s[I] - [\Lambda])[T(s)] = [\Psi][\Lambda][T(s)] \tag{10.61}$$

Thus, we can construct the following equation:

$$\begin{bmatrix} [\alpha(s)] \\ s[\alpha(s)] \end{bmatrix}_{2N \times P} - \begin{Bmatrix} [0] \\ [\Psi][L] \end{Bmatrix}_{2N \times P} = \begin{Bmatrix} [\Psi] \\ [\Psi][\Lambda] \end{Bmatrix}_{2N \times P} [T(s)] \tag{10.62}$$

This equation can be simplified as:

$$\begin{bmatrix} [\alpha(s)] \\ [Y(s)] \end{bmatrix} = \begin{Bmatrix} [\Psi] \\ [\Psi][\Lambda] \end{Bmatrix} [T(s)] \tag{10.63}$$

Here, $[Y(s)]$ is the mobility transfer function. Now let us derive the solution for the eigenvalue matrix $[\Lambda]$ and eigenvector matrix $[\Psi]$ of the system. The eigenvalue problem is as:

$$[A][\Psi] - [\Psi][\Lambda] = [0] \tag{10.64}$$

where matrix $[A]$ is unknown but can be constructed using measured FRF data. First, let us convert equation (10.64) into:

$$\{[A] - [I]\}_{N \times 2N} \begin{Bmatrix} [\Psi] \\ [\Psi][\Lambda] \end{Bmatrix}_{2N \times N} = [0] \tag{10.65}$$

Post-multiplying the equation by the transformation matrix $[T(s)]$ and using the conclusion given by equation (10.62) leads to:

$$\{[A] - [I]\} \begin{Bmatrix} [\Psi] \\ [\Psi][\Lambda] \end{Bmatrix} [T(s)] = \{[A] - [I]\} \begin{bmatrix} [\alpha(s)] \\ s[\alpha(s)] - [\Psi][L] \end{bmatrix} = [0] \tag{10.66}$$

Denote $[B] = [\Psi][L]$, equation (10.66) becomes:

$$[A][\alpha(s)] = s[\alpha(s)] - [B] \tag{10.67}$$

Let $s = -j\omega$, then equation (10.67) becomes:

$$[A][\alpha(j\omega)] = -j\omega[\alpha(j\omega)] - [B] \tag{10.68}$$

This equation can be used 'm' times for 'm' measured frequency points such that:

$$\begin{cases} [A][\alpha(j\omega_1)] = -j\omega_1[\alpha(j\omega_1)] - [B] \\ [A][\alpha(j\omega_2)] = -j\omega_2[\alpha(j\omega_2)] - [B] \\ \cdots \\ [A][\alpha(j\omega_m)] = -j\omega_m[\alpha(j\omega_m)] - [B] \end{cases} \qquad (10.69)$$

These equations can be combined to form the following matrix equation:

$$[[A] + [B]]_{N\times(N+P)} \begin{bmatrix} [\alpha(j\omega_1)] & [\alpha(j\omega_2)] & \cdots & [\alpha(j\omega_M)] \\ [I] & [I] & \cdots & [I] \end{bmatrix}_{(N+P)\times mP} \quad (10.70)$$

$$= [-j\omega_1[\alpha(j\omega_1)] - j\omega_2[\alpha(j\omega_2)] \ldots - j\omega_m[\alpha(j\omega_m)]]_{N\times mP}$$

From here it is possible to estimate matrices $[A]$ and $[B]$ using pseudo matrix inverse. Matrices $[A]$ and $[B]$ can then be substituted into equation (10.65) so that system eigenvalues and eigenvectors can be calculated.

For a real structure, within the measurement frequency range, the number of modes n_e is usually smaller than the modes derived from analytical modelling. When the number of measurement locations n is larger than $2n_e$, $[A]$ and $[B]$ obtained from equation (10.70) will often yield computational modes later. In addition, when n becomes excessively larger, the process of determining $[A]$ and $[B]$ may be ill-conditioned. To avert these problems, a matrix decomposition can be used to reduce the measured FRF matrix $[\alpha(j\omega)]_{n\times P}$ into $[\bar{\alpha}(j\omega)]_{n_e\times P}$. Assume there exists a matrix transformation such that:

$$[\bar{\alpha}(j\omega)]_{n_e\times P} = [Q]_{n_e\times n}[\alpha(j\omega)]_{n\times P} \qquad (10.71)$$

then we can define a deviation matrix as:

$$[SP] = \sum_{i=1}^{m} \mathrm{Re}\,([\alpha(j\omega_i)][\alpha(j\omega_i)]^H)_{n\times n} \qquad (10.72)$$

The singular decomposition of matrix $[SP]$ leads to:

$$[SP] = [U]^T[\Sigma][U] \qquad (10.73)$$

where the singular matrix will be a diagonal matrix with descending singular values:

$$[\Sigma] = \mathrm{diag}\,[\sigma_1 \quad \sigma_2 \quad \cdots \quad \sigma_{n_e} \quad 0 \quad \cdots \quad 0] \qquad (10.74)$$

The first n_e vector in matrix $[U]$ that corresponds to the first n_e singular values will form matrix $[Q]$ in equation (10.71). Having derived matrix $[Q]$, the equation becomes:

$$[\bar{\alpha}(j\omega)] = [Q][\alpha(j\omega)] = [Q][\Psi]e^{[\Lambda]t}[L] = [\bar{\Psi}]e^{[\Lambda]t}[L] \qquad (10.75)$$

Matrix $[\bar{\Psi}]$ $(= [Q][\Psi])$ is an $n_e \times n_e$ matrix that contains n_e effective modes.

To summarize, the procedure of the frequency domain poly-reference method is as follows:

(1) Select an FRF estimation model and from measurement data construct the FRF matrix $[\alpha(j\omega)]$.

(2) Use equation (10.73) to estimate matrix $[SP]$, find its singular value decomposition (SVD) to determine matrix $[Q]$.
(3) Use $[\bar{\alpha}(j\omega)]$ to replace $[\alpha(j\omega)]$ in equation (10.70) to derive matrices $[A]$ and $[B]$.
(4) Derive eigenvalue matrix $[\bar{\Lambda}]$ and eigenvector matrix $[\bar{\Psi}]$.
(5) From the transformation derive the eigenvector matrix $[\Psi]_{n\times n_e}$.
(6) Compute the natural frequency and damping ratio from the diagonal elements of the eigenvalue matrix $[\bar{\Lambda}]$ as:

$$\omega_r = \sqrt{(Re(\lambda_r))^2 + (Im(\lambda_r))^2} \quad \text{and} \quad \zeta_r = -\frac{Re(\lambda_r)}{\omega_r}$$

(7) Derive the modal participation matrix from $[L] = [B][\Psi]^{-1}$.

10.4 Time domain global modal analysis method

The time domain global modal analysis method makes use of the free vibration or impulse responses of an MDoF system to identify its modal parameters. Since the identification is for the whole system simultaneously, the modal parameters usually show good consistency and the overall accuracy is contented. There are three steps involved in implementing this method. They are respectively: (1) establishment of a mathematical model; (2) estimation of the coefficient matrix; and (3) identification of modal parameters.

10.4.1 Establishment of a mathematical model

The first step is to establish a workable mathematical model from measured data. For an N DoF linear and time-invariant system, assume the response is measured from all N coordinates and independent excitations are applied at P coordinates ($P < N$). The system governing the equation of motion can be written as:

$$\{\ddot{X}\}_{N\times1} + [M]^{-1}_{N\times N}[C]_{N\times N}\{\dot{X}\}_{N\times1}[M]^{-1}_{N\times N}[K]_{N\times N}\{X\}_{N\times1} = [D]_{N\times P}\{F\}_{P\times1}$$

(10.76)

Here, $[D]$ is a load distribution matrix. Sampling all inputs and outputs with time step Δ, we can obtain both the input and output time series $\{F\}_k$ and $\{X\}_k$ ($k = 0, 1, 2, \ldots$). Thus, equation (10.76) can be replaced by a series of finite difference equations:

$$\{X\}_k - [A_1]\{X\}_{k-1} - [A_2]\{X\}_{k-2} = [B_0]\{F\}_k + [B_1]\{F\}_{k-1} \quad (10.77)$$

Here, $[A_1]$ and $[A_2]$ are the output coefficient matrices and $[B_0]$ and $[B_1]$ the input coefficient matrices. To remedy the noise in the measurement, this time series can be expanded into a finite difference equation of $2p$th order (actually a $2p$th order of the ARMA model). The equation will become:

$$\{X\}_k - \sum_{r=1}^{2p} [A_r]\{X\}_{k-r} = [B_0]\{F\}_k + \sum_{r=1}^{2p-1} [B_r]\{F\}_{k-r} \quad (10.78)$$

Here, p is the order of the mathematical model. $[A_r]_{N \times N}$ is the output coefficient matrix and $[B_r]_{N \times P}$ the input coefficient matrix. From time domain modal analysis theory we know that $[A_r]$ contain the eigenvalues of the system and $[B_r]$ the eigenvectors. Once these matrices are found from equation (10.78), the modal parameters (natural frequencies, damping loss factors, mode shapes and modal participation factors) of the system, can be derived.

Equation (10.78) can also be written as:

$$\left([I] - \sum_{r=1}^{2p} [A_r]Z^{-r} \right)\{X\}_k = \left([B_0] - \sum_{r=1}^{2p-1} [B_r]Z^{-r} \right)\{F\}_k \qquad (10.79)$$

Here, Z^{-1} is the time shifting operator such that:

$$Z^{-i}\{X\}_k = \{X\}_{k-i} \qquad (10.80)$$

and

$$Z^{-i}\{F\}_k = \{F\}_{k-i} \qquad (10.81)$$

Series $\{X\}_k$ and $\{F\}_k$ are obtained from time domain measurement. Theoretically, they can be used in equation (10.78) to determine the least-square solutions for $[A_r]$ and $[B_r]$. When the auto-correlation matrices $[R_{XX}(0)]$ and $[R_{FF}(0)]$ of the series $\{X\}_k$ and $\{F\}_k$ are rank efficient, unique solutions can be found for $[A_r]$ and $[B_r]$. In real measurement, however, the number of the measurement location n is usually greater than the number of vibration modes n_e. As a result, equation (10.78) does not have a unique solution for $[A_r]$ and $[B_r]$.

The rank of $[R_{XX}(0)]$ is n_e, which is less than n. Therefore, it is rank deficient. A solution is to find the SVD of matrix $[R_{XX}(0)]$ and then condense the response vector $\{X\}_k$ into $\{X'\}_k$ with a rank efficient auto-correlation matrix $[R_{X'X'}(0)]$. This will enable equation (10.78) to produce a unique solution for matrices $[A_r]$ and $[B_r]$.

The SVD of matrix $[R_{XX}(0)]$ is as:

$$[R_{XX}(0)] = [U]^T \text{ diag } [\delta_1 \ \delta_2 \ldots \delta_{n_e} \ldots \delta_n][U] \qquad (10.82)$$

The rank of matrix $[R_{XX}(0)]$ can be determined from the sudden and dramatic decrease of the singular values (likely after the n_eth one). Once the rank of $[R_{XX}(0)]$ is confirmed to be n_e, the response vector $\{X\}_k$ can be condensed by a linear transformation:

$$\{X'\}_k = [Q]\{X\}_k \qquad (10.83)$$

Here, $\{X'\}_k$ is an $n_e \times 1$ condensed response vector and $[Q]$ an $n_e \times n$ transformation matrix that comprises the first n_e vectors of the orthogonal matrix $[U]$:

$$[Q] = \left[\{U\}_1^T \ \{U\}_2^T \ \ldots \{U\}_{n_e}^T \right]^T \qquad (10.84)$$

Thus, in the finite difference model in equation (10.78), the response vector $\{X\}_k$ can be replaced by vector $\{X'\}_k$ for the solution of $[A_r]$ and $[B_r]$.

A mathematical model can also be established using the impulse response functions of a system. As the inverse Fourier transform of the frequency response function, the impulse response function characterizes the input and output relationship of a dynamic system.

Assume there are 'N' response points and 'P' input points on a test structure. The impulse response function between the ith response point and the jth excitation point is $h_{ij}(t)$. An impulse response function matrix can be defined as:

$$[H(t)] = \begin{bmatrix} h_{11}(t) & h_{12}(t) & \ldots & h_{1N}(t) \\ h_{21}(t) & h_{22}(t) & \ldots & h_{2N}(t) \\ \ldots & \ldots & \ldots & \ldots \\ h_{N1}(t) & h_{N2}(t) & \ldots & h_{NP}(t) \end{bmatrix} \tag{10.85}$$

A time series $[H]_k$ $(k = 1, 2, \ldots)$ can be formed by sampling impulse response $[H(t)]$. Thus, a finite difference model based on time series $[H]_k$ is a multi-variable auto-regression process. It should satisfy:

$$[H]_k - [A_1][H]_{k-1} - \ldots - [A_{2p}]\,[H]_{k-2p} = 0 \quad (k = 2p, 2p + 1, \ldots) \tag{10.86}$$

or simply

$$[H]_k = \sum_{r=1}^{2p} [A_r][H]_{k-r} \tag{10.87}$$

The discretized version of the inverse Fourier transform of matrix $[H]_k$ can be written as:

$$[H]_k = [\Psi]e^{[\Lambda]\Delta k}\,[L] = [\Psi][Z]^k\,[L] \tag{10.88}$$

Here, $[\Psi]_{n_e \times 2n_e}$ is the mode shape matrix. $[\Lambda]_{2n_e \times 2n_e}$ is the eigenvalue matrix. $[L]_{2n_e \times P}$ is the mode participation matrix. Δ is the sampling resolution so that Δk is the time for the kth sample. $[Z]_{2n_e \times 2n_e}$ is the z-transformation matrix such that

$$[Z] = e^{[\Lambda]\Delta k} \tag{10.89}$$

With equation (10.88), it is possible to rewrite equation (10.87) as:

$$[H]_k = \sum_{r=1}^{2p} [A_r][\Psi][Z]^{k-r}[L] \tag{10.90}$$

or

$$\sum_{r=0}^{2p} [A_r][\Psi][Z]^{k-r}[L] = [0] \quad \text{where} \quad [A_0] = [I] \tag{10.91}$$

Using equations (10.87) and (10.91) together, a new matrix equation can be formed for coefficient matrix $[A_r]$:

$$\begin{bmatrix} [H]_k \\ [H]_{k-1} \\ \ldots \\ [H]_{k-2p+1} \end{bmatrix} = \begin{bmatrix} [A_1] & [A_2] & \ldots & [A_{2p}] \\ [I] & [0] & \ldots & [0] \\ \ldots & \ldots & \ldots & \ldots \\ [0] & \ldots & [I] & [0] \end{bmatrix} \begin{bmatrix} [H]_{k-1} \\ [H]_{k-2} \\ \ldots \\ [H]_{k-2p} \end{bmatrix} \tag{10.92}$$

or simply as:

$$[\overline{H}]_k = [\overline{A}][\overline{H}]_{k-1} \tag{10.93}$$

If matrix $[H]_k$ is ill-conditioned, it is possible to condense it using SVD. Assume a linear transformation exists:

$$[H']_k = [B][H]_k \tag{10.94}$$

where $[B]$ can be formed from the SVD of matrix $[H]_k$. When the auto-correlation matrix of the impulse response $[R_{HH}(0)]$ has a rank of n_e that is less than N, response matrix $[H]_k$ can be replaced by $[H']$ with a dimension of $n_e \times P$.

10.4.2 Estimation of coefficient matrix [A]

The second step of the time domain global modal analysis method is to estimate the coefficient matrix $[A]$ presented in the formulated mathematical model. Assume data samples are available when $k = 2p, 2p + 1, \ldots, 2p + 2p - 1$. Equation (10.87) will form a series of equations:

$$
\left\{
\begin{array}{l}
[H]_{2p} = [A_1][H]_{2p-1} + [A_2][H]_{2p-2} + \ldots + [A_{2p}][H]_0 \\[4pt]
[H]_{2p+1} = [A_1][H]_{2p} + [A_2][H]_{2p-1} + \ldots + [A_{2p}][H]_1 \\[4pt]
\qquad\qquad\qquad \ldots \\[4pt]
[H]_{2p+2p-1} = [A_1][H]_{2p+2p-2} + [A_2][H]_{2p+2p-3} + \ldots + [A_{2p}][H]_{2p-1}
\end{array}
\right.
\tag{10.95}
$$

Using the same approach that led to equation (10.92) from equation (10.87), a new augmented matrix equation can be derived:

$$
\begin{bmatrix}
[H]_{2p} & [H]_{2p+1} & [H]_{2p+2} & \cdots & [H]_{2p+2p-1} \\
[H]_{2p-1} & [H]_{2p} & [H]_{2p+1} & \cdots & [H]_{2p+2p-2} \\
[H]_{2p-2} & [H]_{2p-1} & [H]_{2p} & \cdots & [H]_{2p+2p-3} \\
\cdots & \cdots & \cdots & \cdots & \cdots \\
[H]_{2p-2p+1} & [H]_{2p-2p+2} & [H]_{2p-2p+3} & \cdots & [H]_{2p-2p+2p}
\end{bmatrix}
$$

$$
=
\begin{bmatrix}
[A]_1 & [A]_2 & [A]_3 & \cdots & [A]_{2p} \\
[I] & [0] & [0] & \cdots & [0] \\
[0] & [I] & [0] & \cdots & [0] \\
\cdots & \cdots & \cdots & \cdots & \cdots \\
[0] & [0] & \cdots & [I] & [0]
\end{bmatrix}
\tag{10.96}
$$

$$
\begin{bmatrix}
[H]_{2p-1} & [H]_{2p} & [H]_{2p+1} & \cdots & [H]_{2p+2p-2} \\
[H]_{2p-2} & [H]_{2p-1} & [H]_{2p} & \cdots & [H]_{2p+2p-3} \\
[H]_{2p-3} & [H]_{2p-2} & [H]_{2p-1} & \cdots & [H]_{2p+2p-4} \\
\cdots & \cdots & \cdots & \cdots & \cdots \\
[H]_{2p-2p} & [H]_{2p-2p+1} & [H]_{2p-2p+2} & \cdots & [H]_{2p-2p+2p-1}
\end{bmatrix}
$$

or simply:

$$
[\overline{H}]_{2p} = [\overline{A}][\overline{H}]_{2p-1}
\tag{10.97}
$$

The coefficient matrices $[A_r]$ can be found from the augmented matrix $[\overline{A}]$.

10.4.3 Identification of modal parameters

The modal parameters can be identified from the coefficient matrices $[A_r]$. The method to do so is to formulate an eigenvalue problem using the coefficient matrices $[A_r]$. The solution to this eigenvalue problem will produce natural frequencies, damping ratios and mode shapes.

Substituting equations (10.90) into (10.86) can lead to an eigenvalue problem. This substitution yields:

$$[\Psi][Z]^k[L] = [A_1][\Psi][Z]^{k-1}[L] + [A_2][\Psi][Z]^{k-2}[L] + \ldots + [A_{2p}][\Psi][Z]^{k-2p}[L] \quad (10.98)$$

Like before, this equation, when used several times for individual sampling points, can form an augmented matrix equation:

$$[Z]\begin{bmatrix} [\Psi][Z]^{2p-1} \\ [\Psi][Z]^{2p-2} \\ \ldots \\ [\Psi][Z]^{2p-2p} \end{bmatrix} = \begin{bmatrix} [A_1] & [A_2] & \ldots & [A_{2p}] \\ [I] & [0] & \ldots & [0] \\ \ldots & \ldots & \ldots & \ldots \\ [0] & \ldots & [I] & [0] \end{bmatrix}\begin{bmatrix} [\Psi][Z]^{2p-1} \\ [\Psi][Z]^{2p-2} \\ \ldots \\ [\Psi][Z]^{2p-2p} \end{bmatrix} \quad (10.99)$$

or
$$[Z][\Theta] = [\bar{A}][\Theta] \quad (10.100)$$

This is an eigenvalue problem. The solutions are:

$$[Z] = e^{[\Lambda]\Delta} \quad (10.101)$$

and
$$[\Theta] = \begin{bmatrix} [\Psi][Z]^{2p-1} \\ [\Psi][Z]^{2p-2} \\ \ldots \\ [\Psi][Z]^{2p-2p} \end{bmatrix} \quad (10.102)$$

Each eigenvalue corresponds to the modal parameters of a vibration mode such that the rth eigenvalue z_r can be expressed as:

$$z_r = e^{\lambda_r \Delta} \quad (10.103)$$

and the whole matrix is:

$$[Z] = \text{diag} [e^{\lambda_r \Delta}] \quad (10.104)$$

This leads to the estimation of the complex natural frequency:

$$\lambda_r = \frac{1}{\Delta} \ln z_r = -\xi_r \omega_r + j\omega_r \sqrt{1-\xi_r^2} \quad (10.105)$$

The diagonal natural frequency matrix can be constructed as:

$$[\Lambda] = \begin{bmatrix} \lambda_1 & & & \\ & \lambda_2 & & \\ & & \ldots & \\ & & & \lambda_{n_e} \end{bmatrix} \quad (10.106)$$

Once the eigenvalues are estimated, the mode shape matrix [Ψ] can be found from equation (10.102).

The last quantity to be determined is the mode participation factor matrix [L]. Equation (10.88) provides a means of estimating [L]. Once the eigenvalue matrix [Z] and the mode shape matrix [Ψ] are determined, the least-square method can be used to minimize the error function defined as:

$$[e]^T[e] = ([H_k] - [\Psi][Z^k][L])^T ([H_k] - [\Psi][Z^k][L]) \qquad (10.107)$$

This will lead to the least square estimate of the participation factor matrix:

$$[L] = (([\Psi][Z^k])([\Psi][Z^k])^T)^{-1} [H_k] \qquad (10.108)$$

To summarize, the time domain global modal analysis method involves the following steps in its implementation:

(1) From 'P' excitation signals and 'N' response signals of a test structure, construct the frequency response function matrix. Take the inverse Fourier transform from the matrix to form the impulse response matrix $[H(t)]$. This matrix, when sampled with an equal time interval, produces a time series $[H_k]$.

(2) Determine the auto-correlation matrix $[R_{HH}(0)]$ from the time series $[H_k]$ and use the SVD method to identify its rank. If the rank N_e is less than the number of response signals N, then use a linear transformation to reduce the dimension of the time series $[H_k]$ from $N \times P$ to $N_e \times P$.

(3) Find the initial estimate of the order of the system. The mode number can be determined from the response spectrum. The order of the system can be finalized after a few iteration processes.

(4) Determine the coefficient matrices $[A_r]$.

(5) Identify the natural frequencies, damping ratios and mode participation factors.

(6) Use the estimated modal parameters to regenerate the time series $[\tilde{H}_k]$ and compare it with the measured one $[H_k]$. If the two series agree with each other, then the identified modal parameters are accurate. Otherwise, increase the order of the model and repeat the process from steps 2 to 5.

The number of modes is a critical parameter for the method. It is not always obvious how many modes should be considered. Too few means missing out genuine modes and too many will result in computational modes. One simple criterion is to define a reasonable range of damping ratio so that if some 'modes' have greater damping ratio estimates, then they are deemed to be noise or computational modes. Mode participation factors of those modes should also testify their minute contribution to the response of the structure. In addition, noise or computational modes show poor reciprocity on the residue matrices. This information can be used to sieve analysed modes.

10.5 Eigensystem realization algorithm method

The eigensystem realization algorithm (ERA) is another global time domain method for modal parameter identification. The basic idea of ERA is to use the minimum realization concept from control theory to identify eigenproperties of a test structure from its impulse response data. The mathematical process is SVD. Once the eigenproperties are identified, the modal parameters can be derived accordingly. Since its inception, ERA has been used as an effective modal analysis method for real engineering structures, including the Galileo Space Vehicle.

The essence of ERA is to apply the SVD on the measured impulse response data in order to determine the number of prominent DoFs of a test structure and to determine the system matrix [A], input matrix [B] and output matrix [C] in the state space representation of the structure. The system matrix [A] can then be used for an eigenvalue solution to determine the modal parameters of the structure. When the dimensions of [A], [B] and [C] are minimum, the minimum realization has been achieved. This renders a controllable and observable system.

10.5.1 State–space equation of an NDoF system

An NDoF system with 'P' excitation forces is governed by the following differential equation:

$$[M]_{N\times N}\{\ddot{x}\}_{N\times1} + [D]_{N\times N}\{\dot{x}\}_{N\times1} + [K]_{N\times N}\{x\}_{N\times1} = [L]_{N\times P}\{F\}_{P\times1} \quad (10.109)$$

Combining the equation with an identity $\{\dot{x}\} = [I]\{\dot{x}\}$ and define the state–space variable $\{X\} = \begin{Bmatrix} \{x\} \\ \{\dot{x}\} \end{Bmatrix}$, we can form the following state–space equation:

$$\{\dot{X}\} = \begin{bmatrix} [0] & [I] \\ -[M]^{-1}[K] & -[M]^{-1}[D] \end{bmatrix}_{2N\times2N} \{X\}_{2N\times1} + \begin{bmatrix} [0] \\ -[M]^{-1}[L] \end{bmatrix}_{N\times2N} \{F\}_{2N\times1}$$

$$(10.110)$$

or
$$\{\dot{X}\} = [A]\{X\} + [B]\{F\} \quad (10.111)$$

If the system has 'L' output locations, the output vector $\{Y\}$ is related to the state variable through an output matrix [C] as:

$$\{Y\}_{L\times1} = [C]_{L\times2N}\{X\}_{2N\times1} \quad (10.112)$$

By sampling the response of the system with a time resolution Δ, we have the discretized version of equations (10.111) and (10.112) as:

$$\begin{cases} \{X(K+1)\} = [A]\{X(K)\} + [B]\{F(K)\} \\ \{Y(K)\} = [C]\{X(K)\} \end{cases} \quad (10.113)$$

Here, $\{X(K)\}$ is the state–space variable at time moment $K\Delta$ and $\{Y(K)\}$ the response output at the time. For a linear time-invariant system, the minimum realization for measured responses can be sought from the impulse responses.

10.5.2 Impulse response matrix

The impulse response of a system can be derived from the inverse Laplace transform of the transfer functions. By sampling the impulse response function $h(t)$, we have a time series for the function $h(K)$ ($K = 1, 2, \ldots$). At moment $K\Delta$, the impulse responses from all measured coordinates form the following impulse response matrix:

$$[\bar{H}(K)] = \begin{bmatrix} h_{11}(K) & h_{12}(K) & \ldots & h_{1P}(K) \\ h_{21}(K) & h_{22}(K) & \ldots & h_{2P}(K) \\ \vdots & \vdots & \vdots & \vdots \\ h_{L1}(K) & h_{L2}(K) & \ldots & h_{LP}(K) \end{bmatrix}_{L \times P} \tag{10.114}$$

The minimum realization becomes a mathematical task of determining matrices $[A]$, $[B]$ and $[C]$ from $[\bar{H}(K)]$ such that the order of these matrices is kept minimum. From the eigenvalue solutions of the determined matrix $[A]$ we can then derive the modal properties of the system.

10.5.3 Construction of a Hankel matrix

The minimum realization of impulse responses commences from the partitioned Hankel matrix. A Hankel matrix is formed in the following equation:

$$[\tilde{H}(K-1)] = \begin{bmatrix} \bar{H}(K) & \bar{H}(K+t_1) & \ldots & \bar{H}(K+t_{\beta-1}) \\ \bar{H}(j_1+K) & \bar{H}(j_1+K+t_1) & \ldots & \bar{H}(j_1+K+t_{\beta-1}) \\ \vdots & \vdots & \vdots & \vdots \\ \bar{H}(j_{\alpha-1}+K) & \bar{H}(j_{\alpha-1}+K+t_1) & \ldots & \bar{H}(j_{\alpha-1}+K+t_{\beta-1}) \end{bmatrix} \tag{10.115}$$

Here, j_r ($r = 1, 2, \ldots, \alpha-1$) and t_r ($r = 1, 2, \ldots, \beta-1$) are arbitrary positive integers. This matrix is constructed from the measured impulse response data.

Theoretically, the rank of matrix $[\tilde{H}(K-1)]$ is a constant and equal to the order of the system. Due to measurement noise, the rank varies until the α and β in equation (10.115) reach certain values. Therefore, the final settlement of j_r, t_r, α and β values occur when the rank of $[\tilde{H}(K-1)]$ approaches a constant and the dimension of the matrix is at a minimum. A common tactic is to let $j_r = t_r = r\Delta$ ($r = 1, 2, \ldots$). Matrix $[\tilde{H}(K-1)]$ will become:

$$[\tilde{H}(K-1)] = \begin{bmatrix} \bar{H}(K) & \bar{H}(K+1) & \ldots & \bar{H}(K+\beta-1) \\ \bar{H}(K+1) & \bar{H}(K+2) & \ldots & \bar{H}(K+\beta) \\ \vdots & \vdots & \vdots & \vdots \\ \bar{H}(K+\alpha-1) & \bar{H}(K+\alpha) & \ldots & \bar{H}(K+\alpha+\beta-2) \end{bmatrix}_{L\alpha \times P\beta} \tag{10.116}$$

$$\text{and} \quad [\tilde{H}(K)] = \begin{bmatrix} \overline{H}(K+1) & \overline{H}(K+2) & \cdots & \overline{H}(K+\beta) \\ \overline{H}(K+2) & \overline{H}(K+3) & \cdots & \overline{H}(K+\beta+1) \\ \vdots & \vdots & \vdots & \vdots \\ \overline{H}(K+\alpha) & \overline{H}(K+\alpha+1) & \cdots & \overline{H}(K+\alpha+\beta-1) \end{bmatrix}_{L\alpha \times P\beta}$$

(10.117)

10.5.4 Impulse response matrix and system matrices

Establishment of the link between the system matrices $[A]$, $[B]$ and $[C]$ and the impulse response matrix $[\overline{H}(K)]$ is the basis of the eigensystem realization algorithm. Using the z-transform, we find that the transfer function matrix of a system is related to its impulse response matrix:

$$[H(Z)] = [\overline{H}(0)] + [\overline{H}(1)]Z^{-1} + [\overline{H}(2)]Z^{-2} + \ldots \qquad (10.118)$$

We also know from Chapter 2 that the transfer function can be written as:

$$[H(Z)] = [C](Z[I] - [A])^{-1}[B] = Z^{-1}[C]([I] - Z^{-1}[A] - (Z^{-1}[A])^2 - \ldots)[B] \quad (10.119)$$

Comparing these equations we establish:

$$[\overline{H}(K)] = [C][A]^{k-1}[B] \qquad (10.120)$$

from here we deduct that:

$$[\overline{H}(K+1)] = [C][A]^K [B] = [C][A][A]^{K-1}[B]$$
$$[\overline{H}(K+2)] = [C][A]^K [B] = [C][A]^2 [A]^{K-1}[B] \qquad (10.121)$$
$$\ldots$$

Therefore, we can decompose the Hankel matrix $[\tilde{H}(K-1)]$ into:

$$[\tilde{H}(K-1)] = \begin{bmatrix} [C] \\ [C][A] \\ [C][A]^2 \\ \vdots \\ [C][A]^{\alpha-1} \end{bmatrix}_{L\alpha \times 2N} [A]^{K-1}[[B]\ [A][B]\ [A]^2[B] \ldots [A]^{\beta-1}[B]]_{2N \times P\beta}$$

or

$$[\tilde{H}(K-1)] = [P]_{L\alpha \times 2N} [A]^{K-1} [Q]_{2N \times P\beta} \qquad (10.122)$$

Matrix $[P]$ is called the observability matrix and $[Q]$ the controllability matrix.

10.5.5 Eigensystem realization algorithm

The derivation of the algorithm can begin from equation (10.122) with the initial impulse response matrix:

$$[\tilde{H}(0)] = [P][Q] \qquad (10.123)$$

Using SVD, we have:

$$[\tilde{H}(0)] = [U]^T[\Sigma][V] \qquad (10.124)$$

This decomposition can be seen physically as passing measured impulse response data through a Wiener filter. The signal filtered out, which is responsible for the zero singular values, corresponds to the random noise which does not correlate to either system input or its output.

The descending singular values are real and form a diagonal matrix:

$$[\Sigma] = \text{diag} [\sigma_1 \quad \sigma_2 \ldots \sigma_r \ldots 0 \ldots] \qquad (10.125)$$

and the vectors are normalized such that:

$$[U]^T[U] = [V]^T[V] = [I] \qquad (10.126)$$

If there exist a matrix $[H^\#]$ such that:

$$[Q][H^\#][P] = [I], \qquad (10.127)$$

then from

$$[\tilde{H}(0)][H^\#][\tilde{H}(0)] = [\tilde{H}(0)] \qquad (10.128)$$

we know that $[H^\#]$ is a general inverse of matrix $[\tilde{H}(0)]$ and, from the SVD theory, it is estimated as:

$$[H^\#] = [V]^T [\Sigma][U] = [V]^T \left[\frac{1}{\sigma_1} \; \frac{1}{\sigma_2} \; \ldots \; \frac{1}{\sigma_r} \ldots 0 \ldots \right][U] \qquad (10.129)$$

Now let us determine the relationship between the system matrices $[A]$, $[B]$ and $[C]$ and the matrices derived from the SVD $[U]$, $[\Sigma]$ and $[V]$. To do so, we define two matrices below which will be used in the analysis:

$$[E_L]^T = [[I_L]_{L \times L} \; [0]_{L \times L} \; \ldots \; [0]_{L \times L}]_{L \times L\alpha} \qquad (10.130)$$

$$[E_P]^T = [[I_P]_{P \times P} \; [0]_{P \times P} \; \ldots \; [0]_{P \times P}]_{P \times P\beta} \qquad (10.131)$$

From equation (10.117) we have:

$$[E_L]^T [\tilde{H}(K)][E_P] = [E_L]^T [P][A]^K [Q][E_P] = [\overline{H}(K+1)] \qquad (10.132)$$

On both sides of matrix $[A]$ we multiply a unity matrix $[Q]\lfloor H^\# \rfloor[P]$, equation (10.132) becomes:

$$[\overline{H}(K+1)] = [E_L]^T ([P][Q])[H^\#]([P][A]^K [Q])[H^\#]([P][Q])[E_P] \qquad (10.133)$$

or $[\overline{H}(K+1)] = [E_L]^T [\tilde{H}(0)][H^\#][\tilde{H}(1)]^K [H^\#][\tilde{H}(0)][E_P] \qquad (10.134)$

Using the SVDs of matrices $[\tilde{H}(0)]$ and $\lfloor H^\# \rfloor$ in equations (10.124) and (10.129), we can rewrite equation (10.134) as:

$$[\overline{H}(K+1)] = ([E_L]^T[U]^T[\Sigma]^{\frac{1}{2}})\,([\Sigma]^{-\frac{1}{2}}[U][\tilde{H}(1)][V]^T[\Sigma]^{-\frac{1}{2}})^K\,([\Sigma]^{\frac{1}{2}}[V][E_P])$$

(10.135)

Comparing it with equation (10.120), we have the derivation for the system matrices as:

$$\begin{cases} [A] = [\Sigma]^{-\frac{1}{2}}[U][\tilde{H}(1)][V]^T[\Sigma]^{-\frac{1}{2}} \\[2mm] [B] = [\Sigma]^{\frac{1}{2}}[V][E_P] \\[2mm] [C] = [E_L]^T[U]^T[\Sigma]^{\frac{1}{2}} \end{cases}$$

(10.136)

Taking these matrices back to equation (10.113), we have the realization. Equation (10.136) shows that the dimension of $[A]$ is dictated by that of $[\Sigma]$ whose dimension is $2N \times 2N$, even $[\tilde{H}(1)]$ has a much bigger dimension of $L\alpha \times P\beta$. For a system with N DoFs, a $2N \times 2N$ $[A]$ is indeed a minimum realization.

The eigenvalue solution of matrix $[A]$ provides the system modal properties. Its eigenvalue decomposition is:

$$[\Psi]^{-1}[A][\Psi] = [Z]$$

(10.137)

Here, $[\Psi]$ contains $2N$ complex mode shapes. The eigenvalues in the diagonal matrix $[Z]$ are related to the system natural frequencies and damping ratios:

$$Z_r = e^{\lambda_r \Delta} = e^{(\lambda_r^R + i\lambda_r^I)\Delta}$$

(10.138)

$$\omega_r = \sqrt{(\lambda_r^R)^2 + (\lambda_r^I)^2}$$

(10.139)

$$\zeta_r = -\frac{\lambda_r^R}{\sqrt{(\lambda_r^R)^2 + (\lambda_r^I)^2}}$$

(10.140)

To summarize, the eigensystem realization algorithm involves the following steps:

(1) Select measured data to construct matrices $[\tilde{H}(0)]$ and $[\tilde{H}(1)]$.
(2) Decompose matrix $[\tilde{H}(0)]$ using SVD and reconstruct it using the non-zero singular values and corresponding vectors.
(3) Derive the system matrices $[A]$, $[B]$ and $[C]$.
(4) Derive modal data from the realized matrix $[A]$.

Literature

1. Allemang, R.J. 1984: Multiple input estimation of frequency response functions. *Proceedings of the 3rd International Seminar on Modal Analysis*, Leuven, Belgium, 710–719.
2. Bendat, J.S. 1980: Modern Analysis Procedures for Multiple Input/Output Problems. *Journal of the Acoustical Society of America*, **68**(2), 498–503.
3. Bendat, J.S. and Piersol, A.G. 1984: *Random Data: Analysis and Measurement Procedures*. New York, John Wiley.
4. Craig Jr., R.R. and Blair, M.A. 1985: A Generalized Multiple-Input Multiple-Output Modal Parameter Estimation Algorithm. *AIAA Journal*, **23**(6), 931–937.

5. Deblauwe, F. and Allemang, R.J. 1985: The Polyreference Time-Domain Technique. *Proceedings of the 10th International Seminar on Modal Analysis*, Part IV, K.U. Leuven, Belgium.
6. Dobbs, C.J. and Robson, J.F. 1975: Partial Coherence in Multivariate Random Processes. *Journal of Sound and Vibration*, **42**, 243–249.
7. Gersch, W. 1974: On the Achievable Accuracy of Structural System Parameter Estimates. *Journal of Sound and Vibration*, **34**(1), 63–79.
8. Gersch, W. and Loo, S. 1972: Discrete Time Series Synthesis of Randomly Excited Structural System Response. *Journal of the Acoustical Society of America*, **51**(1), 402–408.
9. Gersch, W., Nielsen, N.N. and Akaike, H. 1973: Maximum Likelihood Estimation of Structural Parameters from Random Vibration Data. *Journal of Sound and Vibration*, **31**(3), 295–308.
10. Juang, J.N. 1987: Mathematical correlation of modal-parameter-identification methods via system realisation theory. *The International Journal of Theoretical and Experimental Modal Analysis*, **2**(1), 1–18.
11. Juang, J.-N. and Pappa, R.S. 1985: An eigensystem realisation algorithm (ERA) for modal parameter identification and modal reduction. *Journal of Guidance, Control and Dynamics*, **8**(5), 620–627.
12. Juang, J.-N. and Suzuki, H. 1988: An Eigensystem Realization Algorithm in Frequency Domain for Modal Parameter Identification. *Transactions of the ASME (Journal of Vibration, Acoustics, Stress and Reliability in Design)*, **110**, 24–29.
13. Lembregts, F., Snoeys, R. and Leuridan, J. 1987: Application and Evaluation of Multiple-Input Modal Parameter Estimation. *Journal of Modal Analysis*, 19–31.
14. Leuridan, J. *et al.* 1986: Global modal parameter estimation methods: an assessment of time versus frequency domain implementation. In: *Proceedings of the 4th International Modal Analysis Conference*, Los Angeles, CA, 589–598.
15. Leuridan, J., Brown, D.L. and Allemang, R.J. 1986: Time domain parameter identification methods for linear modal analysis: a unifying approach. *Journal of Vibration, Acoustics, Stress and Reliability in Design*, **108**(1).
16. Leuridan, J. and Vold, H. 1983: A Time Domain Linear Model Estimation Technique for Multiple Input Modal Analysis. *The Winter Annual Meeting of the American Society of Mechanical Engineers*, Boston, Massachusetts, 51–62.
17. Maia, N.M.M. *et al.* 1997: *Theoretical and Experimental Modal Analysis*. Research Studies Press, UK and John Wiley & Sons, USA.
18. Mergeay, M. 1983: Least squares complex exponential method and global system parameter estimation used by modal analysis. In: *Proceedings of the 8th International Modal Analysis Seminar*, Leuven, Belgium, **3**.
19. Mitchell, L.D., Cobb, J.C. and Luk, Y.W. 1988: An unbiased frequency response function estimator. *The International Journal of Theoretical and Experimental Modal Analysis*, **3**(1).
20. Mitchell, L.D. 1982: Improved method for FFT calculation of the frequency response function. *Journal of Mechanical Design*, **104**.
21. Schmerr, L.W. 1982: A new complex exponential frequency domain technique for analysing dynamic response data. In: *Proceedings of the 1st International Modal Analysis Conference*, Orlando, FL, 183–186.
22. Shih, C.Y. *et al.* 1988: A frequency domain global parameter estimation method for multiple reference frequency response measurements. *Mechanical Systems and Signal Processing*, Vol. 4, No. 2, 349–365.
23. Van Der Auweraer, H. and Leuridan, J. 1987: Multiple Input Orthogonal Polynomial Parameter Estimation. *Mechanical Systems and Signal Processing*, **1**(3), 259–272.
24. Vold, H. *et al.* 1982: A Multi-Input Modal Estimation Algorithm for Mini-Computers. *SAE Technical Paper Series*, No. 820–194.
25. Vold, H. and Rocklin, G.T. 1986: The numerical implementation of a multi-input modal estimation method for mini computers. In: *Proceedings of the 4th International Modal Analysis Conference*, Orlando, FL.
26. Wicks, A.L. and Vold, H. 1986: The Hs frequency response function estimation. In: *Proceedings of the 4th International Modal Analysis Conference*, Los Angeles, CA.

11

Local structural modification

11.1 The objectives of structural modification

Structural modification as an application of modal analysis is a technique to study the effect of physical parameter changes of a structural system on its dynamic properties which are in the forms of natural frequencies and mode shapes. Because the theory of structural modification is developed using system mass and stiffness matrices, the physical parameter changes are projected into mass and stiffness property changes of the system. For a simple mass–spring system, the physical parameters are actually mass and stiffness values of physical elements. For a continuous system such as a cantilever, these parameters can be the thickness or length of a section of the beam. They are not strictly mass or stiffness quantities but their changes affect both mass and stiffness properties. For a real structure such as a car chassis, physical parameters can be local mass and stiffness changes caused by cross-sectional area changes or by an additional stiffener welded between two locations. Nevertheless, we can find later that the final derivation of structural modification may not have to rely on knowing the full mass and stiffness matrices.

The objective of structural modification is to improve the dynamic properties of a structure. Changing a natural frequency is perhaps the most common objective of structural modification. A natural frequency coinciding with, or at the vicinity of, the ambient vibration frequency is a common source of excessive vibration that may lead to structural failure. Assuming no change can be made to the ambient vibration, an obvious solution for reducing vibration would be to relocate the natural frequency away from the excitation frequency using structural modification. Alternatively, structural modification can help to move an anti-resonanace to match the excitation frequency so as to create a nodal point at a selected location on the system. If the excitation frequency is a narrow-banded range rather than an individual frequency, structural modification can be used to rearrange the natural frequencies so that no natural frequency falls within the band.

11.2 Direct and inverse problems

Structural modification can be a direct problem or an inverse problem. The direct problem is the one that predicts dynamic property changes once mass, stiffness or a physical parameter changes are given. Physical parameter changes usually cause both mass and stiffness changes. This problem requires a straightforward analytical solution. The difficulty of solving a direct problem lies on getting system matrices or sufficient and accurate data from modal testing. In the calculation, we may require either measured FRF data or modal data from selected DoFs. Not all of these data can be provided easily from measurement. However, if structural modification is used as a design tool, then a finite element model of the designed structure of component is normally available. In this case, FRF or modal data can be determined numerically.

The inverse problem takes the opposite path to the direct problem. It seeks to determine necessary structural changes needed to create a targeted dynamic property change such as the change of a natural frequency. By nature, the solution to this problem can be non-unique or non-existent. This is the main difficulty of the inverse problem. Specifically, the difficulty manifests itself in two ways. The first is having to determine where the necessary structural changes should occur so that the solution, if it exists, can be realistic. The second is the need to formulate the structural modification problem analytically once the locations for modification have been identified, so that the solution, if it exists, can be found.

11.3 Modal properties after structural modification – reanalysis

The modal and FRF properties of an n DoF structural system have been given in Chapter 5. The natural frequencies and mode shapes of this system are the solutions of the following eigenvalue problem:

$$([K] - \omega^2[M])\{\Psi\} = \{0\} \tag{11.1}$$

The FRF matrix of the system $[\alpha(\omega)]$ can be determined from the natural frequency matrix $\lceil\cdot\omega_r^2\cdot\rfloor$ and mass-normalized mode shapes matrix $[\phi]$ as:

$$[\alpha(\omega)] = [\phi][\cdot\omega_r^2 - \omega^2\cdot]^{-1}[\phi]^T \tag{11.2}$$

For an individual load path on the structure between two DoFs, there is a unique set of anti-resonance frequencies. The anti-resonances of a given receptance FRF $\alpha_{ij}(\omega)$, denoted as Ω_r, are the real positive roots of the following eigenvalue equation:

$$([K]_{ij} - \Omega^2[M]_{ij})\{X\} = \{0\} \tag{11.3}$$

where matrix $[\ ]_{ij}$ is obtained by deleting the ith row and the jth column of matrix $[\ ]$. The number of the real positive roots of this eigenvalue equation ranges from zero to n.

Structural modification to a dynamic system can be the results of physical parameter changes as well as mass and stiffness changes. Regardless of the causes, we can denote the modification of the system characterized by (11.1) by a mass modification matrix $[\Delta M]$ and a stiffness modification matrix $[\Delta K]$. Either could be a null matrix if no modification occurs. For the change of a physical parameter of a structure such as a thickness, it is most likely that both mass and stiffness change so that neither $[\Delta M]$ nor $[\Delta K]$ is null. Consider that realistically modification on a structure cannot happen at many locations due to various constraints, $[\Delta M]$ and $[\Delta K]$ are sparse matrices.

The main objective of the direct problem of structural modification, also known as reanalysis, is to predict the FRF or the modal properties of a structure after specified modifications. For a structure with an analytical model, this is done without having to repeat the whole eigenvalue solution. For a structure without one, reanalysis avoids making costly modification prototypes and conducting modal testing on the structure. Reanalysis assumes that modification in the form of $[\Delta M]$ and $[\Delta K]$ has been specified. Thus, the dynamics of the modified structure is governed by the following eigenvalue problem:

$$[([K] + \Delta[K]) - \omega^{*2}([M] + \Delta(M)][\{Y\} = \{0\}] \tag{11.4}$$

Here, ω^* is the natural frequency and $\{Y\}$ the mode shape of the modified structure. For a structure without repeated natural frequencies, we know that the mode shape of the modified structure is a linear combination of those from the unmodified one:

$$\{Y\} = [\phi]\{\eta\} \tag{11.5}$$

With this relationship for mode shapes and the orthogonality properties of an MDoF system, we can recast equation (11.4) into:

$$(([I] + [\phi]^T[\Delta M][\phi]) - \omega^{*2}([\cdot\omega_r^2\cdot] + [\phi]^T[\Delta K][\phi]))\{\eta\} = \{0\} \tag{11.6}$$

or

$$([K^*] - \omega^{*2}[M^*])\{\eta\} = \{0\} \tag{11.7}$$

This equation forms a relationship between the modifications ($[\Delta M]$ and $[\Delta K]$) and the modal properties (ω^* and $\{Y\}$) of the modified structure. The modal data of the original structure (ω_r and $\{\phi\}$) can come from its modal testing and analysis. The significance is that the equation does not rely on the mass and stiffness matrices of the original structure. This is particularly useful for problem solving of a real structure where an analytical model is not available. The significance also lies in the fact that the equation is used to predict the modal properties of the modification. As stated earlier, this can avoid costly prototype making for trial-and-error modifications.

However, closer examination of equation (11.7) suggests that its usefulness may be discounted. Firstly, the number of modes identified for the original structure after modal testing is usually limited. This means the eigenvalue problem presented in equation (11.7) becomes a reduced one. For example, for a structural system with 200×200 matrices, a total number of 12 modes from modal testing, assuming all DoFs are at present in mode shapes, means that matrices $[K^*]$ and $[M^*]$ are 12×12 in dimension. They are excessively condensed to yield inaccurate modal data of the modified structure. Secondly because of test limitations, the modes identified for modal testing of the original structure do not include all DoFs the system matrices have. If the DoFs used in the identified modes cover those used in structural modifications $[\Delta M]$ and $[\Delta K]$, then lack of all other DoFs means numerical inaccuracy for the predicted modal properties of the modified structure. If, however, the DoFs used in the identified modes do not cover all DoFs used in modifications, we would face a difficult stalemate where the solution offered by equation (11.7) cannot proceed. For example, if the identified mode shapes do not include rotational DoFs while a structural modification does by welding a beam between two locations as a stiffner, then reanalysis in equation (11.7) cannot be carried out because of DoF incompatibility.

11.4 Local structural modification by mass and stiffness changes

According to its complexity, we can classify structural modification as a unit rank modification, a mass or stiffness modification, and a physical parameter modification. The complexity is reflected in the analytical formulation of the modification problem. Unit rank modification allows an explicit algebraic solution for a structural modification problem. Mass or stiffness modification can be formulated either as an eigenvalue problem or as a linear simultaneous equation problem. In both cases, mathematical solutions are readily available to solve the structural modification problem. Physical parameter modification is often more realistic for structures. For example, the parameter to be modified can be a design parameter for a structure. Such a modification usually leads to both mass and stiffness changes. The mathematical formulation of the modification needs more rigorous analytical treatment and an accurate solution is not guaranteed.

In the following, we will begin with unit rank modification and expand into mass or stiffness modification. Finally, we will study structural modification caused by physical parameter changes.

11.4.1 Unit rank modification

Unit rank modification involves changing a single mass or stiffness element of a structure only. This is usually realized by adding a concentrated mass to a DoF or linking a spring between two DoFs. Theoretically, a mass or stiffness modification matrix of these changes is a matrix with a unity rank, as shown in equation (11.8).

$$[\Delta M] = \begin{bmatrix} 0 & & & 0 \\ & \ddots & & \\ & & \Delta m & \\ & & & \ddots \\ 0 & & & 0 \end{bmatrix} \quad [\Delta K] = \begin{bmatrix} \cdots & & & \\ & \Delta k & \cdots & -\Delta k \\ & \cdots & \cdots & \cdots \\ & -\Delta k & \cdots & \Delta k \\ & & & \cdots \end{bmatrix} \quad (11.8)$$

Figure 11.1 shows a structure before and after a modification is made. Since the modification involves an elastic spring linking two positions of the structure, we can regard this as a unit rank stiffness modification.

Contrary to its appearance, a unit rank modification is actually quite difficult to achieve. For example, it is impossible in reality to add a concentrated mass which only affects vibration in one direction. It is only when vibrations in other directions are insignificant can we formulate the problem as a unit rank modification. The understanding of unit rank modification helps to provide physical insights into structural modification.

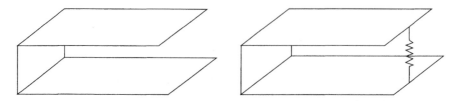

Figure 11.1 A structure with a single spring modification

11.4.2 Mass modification of a lumped system

Structural modification via mass changes is convenient if lumped mass can be applied to a structure which is modelled as a lumped system. Modification by removing distributed mass from an existing structure does not fall into this category since it will usually be accompanied by stiffness reduction. As before, we denote the mass modification matrix as $[\Delta M]$. Since stiffness does not change, we can simplify equation (11.4) to formulate the mass modification case as:

$$([K] - \omega^{*2}[M])\{Y\} - \omega^{*2}[\Delta M]\{Y\} = \{0\} \tag{11.9}$$

or
$$\{Y\} = \omega^{*2}[\alpha(\omega^*)][\Delta M]\{Y\} \tag{11.10}$$

This equation, although it appears to be of full dimension as matrices $[M]$ and $[K]$, is actually in a much reduced size, owing to the fact that matrix $[\Delta M]$ has only a few non-zero elements for modified DoFs. If there are three modified DoFs, $[\Delta M]$ can be reduced to a 3×3 non-zero submatrix and only the relevant portion of other quantities in equation (11.10) will be useful in the analysis. From now on, when we refer to equation (11.10), we actually mean the reduced version of the equation.

There is no obvious mathematical solution to this equation for a prescribed natural frequency ω^*. However, we can find mathematical transformation that will lead us to the solution. First, let us see how the mass modification problem described by equation (11.10) can be turned into a normal eigenvalue problem. Assume the mass modification matrix can be expressed as:

$$[\Delta M] = \zeta_m[\varepsilon] \tag{11.11}$$

where ζ_m is an unknown factor and $[\varepsilon]$ is a given matrix describing the relative mass modifications among selected coordinates. Matrix $[\varepsilon]$ needs to be predetermined to set the relative mass modifications among the selected DoFs. If three masses from the system will be modified, matrix $[\varepsilon]$ prohibits the three mass changes to be independent of each other. Only one change is determined from the solution. The other two are proportionally related to it.

With the mass modification matrix defined, equation (11.10) can now be translated into an eigenvalue problem:

$$\left(\omega^{*2}[\alpha(\omega^*)][\varepsilon] - \frac{1}{\zeta_m}[I]\right)[Y] = \{0\} \tag{11.12}$$

where $1/\zeta_m$ is the eigenvalue. Since the matrix product $[\alpha(\omega^*)][\varepsilon]$ is not a real symmetric matrix, not all the eigenvalues derived from equation (11.12) are real. A real eigenvalue solution can be used to determine the mass modification matrix from equation (11.11) which will lead to the prescribed natural frequency ω^*. The final conclusion of the modification depends on the feasibility of the modified mass matrix $[M] + [\Delta M]$. Obviously any suggestion of negative mass on a structure would be unfeasible. This analysis does not make use of mode shape information before and after modification.

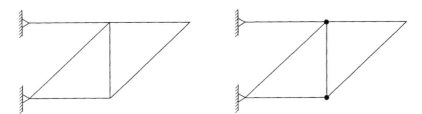

Figure 11.2 A 2-D truss structure without and with mass modification

To fully appreciate this eigenvalue-based structural modification approach, let us study its advantages and limitations. The dynamic data needed in the eigenvalue approach are the FRF data from modified DoFs only. This means that the size of the eigenvalue problem formulated is a much-reduced one in dimension, as explained after equation (11.10). It also means that, without having to rely on the mass and stiffness matrices of the original structure, this approach can be used as a problem solving tool for real structures where analytical models are not available. The approach is, however, limited by the oversimplicity of pure mass modification which may not be realizable on real structures. Inaccuracy of the FRF data may hinder the precision of the resultant natural frequency ω^*.

The eigenvalue formulation of structural modification for a prescribed natural frequency ω^* can be conveniently extended to that for a new anti-resonance Ω^* for the ijth FRF. By replacing matrices $[M]$ and $[K]$ with $[M]_{ij}$ and $[K]_{ij}$ (defined in equation (11.3)), respectively, we can rewrite equation (11.9) as:

$$([K]_{ij} - \Omega^{*2}[M]_{ij})\{Y\} - \Omega^{*2}[\Delta M]_{ij}\{Y\} = \{0\} \qquad (11.13)$$

Assume there is an imaginary dynamic system that consists of a mass matrix $[M]_{ij}$ and stiffness matrix $[K]_{ij}$ (neither one has to be symmetrical). The formulation from equations (11.10) to (11.12) can be applied. The square root of the positive and real eigenvalues are now the 'natural frequencies' of this imaginary system which are actually the same as the anti-resonances of the selected FRF of the original system. The eigenvector of this imaginary system does not possess apparent physical meaning.

The mass modification problem can also be formulated using an approach based on linear simultaneous equations. To derive the solution, we assume that matrix $[\Delta M]$ is diagonal so that all mass modifications are concentrated masses. There is no need to require the mass matrix of the original structure to be diagonal. We will see later that this restriction on [ΔM] is actually not necessary so long as the elemental mass matrix of the modified component can be specified. Back to equation (11.10). For the easiness of the formulation, assume only three DoFs (i, j, k) are going to have mass changes. More mass changes can be treated likewise. If matrix [ΔM] is diagonal, then the equation becomes:

$$
\begin{Bmatrix} Y_i \\ Y_j \\ Y_k \end{Bmatrix} = \omega^{*2} \begin{bmatrix} \alpha_{ii}(\omega^*) & \alpha_{ij}(\omega^*) & \alpha_{ik}(\omega^*) \\ \alpha_{ji}(\omega^*) & \alpha_{jj}(\omega^*) & \alpha_{jk}(\omega^*) \\ \alpha_{ki}(\omega^*) & \alpha_{kj}(\omega^*) & \alpha_{kk}(\omega^*) \end{bmatrix} \begin{bmatrix} \Delta m_i & 0 & 0 \\ 0 & \Delta m_j & 0 \\ 0 & 0 & \Delta m_k \end{bmatrix} \begin{Bmatrix} Y_i \\ Y_j \\ Y_k \end{Bmatrix} \quad (11.14)
$$

With a linear transformation, this equation becomes:

$$
\omega^{*2} \begin{bmatrix} \alpha_{ii}(\omega^*) & \alpha_{ij}(\omega^*) & \alpha_{ik}(\omega^*) \\ \alpha_{ji}(\omega^*) & \alpha_{jj}(\omega^*) & \alpha_{jk}(\omega^*) \\ \alpha_{ki}(\omega^*) & \alpha_{kj}(\omega^*) & \alpha_{kk}(\omega^*) \end{bmatrix} \begin{bmatrix} Y_i & 0 & 0 \\ 0 & Y_j & 0 \\ 0 & 0 & Y_k \end{bmatrix} \begin{Bmatrix} \Delta m_i \\ \Delta m_j \\ \Delta m_k \end{Bmatrix} = \begin{Bmatrix} Y_i \\ Y_j \\ Y_k \end{Bmatrix} \quad (11.15)
$$

This is a set of simultaneous equations for mass modifications. The first matrix in the equation contains FRF data from the modified coordinates. These data can be available from experimental modal testing. The Y vector is for the subset of the mode shape of the modified structure. It needs to be preset for the solution. We can use the mode shape subset of the original structure. Alternatively, equation (11.15) allows us to define the subset of the mode shape for the modified structure. This is an option the eigenvalue approach does not offer.

11.4.3 Stiffness modification of a lumped system

Structural modification via stiffness changes can also be formulated and solved conveniently for a structure modelled as a lumped system. Stiffness changes can be introduced by elastic elements such as springs. Modification by reducing stiffness from an existing part of a structure usually causes simultaneous mass reduction. This is a scenario to be considered later in this chapter. Since mass does not change, we can simplify equation (11.4) to formulate the stiffness modification case as:

$$
([K] - \omega^{*2}[M])\{Y\} + [\Delta K]\{Y\} = \{0\} \quad (11.16)
$$

or

$$
\{Y\} = -[\alpha(\omega^*)][\Delta K]\{Y\} \quad (11.17)
$$

Like the mass modification case, this equation is in a much-reduced size than matrices [M] and [K], owing to the fact that matrix [ΔK] has only a few non-zero elements for modified DoFs. If there are three modified DoFs, [ΔK] can be reduced to a 3×3 non-zero submatrix and other matrix and vectors in equation (11.17) will reduce accordingly. Unlike mass modification, matrix [ΔK] is not diagonal since stiffness changes mainly between DoFs. Typically, for three modified DoFs i, j and k, [ΔK] can be written as:

$$
[\Delta K] = \begin{bmatrix} \Delta k_{ij} + \Delta k_{ik} & -\Delta k_{ij} & -\Delta k_{ik} \\ -\Delta k_{ij} & \Delta k_{ij} + \Delta k_{jk} & -\Delta k_{jk} \\ -\Delta k_{ik} & -\Delta k_{jk} & \Delta k_{jk} + \Delta k_{ik} \end{bmatrix} \quad (11.18)
$$

Assume this modification matrix can be expressed as:

$$[\Delta K] = \gamma_k [\kappa] \tag{11.19}$$

where γ_k is an unknown factor and $[\kappa]$ is a given matrix describing the relative stiffness modifications among selected coordinates. This means that the three stiffness changes are not independent of each other. With the stiffness modification matrix defined, equation (11.17) can now be translated into an eigenvalue problem:

$$\left([\alpha(\omega^*)][\kappa] + \frac{1}{\gamma_k}[I] \right)\{Y\} = \{0\} \tag{11.20}$$

where $-1/\gamma_k$ is the eigenvalue. Since matrix product $[\alpha(\omega^*)][\kappa]$ is not a real symmetric matrix, not all the eigenvalues derived from equation (11.20) are real. A real eigenvalue solution can be used to determine the stiffness modification matrix from equation (11.19) which will lead to the prescribed natural frequency ω^*. The final conclusion of the stiffness modification depends on the feasibility of the modified stiffness matrix $[K] + [\Delta K]$. Obviously any suggestion of negative stiffness elements on a structure would be unfeasible. This analysis does not make use of mode shape information before and after modification. The advantages and limitations outlined for mass modification case also apply here.

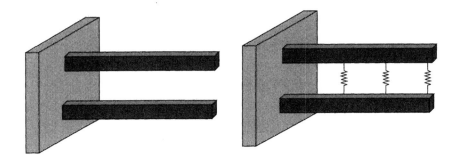

Figure 11.3 Two cantilevers connected by stiffness modifications

The stiffness modification problem can also be formulated using a linear simultaneous equation-based approach. We begin from equation (11.17):

$$\begin{Bmatrix} Y_i \\ Y_j \\ Y_k \end{Bmatrix} = - \begin{bmatrix} \alpha_{ii}(\omega^*) & \alpha_{ij}(\omega^*) & \alpha_{ik}(\omega^*) \\ \alpha_{ji}(\omega^*) & \alpha_{jj}(\omega^*) & \alpha_{jk}(\omega^*) \\ \alpha_{ki}(\omega^*) & \alpha_{kj}(\omega^*) & \alpha_{kk}(\omega^*) \end{bmatrix}$$

$$\begin{bmatrix} \Delta k_{ij} + \Delta k_{ik} & -\Delta k_{ij} & -\Delta k_{ik} \\ -\Delta k_{ij} & \Delta k_{ij} + \Delta k_{jk} & -\Delta k_{jk} \\ -\Delta k_{ik} & -\Delta k_{jk} & \Delta k_{jk} + \Delta k_{ik} \end{bmatrix} \begin{Bmatrix} Y_i \\ Y_j \\ Y_k \end{Bmatrix} \tag{11.21}$$

Through the following transformations, we can convert this equation into a set of linear simultaneous equations for stiffness modifications Δk_{ij}, Δk_{jk} and Δk_{ik}. Equation (11.21) is rewritten as:

$$
\begin{Bmatrix} Y_i \\ Y_j \\ Y_k \end{Bmatrix} = -\begin{bmatrix} \alpha_{ii}(\omega^*) & \alpha_{ij}(\omega^*) & \alpha_{ik}(\omega^*) \\ \alpha_{ji}(\omega^*) & \alpha_{jj}(\omega^*) & \alpha_{jk}(\omega^*) \\ \alpha_{ki}(\omega^*) & \alpha_{kj}(\omega^*) & \alpha_{kk}(\omega^*) \end{bmatrix}
$$

$$
\left\{ \Delta k_{ij}\begin{bmatrix} 1 & -1 & 0 \\ -1 & 1 & 0 \\ 0 & 0 & 0 \end{bmatrix} + \Delta k_{jk}\begin{bmatrix} 0 & 0 & 0 \\ 0 & 1 & -1 \\ 0 & -1 & 1 \end{bmatrix} + \Delta k_{ik}\begin{bmatrix} 1 & 0 & -1 \\ 0 & 0 & 0 \\ -1 & 0 & 1 \end{bmatrix} \right\} \begin{Bmatrix} Y_i \\ Y_j \\ Y_k \end{Bmatrix} \quad (11.22)
$$

$$
\begin{Bmatrix} Y_i \\ Y_j \\ Y_k \end{Bmatrix} = -\begin{bmatrix} \alpha_{ii}(\omega^*) & \alpha_{ij}(\omega^*) & \alpha_{ik}(\omega^*) \\ \alpha_{ji}(\omega^*) & \alpha_{jj}(\omega^*) & \alpha_{jk}(\omega^*) \\ \alpha_{ki}(\omega^*) & \alpha_{kj}(\omega^*) & \alpha_{kk}(\omega^*) \end{bmatrix}
$$

$$
\left\{ \Delta k_{ij}\begin{bmatrix} Y_i - Y_j \\ -(Y_i - Y_j) \\ 0 \end{bmatrix} + \Delta k_{jk}\begin{bmatrix} 0 \\ Y_j - Y_k \\ -(Y_j - Y_k) \end{bmatrix} + \Delta k_{ik}\begin{bmatrix} Y_i - Y_k \\ 0 \\ -(Y_i - Y_k) \end{bmatrix} \right\} \quad (11.23)
$$

Define:

$$
Y_{ij} = Y_i - Y_j, \quad Y_{jk} = Y_j - Y_k, \quad Y_{ik} = Y_i - Y_k,
$$

we have:

$$
-\begin{bmatrix} \alpha_{ii}(\omega^*) & \alpha_{ij}(\omega^*) & \alpha_{ik}(\omega^*) \\ \alpha_{ji}(\omega^*) & \alpha_{jj}(\omega^*) & \alpha_{jk}(\omega^*) \\ \alpha_{ki}(\omega^*) & \alpha_{kj}(\omega^*) & \alpha_{kk}(\omega^*) \end{bmatrix}
$$

$$
\left\{ \Delta k_{ij}Y_{ij}\begin{bmatrix} 1 \\ -1 \\ 0 \end{bmatrix} + \Delta k_{jk}Y_{jk}\begin{bmatrix} 0 \\ 1 \\ -1 \end{bmatrix} + \Delta k_{ik}Y_{ik}\begin{bmatrix} 1 \\ 0 \\ -1 \end{bmatrix} \right\} = \begin{Bmatrix} Y_i \\ Y_j \\ Y_k \end{Bmatrix} \quad (11.24)
$$

$$
\Delta k_{ij}Y_{ij}\begin{bmatrix} \alpha_{ii}(\omega^*) - \alpha_{ij}(\omega^*) \\ \alpha_{ji}(\omega^*) - \alpha_{jj}(\omega^*) \\ \alpha_{ki}(\omega^*) - \alpha_{kj}(\omega^*) \end{bmatrix} + \Delta k_{jk}Y_{jk}\begin{bmatrix} \alpha_{ij}(\omega^*) - \alpha_{ik}(\omega^*) \\ \alpha_{jj}(\omega^*) - \alpha_{jk}(\omega^*) \\ \alpha_{kj}(\omega^*) - \alpha_{kk}(\omega^*) \end{bmatrix} \quad (11.25)
$$

or

$$
+ \Delta k_{ik}Y_{ik}\begin{bmatrix} \alpha_{ii}(\omega^*) - \alpha_{ik}(\omega^*) \\ \alpha_{ji}(\omega^*) - \alpha_{jk}(\omega^*) \\ \alpha_{ki}(\omega^*) - \alpha_{kk}(\omega^*) \end{bmatrix} = -\begin{Bmatrix} Y_i \\ Y_j \\ Y_k \end{Bmatrix}
$$

Define a new FRF quantity:

$$\alpha_{ij,kl}(\omega^*) = \alpha_{ik}(\omega^*) + \alpha_{jl}(\omega^*) - [\alpha_{jk}(\omega^*) + \alpha_{il}(\omega^*)] \tag{11.26}$$

we can convert equation (11.25) into:

$$\Delta k_{ij} Y_{ij} \begin{bmatrix} \alpha_{ij,ij}(\omega^*) \\ \alpha_{jk,ij}(\omega^*) \\ \alpha_{ik,ij}(\omega^*) \end{bmatrix} + \Delta k_{jk} Y_{jk} \begin{bmatrix} \alpha_{ij,jk}(\omega^*) \\ \alpha_{jk,jk}(\omega^*) \\ \alpha_{ik,jk}(\omega^*) \end{bmatrix}$$

$$+ \Delta k_{ik} Y_{ik} \begin{bmatrix} \alpha_{ij,ik}(\omega^*) \\ \alpha_{jk,ik}(\omega^*) \\ \alpha_{ik,ik}(\omega^*) \end{bmatrix} = - \begin{Bmatrix} Y_{ij} \\ Y_{jk} \\ Y_{ik} \end{Bmatrix} \tag{11.27}$$

$$\text{or} \ - \begin{bmatrix} \alpha_{ij,ij}(\omega^*) & \alpha_{ij,jk}(\omega^*) & \alpha_{ij,ik}(\omega^*) \\ \alpha_{jk,ij}(\omega^*) & \alpha_{jk,jk}(\omega^*) & \alpha_{jk,ik}(\omega^*) \\ \alpha_{ik,ij}(\omega^*) & \alpha_{ik,jk}(\omega^*) & \alpha_{ik,ik}(\omega^*) \end{bmatrix} \begin{bmatrix} Y_{ij} & 0 & 0 \\ 0 & Y_{jk} & 0 \\ 0 & 0 & Y_{ik} \end{bmatrix} \begin{Bmatrix} \Delta k_{ij} \\ \Delta k_{jk} \\ \Delta k_{ik} \end{Bmatrix} = \begin{Bmatrix} Y_{ij} \\ Y_{jk} \\ Y_{ik} \end{Bmatrix}$$

$$\tag{11.28}$$

This is a set of simultaneous equations for stiffness modifications. The first matrix in the equation contains processed FRF data from the modified coordinates. These data are available from experimental modal testing. The Y vector and matrix contain elements from the subset of the mode shape for the modified structure. We can use the mode shape subset of the original structure to determine them. Alternatively, we can assign the values of the subset of the mode shape for the modified structure so that both the natural frequency and the mode shape subset will be modification.

11.5 Local structural modification by physical parameter changes

For a real engineering structure, a pure mass or stiffness change is often not practical. It is only realistic to change physical or design parameters such as thickness, diameter and length. For example, for a structure consisting of beams it is more realistic to change cross-sectional area at some locations than to change just stiffness. Such changes will modify both mass and stiffness properties locally. When a physical parameter causes linear changes on both mass and stiffness, we can rely on the theory we have developed so far for structural modification analysis. First, equation (11.4) can be transformed into:

$$([\Delta K] - \omega^{*2}[\Delta M])\{Y\} = -([K] - \omega^{*2}[M])\{Y\} \tag{11.29}$$

Assume here matrices $[\Delta M]$ and $[\Delta K]$ are due to a physical parameter change Δp and they are its linear functions. Once the geometry and physical connectivity are known, it is possible to express the left side of equation (11.29) into:

$$([\Delta K] - \omega^{*2}[\Delta M])\{Y\} = [C]\Delta p \tag{11.30}$$

Then, we can estimate the parameter change Δp as:

$$\Delta p = -([C]^T[C])^{-1}[C]^T([K] - \omega^{*2}[M])\{Y\} \tag{11.31}$$

Use a 2-D truss structure as an example. The elemental mass and stiffness matrices of the truss between coordinates i and j are:

$$[m_e^0] = \frac{\rho A_{ij} l}{6} \begin{bmatrix} 2 & 0 & 1 & 0 \\ 0 & 2 & 0 & 1 \\ 1 & 0 & 2 & 0 \\ 0 & 1 & 0 & 2 \end{bmatrix} \tag{11.32}$$

and

$$[k_e^0] = \frac{A_{ij} E}{l} \begin{bmatrix} c^2 & cs & -c^2 & -cs \\ cs & s^2 & -cs & -s^2 \\ -c^2 & -cs & c^2 & cs \\ -cs & -s^2 & cs & s^2 \end{bmatrix} \tag{11.33}$$

Here, c (= cos θ) and s (= sin θ) are related to the orientation angle θ of the element. A_{ij}, E, l and ρ are respectively the cross-sectional area, elasticity modulus, length and mass density of the element. Select A_{ij} as the modification parameter. The mass and stiffness modification matrices for the element become:

$$[\Delta m_{ij}] = \Delta A_{ij} \frac{\rho l}{6} \begin{bmatrix} 2 & 0 & 1 & 0 \\ 0 & 2 & 0 & 1 \\ 1 & 0 & 2 & 0 \\ 0 & 1 & 0 & 2 \end{bmatrix} \tag{11.34}$$

$$[\Delta k_{ij}] = \Delta A_{ij} \frac{E}{l} \begin{bmatrix} c^2 & cs & -c^2 & -cs \\ cs & s^2 & -cs & -s^2 \\ -c^2 & -cs & c^2 & cs \\ -cs & -s^2 & cs & s^2 \end{bmatrix} \tag{11.35}$$

Substitute the mass and stiffness modification matrices into equation (11.29), we will have

$$\left([\alpha^{*R}(\omega)]_{4\times4} \left(\omega^{*2} \frac{\rho l}{6} \begin{bmatrix} 2 & 0 & 1 & 0 \\ 0 & 2 & 0 & 1 \\ 1 & 0 & 2 & 0 \\ 0 & 1 & 0 & 2 \end{bmatrix} - \frac{E}{l} \begin{bmatrix} c^2 & cs & -c^2 & -cs \\ cs & s^2 & -cs & -s^2 \\ -c^2 & -cs & c^2 & cs \\ -cs & -s^2 & cs & s^2 \end{bmatrix} \right) - \frac{1}{\Delta A_{ij}}[I] \right)$$

$$\times \{y_{ij}\}_{4\times1} = \{0\} \tag{11.36}$$

Here, ω^* is the repositioned natural frequency. The real eigenvalue solution of this equation suggests those modification values ΔA_{ij} which, when applied to the truss structure, may result in the natural frequency ω^*.

When a physical parameter change results in nonlinear changes on mass and stiffness, we will be facing a problem which requires new formulation. For example, for a section of a structure shown in Figure 11.4, let us assume the diameter D is allowed to be changed.

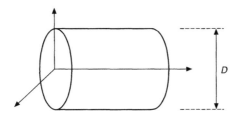

Figure 11.4 A part of a structure with circular cross-sectional area

We can identify that the change of D, denoted as Δ, will cause quadratic mass change and quadruple bending stiffness change. Knowing the elemental matrices, the modification matrices will become respectively:

$$[\Delta m] = (D + \Delta)^2 [\varepsilon] - D^2 [\varepsilon] = \Delta^2 [\varepsilon] + 2D\Delta[\varepsilon] \tag{11.37}$$

$$[\Delta k] = (D + \Delta)^4 [\kappa] - D^4 [\kappa] = \sum_{r=1}^{4} C_4^r D^{4-r} \Delta^r [\kappa] \tag{11.38}$$

Here, C_4^r is a factorial function. Matrix $[\varepsilon]$ is the elemental mass matrix excluding diameter square and matrix $[\kappa]$ the elemental mass matrix excluding diameter quadrature. In general, nonlinear mass and stiffness changes due to a physical parameter change, when substituted into equation (11.4), will lead to an equation in the following form:

$$\Delta^p [A]_p \{Y\} + \Delta^{p-1} [A]_{p-1} \{Y\} + \ldots + \Delta [A]_1 \{Y\} + [I]\{Y\} = \{0\} \tag{11.39}$$

Here, matrices $[A]$ are made of both structural elemental matrices and FRF data. Using a state–space transformation, we can convert the equation into a conventional eigenvalue equation. Assume that:

$$\{Y\} = \{\Theta\} e^{\Delta \tau} \tag{11.40}$$

then

$$\{Y\}^{(r)} = \frac{\partial^r}{\partial \tau^r} \{Y\} = \Delta^r \{Y\} \tag{11.41}$$

Define a state–space vector as:

$$\{V\} = \{\{Y\}^{(p-1)} \{Y\}^{(p-2)} \ldots \{Y\}^{(0)}\}^T \tag{11.42}$$

equation (11.39) will become:

$$([A]_p [A]_{p-1} \ldots [A]_1)\{V\}^{(1)} + ([0] [0] \ldots [I])\{V\} = \{0\} \tag{11.43}$$

This equation, when appended with a number of equalities, forms the following eigenvalue equation:

$$\Delta \begin{bmatrix} [A]_p & [A]_{p-1} & [A]_{p-2} & \cdots & [A]_1 \\ [0] & [I] & [0] & \cdots & [0] \\ [0] & [0] & [I] & \cdots & [0] \\ \cdots & \cdots & \cdots & \cdots \cdots \\ [0] & [0] & [0] & \cdots & [0] \end{bmatrix} \{V\} + \begin{bmatrix} [0] & [0] & [0] & \cdots & [I] \\ -[I] & [0] & [0] & \cdots & [0] \\ [0] & -[I] & [0] & \cdots & [0] \\ \cdots & \cdots \cdots & \cdots \cdots \\ [0] & [0] & \cdots & -[I] & [0] \end{bmatrix} \{V\} = \{0\}$$

(11.44)

The physical parameter change Δ is the eigenvalue of this solution. Only those eigenvalues which result in realistic diameter changes are acceptable.

11.6 Optimization of structural dynamic characteristics

The structural modification theory so far has served mainly for a simple purpose, i.e. to reposition a natural frequency of a structure or an anti-resonance of one of its selected load paths. As structural modification is used primarily for remedying excessive vibration of a dynamic structure, the theory suffices so long as the excitation force has a single frequency or a very narrow frequency band. The principle is either to move natural frequencies of a structure away from the excitation, or to move an anti-resonance to 'match' it. If excitation frequency is more than a very narrow range, moving a natural frequency does not help. A more creative application of the structural modification theory is needed in order to avoid excessive vibration.

11.6.1 Fixing a resonance and an anti-resonance during modification

It is possible to fix a natural frequency of a structure or an anti-resonance of one of its FRFs. While being able to do so itself does not bear apparent significance, it enables us to proceed with some creative applications of structural modification described later in this section.

To understand the mechanism of how a natural frequency can be fixed during structural modification, we need to rewrite equation (11.4) as:

$$([K] - \omega^{*2}[M])\{Y\} + ([\Delta K] - \omega^{*2}[\Delta M])\{Y\} = \{0\}$$

(11.45)

It is obvious that if structural modifications $[\Delta K]$ and $[\Delta M]$ satisfy:

$$([\Delta K] - \omega^{*2}[\Delta M])\{Y\} = \{0\}$$

(11.46)

then from equation (11.45) we know that this combination of modifications has no effect on natural frequency ω^* of the original system and the corresponding mode shape $\{Y\}$. Therefore, after the structure is modified by $[\Delta K]$ and $[\Delta M]$, the natural frequency ω^* is fixed while other natural frequencies have changed.

As a simple example, let us consider a structure modified by attaching the sub-system illustrated in Figure 11.5.

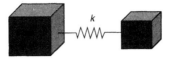

Figure 11.5 A sub-system consisting of two masses connected by a spring

If this sub-system is to attach to two DoFs of the structure whose mode shape subset for the natural frequency ω^* is $\{y_1\ y_2\}^T$, then equation (11.46) becomes:

$$\left(\begin{bmatrix} k & -k \\ -k & k \end{bmatrix} - \omega^{*2} \begin{bmatrix} m_1 & \\ & m_2 \end{bmatrix} \right) \begin{Bmatrix} y_1 \\ y_2 \end{Bmatrix} = \begin{Bmatrix} 0 \\ 0 \end{Bmatrix} \tag{11.47}$$

or $\quad \left\{ \begin{bmatrix} 1 & -1 \\ -1 & 1 \end{bmatrix} \begin{Bmatrix} y_1 \\ y_2 \end{Bmatrix} - \omega^{*2} \begin{bmatrix} 1 & 0 \\ 0 & 0 \end{bmatrix} \begin{Bmatrix} y_1 \\ y_2 \end{Bmatrix} - \omega^{*2} \begin{bmatrix} 0 & 0 \\ 0 & 1 \end{bmatrix} \begin{Bmatrix} y_1 \\ y_2 \end{Bmatrix} \right\} \begin{bmatrix} k \\ m_1 \\ m_2 \end{bmatrix} = \begin{Bmatrix} 0 \\ 0 \end{Bmatrix}$

$$\tag{11.48}$$

This equation can only define the ratios among the unknowns (m_1/k and m_2/k), leaving a scaling factor, γ, to be determined. The defined ratios ensure that once this sub-system is attached to the original system, it will change all natural frequencies but ω^*. The factor γ will be determined in the next step of structural modification analysis such as positioning a natural frequency.

Having explained the principle behind which a natural frequency can remain unchanged during modification, the question is now how to determine the structural modifications that will deliver such an outcome. Equation (11.46) can be transformed into:

$$[A] \begin{Bmatrix} \Delta k_1 \\ \vdots \\ \Delta m_1 \\ \vdots \end{Bmatrix} = \{0\} \tag{11.49}$$

Here matrix $[A]$ is constructed from ω^*, $\{Y\}$ and the connectivity matrices of the elements involved in $[\Delta K]$ and $[\Delta M]$. The non-trivial solution defines the mass and stiffness ratios of the modification system which will satisfy the requirement of 'fixing' the natural frequency ω^* of the original system during modification. Therefore, we have:

$$\begin{Bmatrix} \Delta k_1 \\ \vdots \\ \Delta m_1 \\ \vdots \end{Bmatrix} = \gamma \begin{Bmatrix} \kappa_1 \\ \vdots \\ \varepsilon_1 \\ \vdots \end{Bmatrix} \tag{11.50}$$

or $$[\Delta K] = \gamma [\kappa] \tag{11.51}$$

$$[\Delta M] = \gamma [\varepsilon] \tag{11.52}$$

where γ is unknown. Equation (11.45) can be recast as:

$$-[\alpha(\omega^*)]\,\gamma([\kappa] - \omega^{*2}[\varepsilon])\{Y\} = \{Y\} \tag{11.53}$$

or
$$\left([\alpha(\omega^*)]([\kappa] - \omega^{*2}[\varepsilon]) - \frac{1}{\gamma}[I]\right)\{Y\} = \{0\} \tag{11.54}$$

Here, ω^* is a natural frequency of the modified system. The real eigenvalues of $\frac{1}{\gamma}$ can be used in equations (11.51) and (11.52) to fully determine the modification matrices $[\Delta K]$ and $[\Delta M]$. Only those matrices which do not cause unrealistic changes to the original system are deemed to be feasible modifications. If no real eigenvalues are found from equation (11.54), then it suggests that no mass and stiffness combinations for the selected modification elements (such as the one in Figure 11.5) is able to fix the natural frequency ω^* during modification.

11.6.2 Pole–zero cancellation

A pole–zero cancellation is a term originated from control engineering. In vibration engineering, the poles of a frequency response function are the natural frequencies of the system and the zeros are its anti-resonances. A pole–zero cancellation here means the neutralization of a resonance by an anti-resonance thus creating a nodal point for a particular FRF. For many dynamic structures, the real interest of reducing vibration lies only on selected locations rather than the whole structure itself. For example, for a lathe, vibration of the tool carriage and the chattering during the cutting operation is more important than vibration of other parts of the machine. A successful pole–zero cancellation can reduce dramatically vibration levels in critical parts of a structure.

Since the analysis in this chapter has enabled us to reposition a resonance or an anti-resonance, it seems that pole–zero cancellation can be realized either by fixing a resonance and moving an anti-resonance to meet it, or by the reversed process. However, the former is a fallacy. Fixing a resonance during structural modification means physically preserving both the natural frequency and the mode shape. If by fixing a resonance we can succeed in creating a pole–zero cancellation, then we have made a nodal point on the mode shape. This contradicts to the assumption of preserving the mode shape. Therefore, the latter strategy of moving a resonance to cancel a fixed anti-resonance should be used.

11.6.3 Creation of a frequency range of no resonance

A frequency range with no resonance can be very useful for a real structure with a constant frequency range of excitation. By moving all natural frequencies of the structure outside of the excitation frequency range, we can be assured that excessive resonant vibration does not occur.

Literature

1. Belle, H.V. 1982: Higher Order Sensitivities in Structural Systems. *AIAA Journal*, **20**, 286–288.
2. Bucher, I. and Braun, S. 1993: The structural modification inverse problem: an exact solution. *Mechanical Systems and Signal Processing*, **7**(3), 217–238.
3. Dossing, O. 1990: The Enigma of Dynamic Mass. *Sound and Vibration*.
4. Fox, R.L. and Kapoor, M.P. 1968: Rates of Changes of Eigenvalues and Eigenvectors. *AIAA Journal*, **6**, 2426–2429.
5. Frazer, R.A., Duncan, W.J. and Collar, A.R. 1935: *Elementary Matrices*. Cambridge University Press, London.
6. He, J. and Li, Y. 1995: Relocation of anti-resonances of a vibratory system by local structural changes. *The International Journal of Analytical and Experimental Modal Analysis*, **10**(4), 224.
7. Hallquist, J.O. 1976: An efficient method for determining the effects of mass modifications in damped systems. *Journal of Sound and Vibration*, **44**(3), 449–459.
8. Lancaster, P. 1966: *Lambda-Matrices and Vibrating Systems*. Pergamon Press, New York.
9. Li, T. and He, J. 1999: Local Structural Modification Using Mass and Stiffness Changes. *Engineering Structures*, **21**, 1028–1037.
10. Lim, K.B., Junkins, J.L. and Wang, B.P. 1987: Re-examination of Eigenvector Derivatives. *AIAA Journal of Guidance*, **10**, 581–587.
11. Lord Rayleigh 1945: *Theory of Sound*. Dover Publications, 2nd edition, New York.
12. Maia, N. *et al.* 1997: *Theoretical and Experimental Modal Analysis*. Research Studies Press, UK and John Wiley & Sons, USA, ISBN 0-471-97067-0.
13. Morgan, B.S 1966: Computational Procedure for the Sensitivity of an Eigenvalue. *Electronics Letters*, **2**, 197–198.
14. Mottershead, J.E. 1998: On the zeros of structural frequency response functions and their application to model assessment and updating, *Proceedings of the 16th International Modal Analysis Conference*, Santa Barbara, CA, 500–503.
15. Noor, A.K. and Whitworth. S. 1988: Reanalysis Procedure for Large Structural Systems. *International Journal of Numerical Methods in Engineering*, **26**, 1729–1748.
16. Pomazal, R.J. and Snyder, V.W. 1971: Local modifications of damped linear systems. *AIAA Journal*, **9**, 2216–2221.
17. Ram, Y.M. and Blech, J.J. 1991: The dynamic behaviour of a vibratory system after modification. *Journal of Sound and Vibration*, **150**(3), 357–370.
18. Rogers, L.C. 1970: Derivatives of Eigenvalues and Eigenvectors. *AIAA Journal*, **8**, 943–944.
19. Rosenbrock, H.H. 1965: Sensitivity of an Eigenvalue to Changes in the Matrix. *Electronics Letters*, **1**, 278–279.
20. Rudisill, C.S. 1974: Derivatives of Eigenvalues and Eigenvectors for a General Matrix. *AIAA Journal*, **12**, 721–722.
21. Skingle, G.W. and Ewins, D.J. 1989: Sensitivity Analysis using Resonance and Anti-resonance Frequencies – A Guide to Structural Modification. *Proceedings of the 7th International Modal Analysis Conference*, Las Vegas, Nevada.
22. Smiley, R.G. 1984: Rotational Degrees-of-freedom in Structural Modification. *Proceedings of the 2nd International Modal Analysis Conference*, Orlando, FL, 937–939.
23. Stewart, G.W. 1972: On the sensitivity of the eigenvalue problem $Ax = \lambda Bx$. *SIAM Journal of Numerical Analysis*, **9**, 669–686.
24. Tusei, Y.G. and Yee, Eric K.L. 1987: A method to modify dynamic properties of undamped mechanical systems. *Journal of Dynamic Systems, Measurement, and Control, ASME Transactions*, **111**(9), 403–408.
25. Vanhonacker, P. 1980: Differential and Difference Sensitivities of Natural Frequencies and Mode Shapes of Mechanical Structures. *AIAA Journal*, **18**, 1511–1514.

26. Wang, B.P. 1987: Structural Dynamic Optimisation Using Re-analysis Techniques. *International Journal of Modal Analysis*, **2**(1), 50–58.

27. Wang, B.P. and Chu, F.H. 1987: Effective dynamic reanalysis of large structures. *Shock and Vibration Bulletin*, **51**, Part 3, 73–79.

28. Wang, J., Heylen, W. and Sas, P. 1987: Accuracy of Structural Modification Techniques. *Proceedings of the 5th International Modal Analysis Conference*, London, UK, 65–71.

29. Weissenburger, J.T. 1968: Effects of local modification on the vibration characteristics of linear systems. *Journal of Applied Mechanics*, **35**, 327–332.

30. Yee, Eric K.L. and Tusei, Y.G. 1991: Method of shifting natural frequencies of damped mechanical systems. *AIAA journal*, **29**(11), 1973–1977.

12

System identification using neural network

12.1 Introduction

An artificial neural network is a network made of a large quantity of simple inter-connected processing elements analogous to neurons in the human brain. This network is based on the understanding of the neuromorphic structure, functioning and information process capability of the brain. The research of neural network can be dated back to the 1940s when a psychologist and a mathematician summarized the basic neuromorphic characteristics of biological neurons and mathematically described a structural model of a neuron known as the M–P (McCulloch and Pitts) model. This effort heralded a new era of neural science research. The study of artificial neural network and its engineering application can be traced back respectively to the 1960s and 1980s. Neural network is now beginning to permeate through many engineering disciplines, including structural dynamics and modal analysis. It has found applications in pattern recognition, intelligence control, signal and graphics process, system identification, nonlinear optimization, vibration control, information process and market prediction.

A neural network imitates biological nervous systems. It achieves computational performance by dense interconnection of very simple computational elements. It is nevertheless not a replica of the nervous system in the human brain. It is generally thought that a neural network is a highly sophisticated nonlinear dynamic system. Although each neuron is primitive both in architecture and in function, a network comprising many neurons is intricate. In addition to its nonlinear nature, neural network is a signal processing system. The inherent dynamic process can be classified as a fast process and a slow process. The former is a numerical process to evolve to an equilibrium status with given inputs. The latter is a learning process where the values of the connective weights between neurons are initialized and adjusted according to the environment. After learning, environmental information is stored on the connective weights which have memory capacity.

Neural network can be divided into two types: supervised and unsupervised. A supervised network is given both inputs and desired outputs and adjusts its weights until the errors between its outputs and the desired reach a predefined bound. An unsupervised network is used for clustering problems when the correct output is known. Its weights are adjusted using predefined criteria until the network has performed a classification.

A neural network has dual functions of information storage and information process. Its operation is a process of transferring data from the input to the output. Through that, storage and calculation are performed.

System identification is an essential means of studying the dynamic characteristics of a system. For a linear dynamic system, there have been a number of effective identification methods. For a nonlinear system, however, few effective methods are available. Some existing methods which require prior knowledge or rely on assumptions, may only work for specific types of nonlinearities. A main difficulty in dealing with a nonlinear system lies in finding a reliable mathematical model for it. Without a model, it is usually impossible to proceed with system identification. Neural network does not rely on a preconceived mathematical model or even the number of parameters. Owing to its learning ability and its nonlinear nature, neural network is well suited for nonlinear system identification. The network trains itself from the input–output data of the system to be identified and adjust the weight values. Thus, the trained network reflects the input–output characteristics of the system it intends to represent.

12.2 Neural network model

The neuron model, topology and network algorithm together determine the characteristics of a neural network. They are known as the three elements of a neural network. A neuron is the basic element of a network. The topology of the network reflects the interconnection of the neurons in it. The algorithm determines the way different parts of a network learn and process information.

12.2.1 Neuron

A neuron is the basic constituent and process element in a neural network. Depending on the nature of the problem, the neuron can be given different meanings, such as property, symbol, word, or concept. A basic structure of a neuron is shown in Figure 12.1.

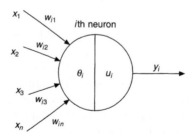

Figure 12.1 Model of a single neuron

Signals $x_1, x_2, \ldots, x_n$ are the inputs of the neuron. They can be the outputs from other neurons. Signals $w_{i1}, w_{i2}, \ldots, w_{in}$ are the connective weights between the ith neuron and other neurons. They reflects the extent of information exchange among neurons. Connections between neurons are not merely a channel of information transfer. There is a weighting coefficient between a pair of neurons that is able to provide synapse strength, similar to what happens to the neurons in a biological nerve system. It can strengthen or weaken the impact of the output of one neuron to the downstream one. This weighting coefficient is called weight (or connection strength, synapse strength). The total input to a neuron is the weighted outputs from other neurons to it, namely:

$$I_i = \sum_{j=1}^{n} x_j w_{ij} \tag{12.1}$$

The neuron cannot produce output until the total input exceeds a certain value, vitalizing the neuron. This value is the threshold of the neuron and is denoted as θ_i. Let:

$$s_i = \sum_{j=1}^{n} x_i w_{ij} - \theta_i \tag{12.2}$$

Then the active value of a neuron is $u_i = g(s_i)$ where g is a continuous function. Usually, we let $g(s_i) = s_i$. Assume y_i is the output of the ith neuron. It is a nonlinear function of the neuron's active value and can be expressed as:

$$y_i = f(u_i) = f\left(\sum_{j=1}^{n} x_i w_{ij} - \theta_i\right) \tag{12.3}$$

The function f in this equation is the activation function of the neuron. It is commonly seen in three forms; threshold form, piecewise form or S form, as shown in Figure 12.2.

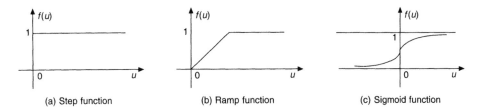

(a) Step function   (b) Ramp function   (c) Sigmoid function

Figure 12.2 Activation functions of neurons

A special type of S form is sigmoid function. Analytically, it is defined as:

$$f(u) = \frac{1}{1 + e^{-u}} \tag{12.4}$$

This is a continuous and differentiable function. It emulates the saturation characteristics of a neuron. It enables a neuron to possess a strong capacity of nonlinear mapping.

12.2.2 Topological structure of neural network

The interconnection pattern of neurons in a neural network demonstrates the architecture of the network and dictates its functionality. The architecture has been evolving and improving. The original pattern had only input and output layers. This type of network is primitive in structure and limited in functionality. A hidden layer (or even several hidden layers) was later introduced. This greatly enhanced the functionality of the neural network. The connection among different neuron layers in a neural network can take various forms. Depending on the connection, a neural network's architecture can be categorized as:

(1) Feedforword multi-layer neural network (Figure 12.3a): the information is transferred unidirectionally from input layer, hidden layer to output layer. This type of network is a static mapping network. It can be used to emulate nonlinear mapping and to facilitate identification of nonlinear systems.
(2) Feedforward network with feedback between output and input layers (Figure 12.3b): this network is a dynamic one because of its built-in feedback character. The information transfer within the network is a dynamic process. This network can be used to store certain pattern series.
(3) Feedforward network with connections among neurons on the same layer (Figure 12.3c).
(4) Network with possible connections among neurons (Figure 12.3d).

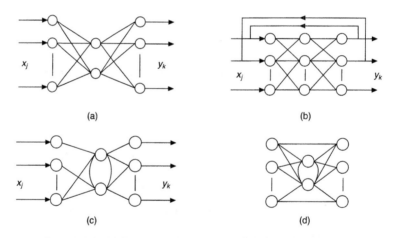

Figure 12.3 Neural network with different architectures

12.2.3 Learning algorithm of a neural network

A neural network can become a useful system only after learning. The learning is a process of satisfying the weightings of the network. It relies on a set of sample data (usually a set of measurement data) to adjust the weightings in order to satisfy given output error criteria. When the output error is below a specified value, learning is completed. The network is now seen as best representing the characteristics of the system under study.

There are different algorithms a neural network can use to learn. One learning algorithm is based on the Hebb rule of weighting adjustment. The Hebb rule basically states that if two neurons are both highly excited, then the weighting for the mutual connection should be strengthened. The Hopfield network adopts this rule. Another algorithm is based on steepest decent method. The weighting adjustment is a gradual process of optimization. This category includes the δ rule, the generalized δ rule and the Boltzman machine learning rule. The δ rule follows that the weighting adjustment should be proportional to the error. The generalized δ rule dilates the δ rule. The back propagation (BP) network adopts a learning algorithm based on the generalized δ rule. The Boltzman random learning rule replaces the generalized δ rule with a statistical method called simulated annealing. It is more suitable for multi-layer neural networks. The steepest decent method is a slow learning method and is susceptible to being trapped on local minima. Another learning method is called self-regulated learning. The network self-adjusts itself during the learning and approaches the characteristics encapsulated within data samples.

Presently, neural networks used for system identification are often multi-layer feedforward networks. The learning method is largely based on the generalized δ rule. The BP network is a widely used network of this type. Its algorithm is simple and effective. The BP network is introduced below. Other types of networks can be found in the literature.

12.3 BP network and its algorithm

A BP network is a back propagation, feedforward, multi-layer network. Its weighting adjustment is based on the generalized δ rule. In the following, details of a BP network, back propagation and the generalized δ rule will be studied.

The structure of a BP network is shown in Figure 12.4. The network consists of an input layer, several hidden layers (only two shown in the figure) and an output layer. The input layer comprises n neurons and the output layer m neurons. The two hidden layers have r and s neurons, respectively. This network is a mapping from R^n space to R^m space. The complexity of the mapping depends on the hidden layers. Neurons on adjacent layers may be connected but neurons on the same layer do not connect. This network is capable of nonlinear mapping. It can be used to approach a continuous mathematical function on a compact set with any given precision.

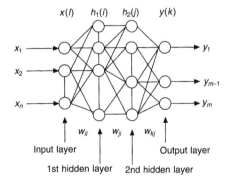

Figure 12.4 Structure of BP network

As shown in Figure 12.4, $x(l)$, $l = 1, 2, \ldots, n$, are the inputs of the network. It is customary to think that the inputs and outputs of the neurons on the network input layer are the same. $h_1(i)$, $i = 1, 2, \ldots, r$, are the outputs of the neurons on the first hidden layer. $\theta(i)$ are corresponding thresholds of these neurons. w_{li} is the weighting between neurons l and i. Let $y(k)$, $k = 1, 2, \ldots, m$, be the outputs of the neurons on the output layer, $\theta(k)$ be their thresholds and w_{kj} the weighting between neurons k and j. Assume the nonlinear functions of layers (excluding the output layer) are S functions, as given in equation (12.4).

First, the network needs to be trained. Training needs samples of measurement input and output data from a real system. Assume the learning samples are $\{(X_p, \bar{Y}_p), X \in R^n, \bar{Y} \in R^m, p = 1, 2, \ldots, N\}$. There are then N input–output data samples for training the network. Take one sample $(X_p, \bar{Y}_p)$ where $X_p = (x_1^p, x_2^p, \ldots, x_n^p)$ and $\bar{Y}_p = (\bar{y}_1^p, \bar{y}_2^p, \ldots, \bar{y}_m^p)$. If the output of the network due to input X_p are $Y_p = (y_1^p, y_2^p, \ldots, y_m^p)$, then error $Y_p - \bar{Y}_p$ exists between the anticipated output Y_p and the actual output $\bar{Y}_p$. The total mean square errors will be:

$$e_p = \frac{1}{2} \sum_{k=1}^{m} (y_k^p - \bar{y}_k^p)^2 \tag{12.5}$$

For the whole sample space, the total error will be

$$E = \sum_{p=1}^{N} e_p \tag{12.6}$$

The learning process of the network is to adjust the weightings in order to minimize the error E. In other words, the aim is to make the system outputs approach the actual outputs. As the error becomes less than the allowable level, network learning is completed. After this process, the network is at its best to represent the real system. Subsequent adjustment of weightings is then based on the steepest decent method. The gradient of a weighting to the error is defined as:

$$G = \frac{\partial E}{\partial w} = \sum_{p=1}^{N} \frac{\partial e_p}{\partial w} \tag{12.7}$$

Here w is one of the network weightings w_{il}, w_{ji}, w_{kj} or one of their thresholds $\theta(i)$, $\theta(j)$ or $\theta(k)$. In the following, the analysis of gradients with respect to different weightings is given. For the sake of simplicity, the subscript 'p' is omitted.

The output of neuron k on the output layer is given as:

$$y_k = f(u(k)) \quad u(k) = \sum_{j=1}^{s} w_{kj} h_2(j) - \theta(k) \quad k = 1, 2, \ldots, m \tag{12.8}$$

The output of neuron j on the second hidden layer is given as:

$$h_2(j) = f(u(j)) \quad u(j) = \sum_{i=1}^{r} w_{ji} h_1(i) - \theta(j) \quad j = 1, 2, \ldots, s \tag{12.9}$$

The output of neuron i on the first hidden layer is given as:

$$h_1(i) = f(u(i)) \quad u(i) = \sum_{l=1}^{n} w_{il} x(l) - \theta(i) \quad i = 1, 2, \ldots, r \tag{12.10}$$

From equation (12.8) we have:

$$\frac{\partial e}{\partial w_{kj}} = \frac{\partial e}{\partial y(k)} \frac{\partial y(k)}{\partial u(k)} \frac{\partial u(k)}{\partial w_{kj}} = (y_k - \bar{y}_k) f'_{u(k)} h_2(j) \qquad (12.11)$$

Here, $f'_{u(k)}$ is the first derivative of function $f(u(k))$ with respect to $u(k)$. Substituting equation (12.9) leads to:

$$\frac{\partial e}{\partial w_{ji}} = \frac{\partial e}{\partial h_2(j)} f'_{u(j)} \frac{\partial u(j)}{\partial w_{ji}} = \sum_{k=1}^{m} \frac{\partial e}{\partial y(k)} \frac{\partial y(k)}{\partial h_2(j)} f'_{u(j)} \frac{\partial u(j)}{\partial w_{ji}}$$

$$= \sum_{k=1}^{m} (y_k - \bar{y}_k) f'_{u(k)} w_{kj} f'_{u(j)} h_1(j) \qquad (12.12)$$

Note that
$$\delta_k = (y_k - \bar{y}_k) f'_{u(k)}, \quad \delta_j = \sum_{k=1}^{m} \delta_k w_{kj} f'_{u(j)} \qquad (12.13\text{a, b})$$

Equations (12.11) and (12.12) can then be recast into:

$$\frac{\partial e}{\partial w_{kj}} = \delta_k h_2(j) \qquad (12.14)$$

$$\frac{\partial e}{\partial w_{ji}} = \delta_j h_1(i) \qquad (12.15)$$

From equation (12.10) we have:

$$\frac{\partial e}{\partial w_{il}} = \frac{\partial e}{\partial h_l(j)} \frac{\partial h_l(j)}{\partial u(i)} \frac{\partial u(i)}{\partial w_{il}} = \sum_{j=1}^{s} \frac{\partial e}{\partial h_2(j)} \frac{\partial h_2(j)}{\partial u(i)} \frac{\partial u(i)}{\partial h_1(i)} \frac{\partial h_1(i)}{\partial u(i)} \frac{\partial u(i)}{\partial w_{il}}$$

$$= \sum_{j=1}^{s} \delta_j w_{ji} f'_{u(i)} x(l) \qquad (12.16)$$

Because
$$\delta_i = \sum_{j=1}^{s} \delta_j w_{ji} f'_{u(i)}, \qquad (12.17)$$

equation (12.16) can be rewritten as:

$$\frac{\partial e}{\partial w_{il}} = \delta_i x(l) \qquad (12.18)$$

Here, δ_k, δ_j and δ_i are the output errors of the output layer and hidden layers. From equations (12.13) and (12.17), it is found that errors propagate backwards. This is called the back propagation of errors. The BP network is known from the abbreviation. Equations (12.14), (12.15) and (12.18) are the estimators for error gradients.

Gradients of thresholds $\dfrac{\partial e}{\partial \theta(k)}$, $\dfrac{\partial e}{\partial \theta(j)}$ and $\dfrac{\partial e}{\partial \theta(i)}$ can be estimated likewise, leading to:

$$\frac{\partial e}{\partial \theta(k)} = -\delta_k \qquad (12.19)$$

$$\frac{\partial e}{\partial \theta(j)} = -\delta_j \qquad (12.20)$$

$$\frac{\partial \dot{e}}{\partial \theta(i)} = -\delta_i \tag{12.21}$$

The total error gradient is estimated by equation (12.7). The adjustments of the weightings and their thresholds are given as:

$$\Delta w_{kj} = -\eta G_{kj}, \quad \Delta w_{ji} = -\eta G_{ji}, \quad \Delta w_{il} = -\eta G_{il} \tag{12.22}$$

$$\Delta \theta(k) = -\eta G(k), \quad \Delta \theta(j) = -\eta G(j), \quad \Delta \theta(i) = -\eta G(i) \tag{12.23}$$

Here, η is the learning step. The adjustment outlined in equations (12.22) and (12.23) is known as the generalized δ rule. It is important to select a reasonable step. A fine step leads to a lengthy learning process while a coarse step may be susceptible to instability or diversion. To improve the stability and accelerate learning, a weighted adjustment can be used. This means:

$$\Delta w^{\text{new}} = -\eta G + \alpha \Delta w^{\text{old}} \tag{12.24}$$

Here, Δw^{new} is the present adjustment for weighting or for threshold while Δw^{old} is the immediate past value of its counterpart. α is a dynamic coefficient. It takes value in the range $\alpha = 0 \sim 1$. G is the same as in equation (12.7).

The BP algorithm can be summarized by the steps below:

(1) Initialize all weightings and thresholds. Let sample number p be 1 and gradient G be 0.
(2) Feed sample p to the neural network and estimate the output of each layer.
(3) Calculate total error and update G using (12.13), (12.17), (12.14), (12.15) and (12.18).
(4) Let $p + 1$ sample replace sample p. If $p < N + 1$, repeat from 2.
(5) Properly select η and α. Use equation (12.24) to estimate the adjustment for connective weights and thresholds.
(6) Update connective weights and thresholds and recalculate the network output and total error E.
(7) If $E > \varepsilon$ (allowable error), reset $p = 1$, $G = 0$ and repeat from 2.
(8) If $E < \varepsilon$, stop learning.

The above is the basics of a neural network. In the following section, we will introduce the application of neural network in the nonlinear system identification.

12.4 Application of neural network in nonlinear system identification

The essence of using neural network for system identification is to find a neural network model that closely resembles the real system. As a result, with the same input, the neural network model should produce similar output as the real system does within a defined error range. Figure 12.5 illustrates the principle of system identification using neural network. In the figure, ANN stands for Artificial Neural Network. $x(k)$ and $\bar{y}(k)$ are the input and output of the real system. (k) is the output of the neural network. TDL stands for Tapped Delay Line. It provides a phase

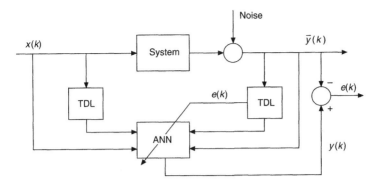

Figure 12.5 System identification with neural network

delay for a signal. With the same input, the difference between the outputs of the real system and the network is the error $e(k)$. According to the introduction of BP network, the network can update itself from a learning process until the error falls within a specified range. On its completion, the input–output characteristics of the network should resemble that of the real system. The neural network model can then be used as the model for the system for further applications.

When using neural network for system identification, the following characteristics should be noted.

(1) There is no need to establish a mathematical model for the real system since it is seen like a 'black box'. Only the input–output relationship is studied. Model parameters are reflected by the connective weights of the network. Network learning is done using the input–output data samples of the real system. No consideration is given to the internal structure of the system. The network learns the overall dynamic characteristics of the system. This is particularly useful for nonlinear system identification since it is usually difficult to establish a mathematical model for such a system.

(2) The convergence rate of the identification does not depend upon the dimension of the system. It is dictated only by the network structure and the learning algorithm adopted.

(3) Neural network as an identification model is a physical realization of a real system. It can be used for online control.

In addition, issues which require special attention for modal analysis related system identification are listed below.

(1) The selection of neural network structure is based on a compromise between the identification accuracy and the complexity of the network. Simple network structure yields low accuracy of identification while complicated network structure leads to lengthy learning time. Usually, several simulative experiments or other means are needed to determine the network of optimal structure that would satisfy identification accuracy.

(2) The excitation of a system should be broad band for the sake of training the network using the measured data. It should ideally excite all modes of the system. This is usually unattainable but it is reasonable to design the input to cover the frequency range of interest so that the trained network is useful for that range.
(3) Always use static neural network for system identification. If the input signals in the network contains feedback components, they will complicate the identification algorithm.

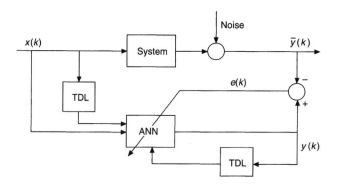

Figure 12.6 Parallel model for system identification

According to the different way of learning, the network model can be classified as the series–parallel model (as shown in Figure 12.5) and the parallel model (shown in Figure 12.6). The series–parallel model does not have feedback inputs. The parallel model does. It uses the real system's input and the output feedback of the network as the inputs of the network. These two models have their pros and cons. The series–parallel model converges fast and the convergence is usually assured. However, after learning, it needs the real system's output as its input. This limits its application. The parallel network relies on itself after learning. The output depends on its own input. However, the convergence is not assured. To take the advantages of both models, it is customary to use the series–parallel model at the learning stage to adjust connective weights and then use these values as the initial values for a serial network to conduct its learning. Experience has shown that this hybrid method can lead to speedy convergence to ideal connective weights.

The mathematical model of a series–parallel model can be outlined by the following equation:

$$y(k + 1) = f(\bar{y}(k), \bar{y}(k - 1), \ldots, \bar{y}(k - n_- + 1), x(k), x(k - 1), \ldots,$$
$$x(k - m + 1)) \tag{12.25}$$

The equation for the mathematical model of a parallel model is given as:

$$y(k + 1) = f(y(k), y(k - 1), \ldots, y(k - n_- + 1), x(k), x(k - 1), \ldots, x(k - m + 1)) \tag{12.26}$$

12.5 Examples of using neural network for system identification

In the following, a number of examples are given to demonstrate the application of neural network on nonlinear system identification.

Example 1

Use neural network to identify the nonlinear dynamic characteristics of an SDoF vibratory system governed by the following equation of motion:

$$m\ddot{x} + c\dot{x} + kx + p(x, \dot{x}) = f(t) \tag{12.27}$$

Here, m, c, k are mass, damping and stiffness of the system. $f(t)$ is the excitation force and $p(x, \dot{x})$ is an unknown nonlinear term.

Solution

Usually, a solution with a nonlinear term involves definition of a nonlinear mathematical model based on some assumptions. For a complicated nonlinear system, it is often impossible to find such a mathematical model. System identification using neural network does not rely on a mathematical model. The system is treated as a 'black box'. The identified overall linear or nonlinear system characteristics are the objectives.

Let $\dot{x} = y$. Equation (12.27) becomes:

$$\begin{cases} \dot{x} = y \\ \dot{y} = \dfrac{1}{m}(f(t) - cy - kx - p(x, y)) \end{cases} \tag{12.28}$$

Using the regressive Euler formula, equation (12.28) can be expressed using finite difference as:

$$\begin{cases} x(n + 1) = x(n) + y(n)\Delta t \\ y(n + 1) = y(n) + \dfrac{\Delta t}{m}(f(t_n) - cy(n) - kx(n)) - \dfrac{\Delta t}{m}p(x(n), y(n)) \end{cases} \tag{12.29}$$

Here, n is the sampling time. Let:

$$\tilde{y}(n + 1) = y(n) + \frac{\Delta t}{m}(f(t_n) - cy(n) - kx(n)) \tag{12.30}$$

Equation (12.29) can be recast as:

$$\begin{cases} x(n + 1) = x(n) + y(n)\Delta t \\ y(n + 1) = \tilde{y}(n + 1) - \dfrac{\Delta t}{m}p(x(n), y(n)) \end{cases} \tag{12.31}$$

Equation (12.31) shows that the system consists of linear and nonlinear parts. Design the structure of the neural network as shown in Figure 12.7. It consists of a linear network and a nonlinear network, as framed by dotted lines in the figure.

If the system parameters m, c, k are known, then the connective weights of the linear network can be computed using equations (12.30) and (12.31). As shown in the figure, the values $x(n + 1)$ and $\tilde{y}(n + 1)$ can be obtained from values at the last

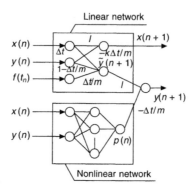

Figure 12.7 Neural network structure for the SDoF system

sampling moment, namely $x(n), y(n), f(t_n)$. The nonlinear network is used to identifiy the nonlinear characteristics of the system. The connective weights of this network are obtained through learning. $y(n + 1)$ can be estimated using equation (12.31). The feedforward algorithm of the nonlinear network can be computed using:

$$h_i = f\left(\sum_{l=1}^{2} O_l w_{il} - \theta_t \right) \tag{12.32}$$

Here, i denotes the ith neuron on the hidden layer, w_{il} is the connective weight between the neuron on the input layer and the neuron on the hidden layer, O_l is the output of the neuron on the output layer. As $i = 1$, $O_l = x(n)$. As $i = 2$, $O_i = y(n)$. The output from the output layer is quantified as:

$$p(n) = f\left(\sum_{i=1}^{r} h_i w_{ki} - \theta_k \right) \quad k = 1 \tag{12.33}$$

Here, r is the number of neurons on the hidden layer. w_{ki} is the connective weight between the neuron on the hidden layer and that on the output layer. Function f is a sigmoidal function.

The data samples for learning are from the input and output relationship of the real system. The learning takes place for data sampled at each moment. For example, at moment n, the sample data are $x(n), y(n), f(t_n)$. Using equations (12.30) to (12.33), the output of the network moment $n + 1$ will be:

$$y(n + 1) = \tilde{y}(n + 1) - \frac{\Delta t}{m} p(n) \tag{12.34}$$

Assume the error between the output of the network and that of the system is E. Then:

$$E = \frac{1}{2} \sum_{n=1}^{M} (y(n + 1) - \bar{y}(n + 1))^2 \tag{12.35}$$

Here, M is the total number of samples. If E is greater than the allowable value, then the connective weights and thresholds are updated using the BP network algorithm. This is repeated with the following formulae:

$$w_{ki}^{\text{new}} = w_{ki}^{\text{old}} - \eta \frac{\partial E}{\partial w_{ki}}$$

$$= w_{ki}^{\text{old}} - \eta \sum_{n=1}^{M} e(n+1)\left(-\frac{\Delta t}{m}\right)p(n)(1-p(n))h_i \quad k=1 \tag{12.36a}$$

$$w_{il}^{\text{new}} = w_{il}^{\text{old}} - \eta \frac{\partial E}{\partial w_{il}}$$

$$= w_{il}^{\text{old}} - \eta \sum_{n=1}^{M} e(n+1)\left(-\frac{\Delta t}{m}\right)p(n)(1-p(n))w_{ki}h_i(1-h_i)O_l(n) \tag{12.36b}$$

$$\theta_{ki}^{\text{new}} = \theta_{ki}^{\text{old}} - \eta \frac{\partial E}{\partial \theta_{ki}}$$

$$= \theta_{ki}^{\text{old}} - \eta \sum_{n=1}^{M} e(n+1)\left(-\frac{\Delta t}{m}\right)p(n)(1-p(n)) \quad k=1 \tag{12.36c}$$

$$\theta_{il}^{\text{new}} = \theta_{il}^{\text{old}} - \eta \frac{\partial E}{\partial \theta_{il}}$$

$$= \theta_{il}^{\text{old}} - \eta \sum_{n=1}^{M} e(n+1)\left(-\frac{\Delta t}{m}\right)p(n)(1-p(n))w_{ki}h_i(1-h_i) \tag{12.36d}$$

Here, $e(n+1) = y(n+1) - \tilde{y}(n+1)$ is the output error at moment $n+1$.

Now let us see the numerical part of this example. Assume the nonlinear term $p(x, \dot{x})$ is a known function. Using the 4th order Ronger–Khutta algorithm, a group of samples can be generated. These samples can then be used to train the nonlinear neural network until it is able to represent the dynamic characteristics. Let $c/m = 1000$ (N.s)/(m.kg), $k/m = 120\,000$ N/(m.kg), $f/m = 100\,000 \sin(314t)$ N/kg. Assume the nonlinear term is $p = 100\,000x^3$. The system has a hardening stiffness. Use five hidden layers. The learning step is set to be $\eta = 0.7$. One hundred samples are used ($M = 100$). After 2000 learning steps, the error range is $E < 0.012$. After completion, the identified nonlinear characteristics are shown in Figure 12.8.

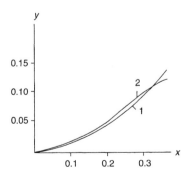

Figure 12.8 Nonlinear stiffness identification: 1 – true value, 2 – identified value

As a different case, assume the nonlinear term is $p = 10\,000\dot{x}^2$ so that the system has nonlinear damping. The identification results are shown in Figure 12.9. The trained neural network can now be used as the dynamic model of the system. With given initial conditions $(x(0), \dot{x}(0))$, the system response is given in Figure 12.10. From Figures 12.8, 12.9 and 12.10, it is evident that the nonlinear characteristics and dynamic response of the neural network match well with that of the system.

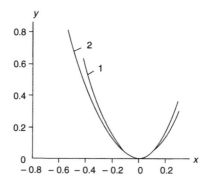

Figure 12.9 Nonlinear damping identification: 1 – true value, 2 – identified value

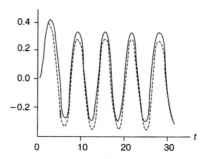

Figure 12.10 Response of the neural network (solid) and that of the system (dotted)

Example 2
Consider the identification of a cantilever plate as a smart structure. The plate consists of ten layers of Fiberite – E9134B. The dimensions are $265 \times 40.5 \times 1$ (mm). A PZT actuator with a dimension of $120 \times 30 \times 0.3$ (mm) is buried between layers 2, 3, 4 and 7, 8, 9 for the purpose of exciting and controlling the vibration of the plate. A piezoelectric sensor with a dimension of $12.5 \times 25 \times 0.3$ (mm) is also buried between layers 7, 8 and 9 to measure the vibration of the plate. Figure 12.11 shows the cross-section of the plate. The dynamic properties of the plate are given in Table 12.1.

This multi-layer composite plate is highly nonlinear. A neural network can be used to establish its nonlinear input–output relationship. Using three layers of a series–parallel network model with neuron numbers being 6, 7 and 2, the identification structure can be shown in Figure 12.11.

In the experiment, the plate was excited by the PZT actuator excited by a random signal of ±9 V. The acceleration signal was measured by the accelerometer and was

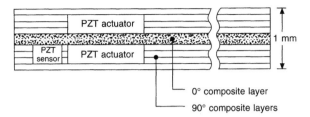

Figure 12.11 The cross-section of the composite laminated smart structure

Table 12.1 Mechanical properties of the unidirectional laminates (S-glass/epoxy)

E_1	55.55 GPa
E_2	25.90 GPa
G_{12}	7.70 GPa
G_{13}	7.70 GPa
γ_{12}	0.26
ρ	1881 kg/m^2

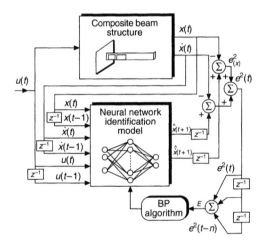

Figure 12.12 System identification of the smart structure by batch learning

integrated once and twice to yield velocity and displacement signals. In order to establish the nonlinear input–output model of the plate for the first two modes at 10 Hz and 51.1 Hz, respectively, the velocity and displacement signals at the free end of the plate and the voltage signal of the PZT actuator were used as the input for the neural network. The outputs will be $x(k-2)$, $x(k-1)$, $\dot{x}(k-2)$, $\dot{x}(k-1)$, $u(k-2)$, $u(k-1)$. The sampling frequency is 100 Hz and duration 10 seconds. The BP algorithm was used for learning. After the learning, the output of the plate matches well with that from measurement (see Figure 12.13).

Using the neural network model, we are able to predict the output of the plate with any given excitation and to design the vibration control for the composite plate.

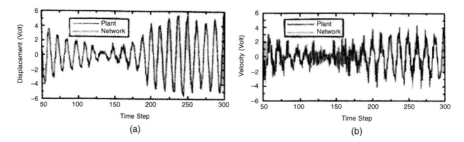

Figure 12.13 Comparison of the time responses between the smart structure and the identified neural network model under random excitation: (a) tip displacement and (b) tip velocity

For more applications of neural network in modal analysis, readers may refer to the literature list at the end of this chapter.

Literature

1. Honik, K., Stincheombe, M. and White, H. 1989: Multilayer Feedforward Networks are Universal Approximators. *Neural Networks,* **2**, 359–366.
2. Hu, S. 1993: Introduction of Neural Networks. *Publishing House of National Defence University of Science and Technology,* Hefei, China (in Chinese).
3. Jiao, L. 1995: System Theory of Neural Networks. *Publishing House of Xi'an University of Electronic Science and Technology* (in Chinese).
4. Jiao, L. 1995: Application and Implementation of Neural Networks. *Publishing House of Xi'an University of Electronic Science and Technology* (in Chinese).
5. Kuschewski, J.G. *et al.* 1993: Application of Feedforward Neural Networks to Dynamical System Identification and Control. *IEEE Trans. on Control Systems Technology,* **1**(1), 37–49.
6. Lieven, N.A.J. 1998: Neural Networks for Modal Analysis. *NATO ASI Series E, Applied Sciences,* 363.
7. Lim, T.W., Cabell, R.H. and Silcox, R.J. 1996: On-Line Identification of Modal Parameters Using Artificial Neural Networks. *Journal of Vibration and Acoustics,* **118,** 649–656.
8. Narendra, K.S. and Pathasarathy, K. 1990: Identification and Control of Dynamic Systems Using Neural Networks. *IEEE Trans. on Neural Networks,* **1**, 4–27.
9. Phan, M. *et al.* 1993: On Neural Networks in Identification and Control of Dynamic Systems. *NASA TM 107702.*
10. Widrow, B. and Steams, S.D. 1985: *Adaptive Signal Processing.* Prentice Hall, Englewood Cliffs, NJ, USA.
11. Wu, C.-J. and Huang, C.-H. 1996: Back-propagation Neural Networks for Identification and Control of a Direct Drive Robot. *Journal of Intelligent and Robotic Systems,* **16**, 45–64.
12. Yang, J. *et al.* 1995: Identification of Nonlinear Characteristics of Vibration Systems Using Structural Neural Networks. *Journal of Vibration Engineering* (in Chinese), **8**(1), 62–66.
13. Yang, S.M. and Lee, G.S. 1997: Vibration Control of Smart Structures by using Neural Networks. *ASME Journal of Dynamic Systems, Measurement and Control,* **119**, 34–39.

Applications of modal analysis on real structures

13.1 Introduction

The development of experimental modal analysis as a new technology is propelled by its ability to offer quick and effective solutions to real life engineering problems. Along with the development of modern computer technology, experimental modal analysis has become the main tool for solving complicated structural vibration problems. For an existing engineering structure, it provides vital information on its dynamic behaviour, thus permitting intelligent solutions to vibration problems the structure may be experiencing.

For product design, the inclusion of modal analysis changes the fundamental philosophy from static design to a combined static and dynamic design. The analysis of the computer model of a product enables us to identify potential dynamic problems of the structure and determine possible design solutions. In so doing, reliability and safety of the manufactured product can be better assured.

The application of modal analysis is not limited to a specific engineering discipline. In recent years, extensive applications have been found in aerospace engineering, acoustical engineering, mechanical engineering, civil engineering, automotive engineering, ship building, mining, manufacturing, nuclear power plants, transportation, weapon systems and building engineering. Modal analysis and finite element analysis have become two pillars in modern structural dynamics.

This chapter presents the reader with a number of practical application cases of using modal analysis to solve vibration problems.

13.2 Modal analysis of a car chassis

The automotive industry has been one of the biggest users of modal analysis technology. The application of modal analysis ranges from simple chassis, suspension systems, driver's seats to whole vehicle analysis and optimal design.

The modal behaviour of a car chassis is essential information for the dynamic analysis of the whole vehicle. This application is to use modal analysis technology to study the vibration behaviour of the chassis. The modal testing was carried out by suspending the chassis with a number of rubber springs, as shown in Figure 13.1.

This was to simulate the free–free boundary conditions of the structure so that the modal information of the chassis alone could be derived. Sinusoidal excitation was used to provide sufficient input energy to excite the chassis. The shaker was also suspended in order to avoid introducing physical constraints to the chassis during the excitation.

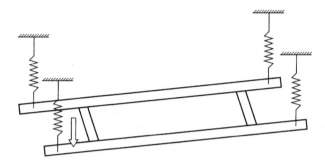

Figure 13.1 A diagram for the boundary conditions of a car chassis for testing

Such a suspension of the structure ensures that the rigid body modes of the supported chassis in both vertical bending and torsional directions are less than 1 Hz, which is much less than the first natural frequency of the chassis. For a modal analysis with sinusoidal excitation, the measurement system can be shown in Figure 13.2.

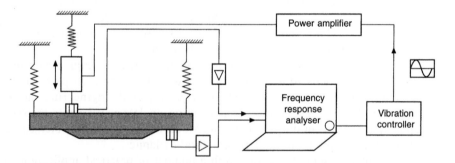

Figure 13.2 Measurement set-up for modal testing of a car chassis

The sinusoidal excitation swept from 5 Hz to 100 Hz. Figure 13.3 shows the amplitude and phase of the point frequency response function at the excitation point. The real and imaginary parts of the accelerance of this point FRF are shown in Figure 13.4.

Within the measured frequency range, we note that vibration modes of the chassis cram together as frequency increases. For example, between 81 Hz and 83 Hz, there appears to have more than one mode. Figure 13.5 shows the FRF data and the Nyquist plot with this frequency range.

The main vibration modes of the chassis are derived from the modal analysis of the measured FRF data and are presented in Table 13.1.

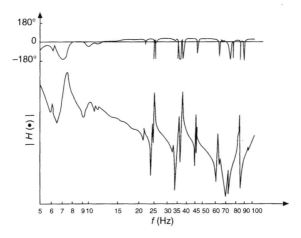

Figure 13.3 Amplitude and phase of the point FRF (receptance)

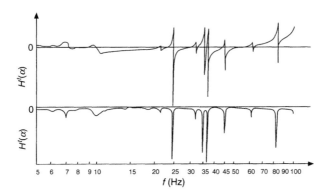

Figure 13.4 The real and imaginary parts of the point FRF (accelerance)

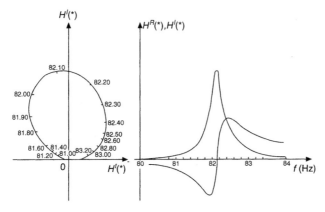

Figure 13.5 Zoomed-in point FRF and the Nyquist plot

Table 13.1 Identified vibration modes of the tested chassis

Mode number	Natural Freq. ω_r (Hz)	Modal damp. $\zeta_r \times 10^3$	Modal mass m_r kg.s²/cm	Modal stiffness $k_r \times 10^{-4}$ kg/cm	Remarks
1	7.23	15.380	0.560	0.116	First torsional mode
2	17.53	1.281	9.438	11.500	First side bending mode
3	24.36	2.581	1.380	3.206	First vertical bending mode
4	31.22	4.959	2.057	7.915	Torsional and side bending mode
5	34.98	4.673	1.202	0.805	Left side bending mode
6	36.60	2.982	1.446	7.646	Second torsional mode
7	60.62	4.926	1.334	19.340	Third bending mode
8	82.10	3.043	1.397	13.970	Third torsional mode

Figures 13.6 and 13.7 show the first torsional mode shape and the first bending mode shape of the chassis.

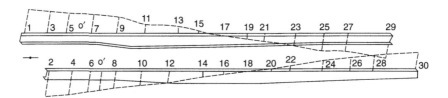

Figure 13.6 First torsional mode shape of the chassis

Figure 13.7 First bending mode shape of the chassis

13.3 Modal analysis of a lathe

The main objective of modal testing and analysis of a lathe was to provide experimental data for improving its dynamic performance and for optimizing future design. Dynamic performance of a lathe impacts directly on the precision, surface quality and efficiency of the machine during operation. Modal testing and analysis can serve as an ideal tool in studying the dynamic performance of a lathe. Modal testing can be carried out when the lathe is idle or when it is in operation. The former is easy to conduct but the results are not truly reflective of the dynamic performance of the machine. A lathe in operation experiences several cutting forces, some being periodic and some random. These forces excite the lathe to vibrate. The vibration of a lathe can be forced or self-excited. The latter is mainly responsible for poor quality of cutting. A lathe with good dynamic performance minimizes forced vibration and eliminates self-excited vibration (chattering).

A lathe is a highly nonlinear structure. Using modal testing to study it will result in linearized modal data of the lathe which are useful to assess mainly qualitatively, and, quantitatively to some extent, its dynamic performance. The modal testing was conducted when the lathe was in operation. The excitation of the lathe was generated from a non-contact electric magnetic exciter installed on the tool carriage to simulate the cutting force. When connected, this exciter simulated the cutting force. The excitation signal was a pseudo random signal shown in Figure 13.8 together with its auto-correlation and auto-spectrum density function. A pseudo random signal can minimize energy leakage in the signal processing and is easily generated numerically. The response of the lathe was measured by accelerometers located at 70 measurement points covering three translational directions.

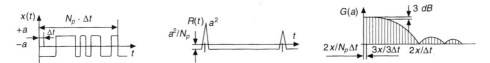

Figure 13.8 The excitation signal and its auto-correlation and auto-spectrum

During the test, the operating lathe did not have feeding motion. The exciter replaced the cutting tool. A rigid workpiece was used so that no vibration was generated from the workpiece to excite the lathe. Table 13.2 shows the modal data of the lathe under idle and operational conditions. The notable difference in modal damping is attributed to the surface friction and damping inside bearings.

Table 13.2 Modal data of the lathe using pseudo excitation

Mode number	Natural frequency (Hz)			Modal damping		
	Operational	Idle	Error %	Operational	Idle	Error %
1	34.95	32.89	5.8	0.1055	0.0985	6.6
2	42.50	44.37	−4.4	0.0711	0.0849	−19.6
3	90.34	85.73	5.1	0.0583	0.0485	16.8
4	154.50	155.38	−0.6	0.0815	0.0884	−8.5
5	198.37	193.84	2.5	0.0573	0.0519	9.4
6	288.38	286.61	0.6	0.0518	0.0426	21.6

For comparison, a sweep sinusoidal signal was also used on the electrical magnetic exciter. The modal data obtained from this excitation were derived from the measure FRF data and are shown in Table 13.3.

The modal analysis of the measured FRF data also revealed the nature of the first six vibration modes of the lathe. They are summarized in Table 13.4.

Table 13.3 Modal data of the lathe using sinusoidal excitation

Mode number	Natural frequency (Hz)			Modal damping		
	Operational	Idle	Error %	Operational	Idle	Error %
1	34.37	35.99	− 4.7	0.0630	0.0555	11.9
2	42.90	45.96	−7.1	0.0558	0.0587	−5.2
3	89.22	85.15	4.6	0.0428	0.0402	6.1
4	154.89	154.81	0.5	0.0735	0.0730	0.7
5	198.17	197.41	0.38	0.0602	0.0590	2.0
6	284.83	284.46	0.13	0.0472	0.0387	22.0

Table 13.4 Vibration modes of the lathe

Mode number	Natural frequency (Hz)	Description of vibration mode shape
1	34.92	First vertical bending mode for the lathe bed. The tailstock and tool carriage move with the lathe bed back and forth.
2	42.50	First torsional mode for the lathe bed. The tailstock moves with the lathe bed. Headstock vibrates along the feeding direction.
3	96.34	First horizontal bending mode for the lathe bed. The headstock moves back and forth.
4	154.50	Second horizontal bending mode for the lathe bed. The headstock rotates on the horizontal plane.
5	198.87	Second horizontal bending mode for the lathe bed. The tool carriage moves out of phase with the workpiece.
6	288.38	The tool carriage moves back and forth. The chuck shows significant torsion.

13.4 Modal analysis of a shaper

A modal testing and analysis was carried out on a shaper. The excitations used were random and impact excitation, respectively. Figure 13.9 shows the measurement set-up.

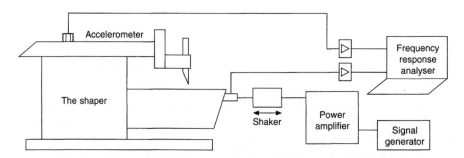

Figure 13.9 Measurement set-up for modal testing of a shaper

The frequency range for the random excitation was selected as 0–500 Hz to cover the main modal activities of the machine. The excitation force was located at the box table of the shaper. A triaxial accelerometer was used at different locations to acquire the responses. When impact excitation was used, the hammer excited the shaper at various locations alternately and response was measured at a fixed location. A hard lead hammer head was used. The frequency range was chiefly 0–600 Hz. The test was conducted with the ram rail being 0.3 m, 0.5 m and 0.7 m, respectively.

Table 13.5 shows the natural frequencies and damping ratios of the first three modes of the shaper with one sliding rail length. The first two mode shapes of the machine are shown in Figure 13.10. The test was conducted when the machine was idle.

Table 13.5 The first three identified vibration modes of the shaper

Mode number	Random excitation		Impact excitation	
	ω_r Hz	ζ_r %	ω_r Hz	ζ_r %
1	48.41	3.06	48.50	2.51
2	78.24	1.96	79.71	4.68
3	113.59	1.65	111.48	1.93

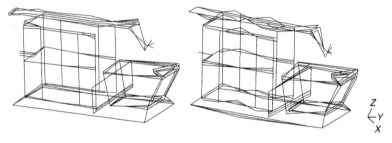

Figure 13.10 First two mode shapes of the shaper

13.5 Modal analysis of a combustion locomotive structure

A combustion locomotive has a dimension of 15 m × 3 m × 2 m and weighs about 20 tonnes. It is made of welded steel parts. A modal testing and analysis was conducted to extract the modal data of the locomotive structure in order to provide information for modifying it. The excitation method was a random impact. This method produces a series of irregular impacts on the structure. Such excitation enables more energy to be inputted into the structure. It produces an excitation time signal with dispersed peaks. Figure 13.11 shows the time history of a random impact signal and its power spectrum. A typical acceleration response of the random impact from the locomotive structure is shown in Figure 13.12.

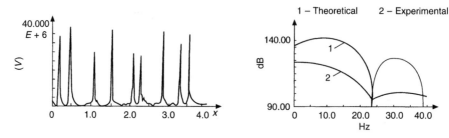

Figure 13.11 Time history and power spectrum of a random impact signal

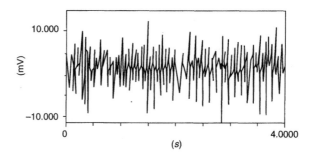

Figure 13.12 An acceleration response of the random impact excitation

The modal testing was carried out in four steps. In step one, a coarse finite element model of the locomotive structure was created. The analysis of this model provided an estimate of the natural frequencies and an insight into the modal density of the structure. In step two, trial impact excitation and measurement was conducted. The objective was to determine the favourite excitation condition for measurement and to locate the optimal measurement sites. Step three of the testing was to conduct the full measurement. The locomotive structure was laid on an elastically supported platform. Altogether there were 120 measurement points. The responses were recorded by a triaxial accelerometer. Both excitation and response signals were recorded on a tape for analysis.

Figure 13.13 shows a typical measured FRF curve and the coherence function. An SDoF frequency domain modal analysis method was used to analyse the recorded FRF data. Modal data of the first seven identified vibration modes are shown in Table 13.6.

The results of modal analysis made available vital experimental data for improving the dynamic characteristics of the locomotive structure. For example, the excessive vibration experienced in the driver's apartment was attributed to the combined effect of the bending and torsional vibration modes below 25.25 Hz and the rigid body modes. This provided a basis for vibration reduction of the driver's seat. In addition, excessive vibration was observed from the side panel at the front. This is detrimental to the fatigue life of the suspension system. Subsequent stiffness enhancement was introduced to the structure to rectify the problem.

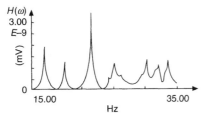
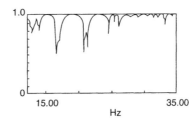

Figure 13.13 A typical measured FRF curve (left) and the coherence function (right)

Table 13.6 The first seven identified vibration modes of the locomotive

Mode number	Natural frequency ω_r Hz	Modal mass $m_r \times 10^{-5}$ Ns^2/m	Modal stiffness $k_r \times 10^{-5}$ N/m	Modal damping $\zeta_r \times 10^{-5}$ N/m
1	13.11	0.0110	89	0.37
2	17.20	0.0093	110	0.29
3	21.37	0.0074	130	0.34
4	25.25	0.0063	160	0.73
5	30.32	0.0053	190	0.42
6	31.92	0.0001	200	0.54
7	33.57	0.0001	210	0.26

13.6 Modal analysis of a power generator

A tidal power station was equipped with a 500 kW hydraulic generator installed within a steel shell structure. The turbine of the generator has four blades. The designed operating speed is 118 rev/min. The whole generator and shell structure is compact and bulky. Traditional external force excitation cannot provide sufficient energy to induce vibration to a significant extent. Instead, the vibration responses at designated locations were measured and time domain modal analysis was used to derive modal data of the structure. The analysis was based on the ARMA(p,q) time series model, as introduced in Chapter 2.

Figure 13.14 shows the diagram of the generator assembly. Altogether 11 locations (marked as ∇ in the figure) were selected for recording vibration responses. Among them, point 1 was located at the top of the bearing pillow block of the first electric motor to observe the vertical vibration of the motor-bearing system. Point 10 was located at the bottom to observe the coupling motion between the shell structure and the armature–bearing system. Points 4 and 9 were located at the top of the bearing pillow block and gear train of the second motor. Point 7 was at the casing of the rear bearing stand. Point 11 was positioned at the entrance chamber of the turbine to record the vibration of the shell structure at the vicinity of the turbine. All responses were measured using accelerometers and recorded on a multi-channel tape recorder.

Table 13.7 shows the time series analysis results from data measured at points 1 and 10, respectively. The results from measurement point 1 reflected the dynamic properties of the armature–bearing–bearing stand assembly. Results from point 10 are indicative of the dynamics of the shell structure. The coupling between the rotor–bearing–bearing stand assembly and the shell structure is also shown in the table.

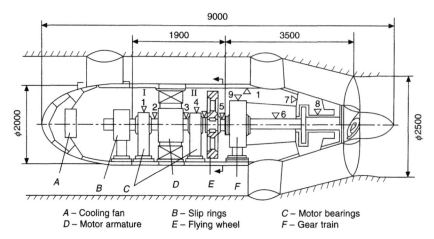

A – Cooling fan   B – Slip rings   C – Motor bearings
D – Motor armature   E – Flying wheel   F – Gear train

Figure 13.14 The diagram of the generator assembly

Table 13.7 Modal analysis results of data measured from points 1 and 10

Time series order No.	Point 1		Point 10	
	Natural frequency ω_r Hz	Modal damping ζ_r	Natural frequency ω_r Hz	Modal damping ζ_r
1	3.30	0.0840		
2	6.60	0.0239	6.200	0.0510
3			24.88	0.275
4		0.0297		
5	57.07	0.0138	74.47	0.0160
6	74.50	0.0075		
7	79.99	0.0037		
8	88.81	0.0087	96.02	0.0297
9	96.50	0.0105		
10	113.39	0.0431	132.7	0.0128
11	175.21	0.0250		
12	213.32	0.0021		

13.7 Modal analysis of a flat flood gate

The flood relief gate of a dam is a flat and multi-layer plate structure. During operation, it is often partially open to control the rate of water relief. The flowing water inevitably impacts on the gate and excites its vibration. If the vibration becomes excessive, the gate risks dramatic reduction of its fatigue life and safety of the operation is in jeopardy. Against this background, a modal testing was conducted in order to investigate the dynamic characteristics of the gate and to predict dynamic response during operation.

The dimension of the flood relief hole is 13.0×12.0 m^2. The maximum flow rate is 800 m^3/sec. The dimension of the gate is 13.7×12.5 m^2. The designed static loading of the gate is 1000 tonnes. The gate is an assembly of three plates in cascade, each being an elastic beam. The gate weighs 70 tonnes and sits in a groove operated

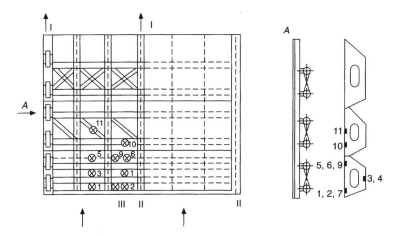

Figure 13.15 The diagram of the flood relief gate and pressure sensor locations

by two lifters which weigh 100 tonnes each. Figure 13.15 shows the diagram of the gate.

The investigation commenced with an experiment to determine the hydraulics of the gate. This was to ascertain the dynamic loading and excitation frequencies the gate has to experience during operation. Eleven pressure sensors were distributed on the gate (as shown in Figure 13.15). Water pressure under different working conditions was recorded and plotted. Figure 13.16 shows a typical water pressure measured by a sensor. Spectral and statistical analyses were carried out on the measured pressure data. Figure 13.17 shows the power spectrum of a pressure signal. Figure 13.18

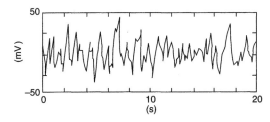

Figure 13.16 A typical water pressure measured by a sensor

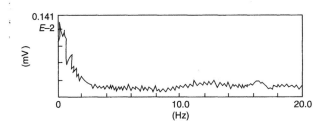

Figure 13.17 Power spectrum of a pressure signal

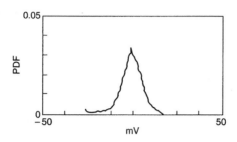

Figure 13.18 Probability density curve of a water pressure signal

shows its probability density curve. It was found that the water pressure load is a stationary random process with zero average. It follows a normal distribution. The frequencies of the transient water load is largely below 1.5 Hz. The triple of the standard deviation of the dynamic load (3σ) is below 90 tonnes, which is smaller than the one-tenth of the designed static load threshold.

The next step was to conduct modal testing. The experiment was carried out on a replica of the real gate, one-tenth of the original size. All material properties and installation were the same as the original. The replica was installed at a small reservoir. The natural frequency of the replica is ten times as great as that of the real gate.

Due to the symmetry of both the structure and load, only those symmetrical vibration modes are of interest. A total number of 84 measurements were selected, 28 for each plate on the gate. All measurement points were distributed on the stiffeners to avoid local flexibility.

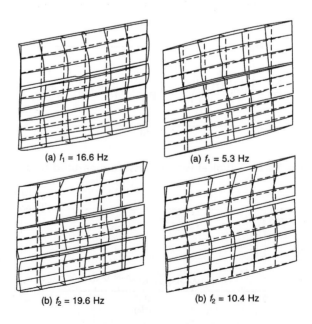

(a) $f_1 = 16.6$ Hz   (a) $f_1 = 5.3$ Hz

(b) $f_2 = 19.6$ Hz   (b) $f_2 = 10.4$ Hz

Figure 13.19 The first two mode shapes of the gate with and without water

Simple sinusoidal sweep was used to excite the structure. The experiment was carried out when the reservoir was empty and full. Table 13.8 shows the modal data of the first three vibration modes (converted to the original gate). The mode shapes are given in Figure 13.19.

Table 13.8 Modal data of the first three modes

Modal data	Empty reservoir			Full reservoir		
	1	2	3	1	2	3
f_r Hz	16.6	19.6	20.0	5.3	10.4	12.8
$2\zeta_r$	0.03	0.026	0.036	0.039	0.045	0.055
K_r t/cm	110.9	154.6	161.0	11.3	43.5	65.9

The natural frequencies of the gate when the reservoir was full are generally lower than those when it was empty. The damping ratios for the former are greater. These are due to the effect of water. Nevertheless, the damping ratios in both cases are reasonably small. The mode shapes are different in these two cases. This reflects the coupling between the gate and water, and the nonlinear effect generated from the supporting mechanisms.

The modal testing and analysis results show that the natural frequencies of the gate are greater than the frequency of the transient water load (which is about 1.5 Hz). Such dynamic properties of the gate are ideal.

13.8 Modal analysis used for stability diagnosis

Pad bearing is often used on machines with a high rotating speed such as compressors and turbines. For a long time, rotors supported by pad bearings are regarded as stable and free of self-excitation. As the rotating speed and operating power increase, self-excitation could occur to cause excessive structural vibration. Theoretically, this type of excitation is caused by gas flow and can be attributed to the skew-symmetric coupling terms in the system matrix. When the rotor of a gas turbine is sealed, the periodic motion of the rotor causes a periodic change of the gap between the rotor and the sealing cubicle. The atmospheric pressure within the sealing cubicle varies accordingly. The total pressure can be separated into a normal pressure component and a tangential pressure component. The latter, being in the same direction of the rotor's rotation, has the potential to cause self-excitation. A critical parameter in the study of the self-excitation, and therefore the stability, of a rotor system is the 'system damping'. It is the real part of the system eigenvalues.

Modal analysis can be used to identify the damping of a rotor system to assist the stability study. When the machine is not operating, external excitation (sinusoidal, random or impact) can be applied and the damping can be identified from measured FRF data. If the rotor is in operation, vibration responses can be acquired and damping extracted using time domain modal analysis methods.

Figure 13.20 shows a rotor rig for stability study. The rotor is supported at each

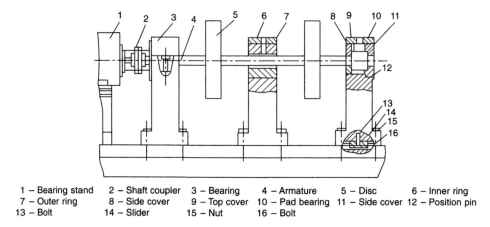

1 – Bearing stand   2 – Shaft coupler   3 – Bearing   4 – Armature   5 – Disc   6 – Inner ring
7 – Outer ring   8 – Side cover   9 – Top cover   10 – Pad bearing   11 – Side cover   12 – Position pin
13 – Bolt   14 – Slider   15 – Nut   16 – Bolt

Figure 13.20 Test rig for stability study of a rotor system

end by a pad bearing. There are two discs on the rotor. The pad bearing is shown in Figure 13.21. The five pads can be tilted to adjust slight angle changes.

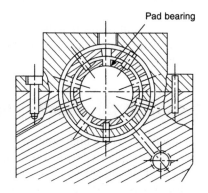

Pad bearing

Figure 13.21 Cross-section of a pad bearing assembly

Figure 13.22 shows the test diagram. The measured vibration signal was analysed to extract the impulse response. Subsequently, the Ibrahim time domain method was used to identify the damping ratios of the rotor system. The test speed was set to 1000, 1500 and 2000 rev/min. The sealing air pressures were 0, 0.25, 0.5, 1.0, 1.5 and 2.0 10^5 Pa. The gravity load of the bearing was 2.54 kg.

According to the stability theory, the more damping, the more stable the rotor system is. When the system damping experiences a dramatic decrease or the value varies significantly, the rotor system may become unstable. Table 13.9 shows the system damping (real part of the eigenvalue) and modal damping from modal analysis.

Figure 13.23 shows the power spectrum of the system's impulse response.

This study led to the following conclusion. The damping of the rotor system varies as the air pressure changes. By studying the modal damping of the first mode of the

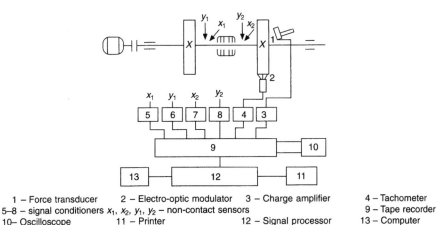

1 – Force transducer   2 – Electro-optic modulator   3 – Charge amplifier   4 – Tachometer
5–8 – signal conditioners x_1, x_2, y_1, y_2 – non-contact sensors   9 – Tape recorder
10– Oscilloscope   11 – Printer   12 – Signal processor   13 – Computer

Figure 13.22 Test diagram for a rotor system

Table 13.9 Modal analysis results of the pad bearing

Sealing air pressure × 10^5 Pa		0.0	0.50	1.00	1.25	1.50	2.00
X direction	System damping σ	13.11	8.063	8.398	7.933	7.219	6.923
	Damping ratio ζ	0.320	0.171	0.182	0.169	0.154	0.148
	Damped nat. freq. ω_d	46.72	47.03	46.21	46.76	46.73	46.79
Y direction	System damping σ	15.08	16.80	16.61	7.421	8.349	11.52
	Damping ratio ζ	0.324	0.345	0.343	0.155	0.175	0.243
	Damped nat. freq. ω_d	46.51	48.71	48.40	47.74	47.58	47.33

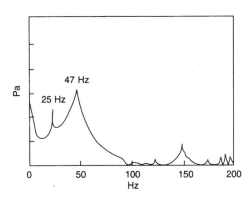

Figure 13.23 Power spectrum of the rotor system's impulse response

rotor system with different air pressure, we can reach a qualitative conclusion on the stability of the system. Figure 13.24 shows the impulse responses at zero air pressure and when pressure is 1.5×10^5 Pa. As the air pressure increases, the free decay rate slows down and the system stability deteriorates.

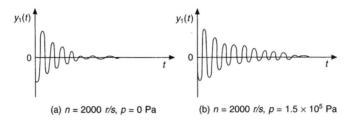

(a) $n = 2000$ r/s, $p = 0$ Pa   (b) $n = 2000$ r/s, $p = 1.5 \times 10^5$ Pa

Figure 13.24 Impulse responses with two different air pressures

13.9 Modal analysis used for stump quality check

Concrete stump forms the foundation for high building, bridges, nuclear plants and many other structures. The significance of its quality cannot be overstated. Once the concrete is poured and set, the quality check of the deeply buried stumps is a formidable task. One engineering practice is to take 2% of total stumps as samples (by using stump pullers) and conduct a static loading test. This method could be costly and the reliability of the outcome is questionable.

A method based on modal testing can assist in checking the quality of stumps. This method uses an impact excitation to measure the power spectrum and FRF data from the top of the stump and analyse the stump quality and loading ability from the measured data. A hammer is used to produce an impulse input to the stump. By analysing the force signal and the acceleration response, we can derive the accelerance of the stump. From the amplitude and phase information of the accelerance, we can derive the quality of the stump. Figure 13.25 shows the measurement set-up of modal testing of a stump.

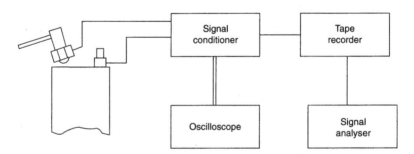

Figure 13.25 Measurement set-up for modal testing of a stump

The theory used in this application is simple. From vibration theory of continuous systems, we know that the natural frequency of a free–fixed stump is given as:

$$f_n = \frac{2n - 1}{4L} c_0 \tag{13.1}$$

Here, L is the total length and c_0 is the wave propagation velocity along the stump. Likewise, the natural frequency of a fixed–fixed stump is given as:

$$f_n = \frac{n}{2L} c_0 \qquad (13.2)$$

We also know that

$$c_0 = \sqrt{\frac{E}{\rho}} \qquad (13.3)$$

Here, E is the elasticity modulus and ρ is the mass density of the stump. In both boundary condition cases, we find that the difference between two consecutive natural frequencies is:

$$\Delta f = \frac{c_0}{4L} \qquad (13.4)$$

This frequency difference is a constant for a uniform stump. From the measured natural frequency difference, we can estimate the wave propagation velocity:

$$c_0 = 4L\Delta f \qquad (13.5)$$

For a uniform stump, we can expect the velocity to be within a range 2800 m/s to 3000 m/s. If the estimated velocity falls outside this range, the stump is susceptible to quality problems. Conversely, from the expected velocity, we can estimate the effective stump length using equation (13.5). If it is shorter than the nominal length, it is indicative of the possible position of a crack on the stump.

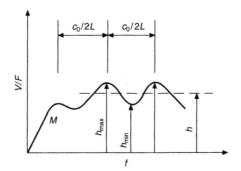

Figure 13.26 A typical mobility FRF of a stump

Another parameter useful for stump quality check is the geometrical average of the mobility function:

$$h = \sqrt{h_{max} h_{min}} \qquad (13.6)$$

Here, h_{max} is the average peak value of the mobility and h_{min} is its average trough value. This value can be compared with the theoretical mobility average h_T which is given by:

$$h_T = \frac{1}{\rho c_0 A} \qquad (13.7)$$

Here, A is the cross-sectional area. The difference between the results from these two equations is used to determine the quality of the stump materials.

The first example, Figure 13.27, shows the mobility curve of a stump 3.5 metres long with a cross-sectional area of 150×150 mm^2. The mobility shows equally spaced natural frequencies. The frequency difference is about 410 Hz. The wave propagation velocity can be estimated to be 2870 m/s. This is within the expected range. Thus, this stump is of a good quality.

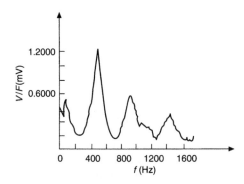

Figure 13.27 Mobility FRF of a 3.5 metre long stump

The second example is a stump 9 metres long with a diameter of 650 mm. The measured mobility FRF is shown in Figure 13.28.

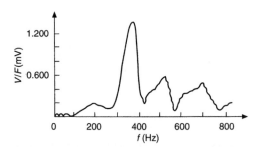

Figure 13.28 Mobility FRF of a 9 metre long stump

From the frequency difference of 180 Hz, the wave velocity is estimated to be 3240 m/s. This indicates that the stump is not uniform throughout its length. The first peak of 360 Hz signals a potential problem at 4.1 metres from the top. This is found to coincide with the crack introduced at 4 metres.

13.10 Modal analysis of ancient bronze bell

This application shows an example of using modal analysis to help to design and manufacture a bronze bell which preserves the acoustic characteristics of the ancient

Chinese bronze bell. The manufacture of bronze bells in ancient China is more a craft than a science. This, however, did not prevent past craftsmen from making acoustically brilliant bronze bells. Some of them survived centuries of wars and turmoil. They provide the opportunity to study bronze bells using modal testing and analysis. The quality of a bell depends on its structural vibration and its properties to propagate sound. Ancient bronze bells in China are renowned for having rich acoustic texture, being able to propagate far and possessing a distinct beating sound. To revive these characteristics in today's manufacture of metal bells, we need to understand the dynamics of ancient bronze bells. Modal testing and analysis has been used in this quest. The objective of this study was to design a bronze bell for a temple in Hawaii that could recapture the brilliance of ancient bells.

The first step of the investigation was to test a number of existing ancient bells. The bronze bell in the Longhua Temple in Shanghai was one of them. Figure 13.29 shows the bell. Figure 13.30 presents the sound spectrum of the bell. It was found that the fundamental frequency of the bell was 122 Hz. Within 1000 Hz, the bell exhibited plenty of clearly defined modal activities. From the sound spectra of a few bells, it was decided that the fundamental frequency was designed as 100 Hz.

Figure 13.29 The bronze bell in the Longhua Temple in Shanghai

The initial geometrical design of the bell was based on knowledge and artistic consideration. Subsequently, a finite element analysis was carried out to derive the modal properties of the designed bell. The bell was 1.70 metres tall with the maximum diameter of 1.10 metres. The thickness increases from the bottom to the top. Figure 13.31 shows the FE model of the designed bell and the manufactured product.

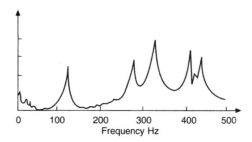

Figure 13.30 The sound spectrum of the bronze bell in the Longhua Temple

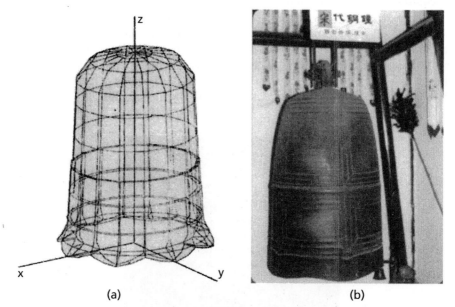

Figure 13.31 (a) The FE model of the designed bell, (b) the manufactured bell

Due to the axial symmetry of the shape, it can be expected that vibration modes will occur in pairs. Table 13.10 shows the first four natural frequencies of the designed bell. The slight non-symmetry of a bell is responsible for splitting a pair of identical natural frequencies and causing the beating sound.

Table 13.10 The first four modes of the designed bell

	Mode			
	1	2	3	4
Frequency Hz	18.24	18.24	134.5	135.0

The next phase of FE analysis was to introduce on the bell model the decorations such as small Buddhist carvings and study their effects on the acoustics, especially on the beating sound. These decorations are additional masses in the FE model. Table 13.11 shows the analysis results. When the bell is completely axial symmetrical (not feasible in reality), there is no beating frequency. The presence of a slight non-symmetry by a Buddhist carving induced beating. By selectively using decorations, we can control the beating frequencies. For example, increasing mass (by adding Buddhist carvings) at anti-nodal points of low frequency modes helps to enhance beating frequencies. Decreasing mass at anti-nodal points of high frequency modes can achieve the same outcome.

Table 13.11 FE analysis results of the bell model with slight asymmetry

Modes	Model		
	Axial symmetrical bell	Bell with rim	Bell with one Buddhist carving
Lower beating frequency Hz	194.6	134.5	194.7
Higher beating frequency Hz	194.6	135.0	195.2
Beating frequency Hz	0	0.5	0.5

The fundamental frequency of the bell was calculated by FE analysis as 134.5 Hz. This did meet the target of 100 Hz. To modify the design more effectively, sensitivity analysis was carried out to determine possible locations for change. The FE model was then adjusted with both mass and stiffness changes according to the sensitivities.

After the design was finalized, the bell was manufactured. Modal testing was then used to study and verify the dynamic characteristics of the bell. Hammer test, single-input random and multi-input random excitations were used in the test. Figure 13.32 shows a typical accelerance FRF of the bell.

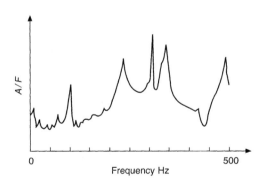

Figure 13.32 A typical FRF of the bronze bell

Figure 13.33 shows the spectrum of the sound. It bears a striking resemblance to the FRF curve. This suggests the intrinsic correlation between the structural dynamics and the acoustics of the bell.

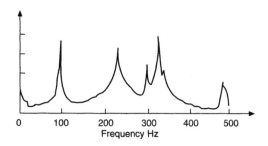

Figure 13.33 A typical sound spectrum of the bronze bell

The circumferential mode shape corresponding to the designed fundamental frequency of 100 Hz is shown in Figure 13.34. It has four nodal points. Theoretically, this mode is not the fundamental mode for a circular structure. It is actually the third vibration one. However, the first two modes require much greater energy to vibrate and, therefore, are not sustainable in reality.

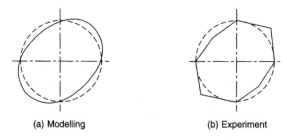

(a) Modelling (b) Experiment

Figure 13.34 The circumferential mode shape of the bronze bell

The beating of the bell was also tested. Figures 13.35 and 13.36 show the measured beating signal of the bell and its spectrum. This confirms that the designed asymmetry has produced the expected beating phenomenon.

Figure 13.35 The measured beating signal of the bronze bell

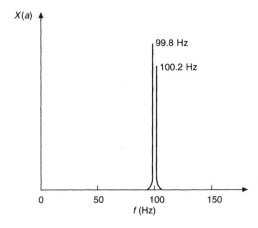

Figure 13.36 The spectrum of the beating signal

Figure 13.57 shows the sound spectrum of the bell at zero second and after 60 seconds. The beating signal based on the fundamental frequency has shown a remarkable longevity.

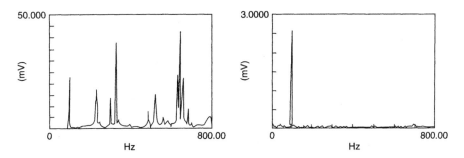

Figure 13.37 The instantaneous sound spectrum of the bronze bell

To determine the free decay of the vibration modes, the instantaneous spectrum of the impulse response was observed. Figure 13.37 shows the instantaneous sound spectrum and the spectrum after 20 seconds. It is evident that the designed frequency is the most lasting one.

13.11 Modal analysis of bus roll cage structure for optimum rollover design

Improvement of vehicle stability and crashworthiness to reduce rollover and to provide increased occupant protection in the event of rollover requires that the effects of design parameters on vehicle rollover propensity be thoroughly understood. Improvement of rollover characteristics of buses currently requires extensive experimentation and considerable resources associated with rollover destruction of experimental bus structures. The developmental process of a bus roll cage structure relies on computer modelling and simulation for efficient outcomes. This process, however, requires verification and sometimes additional information which the process itself cannot provide.

Modal analysis lends much needed assistance to the study of the bus roll cage structure. It can be used to identify the dynamic characteristics of the structure which can be used as the basis for verifying the analytical model. It can also provide correct damping information which modelling itself is incapable of supplying. The case study in this section shows the application of modal analysis in the modelling and testing of a standard roll cage structure which has been designed and manufactured to suit typical two axle single deck bus constructions used for route, school or charter bus services. Figure 13.38 shows the structure. It weighs 846 kg. It has eight front–rear beams and nine left–right inverse U-shaped frames.

Figure 13.38 A standard roll cage structure

FE modelling of the bus frame was carried out. The FE model shown in Figure 13.39 consists of 260 3-D beams. A grounded condition of support has been simulated by constraining the model at each floor stool. Figure 13.40 shows the first and third mode shapes of the bus frame obtained through this analysis.

The role of modal analysis in this study was to determine the structural dynamic characteristics of the standard bus frame with its rollover protection cage. The information would be used to verify the finite element model of the structure and provide the correct modal damping information for it. The results would be stored for future use to identify appropriate structural modification strategies by relating those dynamic characteristics to the design parameters with respect to the desired vehicle stability.

The roll cage structure is welded to the remaining parts of the bus. Due to the relative rigidity of bus chassis and flexibility of the cage structure, it was believed

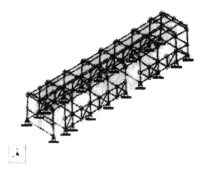

Figure 13.39 The FE model of the roll cage structure

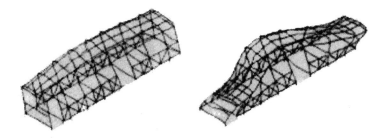

Figure 13.40 The first and third mode shapes of the bus frame

that the grounded boundary conditions for the cage structure were appropriate. The structure was therefore cemented on a concrete floor at each floor stool. This resembled the operational condition of the structure and it was consistent with the condition used in computer modelling. Figure 13.41 shows the boundary conditions applied.

Figure 13.41 The boundary conditions applied during testing

There were in total 131 points selected for modal testing purposes. They constituted all the crossing points between the floor line and side bracings, and between hoops and side bracings. For these points, the main interest was the vibration at y (up and down) direction. In addition, the crossing points of the roof and the roof beams with hoops and door openings were selected. For points on the roof, vibration at the z (fore and aft) direction was of the main concern.

The structure was excited by an impact hammer which provided sufficient excitation force for the cage structure. A dual channel analyser was used to acquire frequency response function data from the measurement. Figure 13.42 shows the measurement set-up.

Figure 13.42 Measurement set-up for the bus roll cage structure

The response point was selected after a trial-and-error attempt. Impact force was then applied in the y direction to points on the sides and the z direction for points on the roof. Figure 13.43 shows the point FRF of this test. Within 40 Hz, there are well-defined modes of the bus frame.

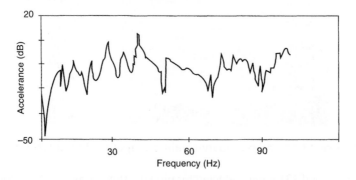

Figure 13.43 The point FRF of this test

The baseband frequency range of the measurement was selected to be 100 Hz. This included the main modal activities computer modelling had dealt with. The measured FRF data were later processed first using the SDoF analysis method. The MDoF method was then used to process all 154 FRFs. Figure 13.44 shows a typical regenerated FRF after the analysis was carried out.

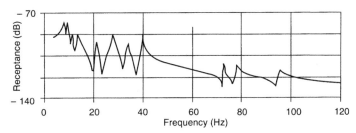

Figure 13.44 A typical regenerated FRF after the analysis

Table 13.12 shows the natural frequencies of the first five modes analysed. All mode shapes have been saved as real modes. The results show very good correlation except for mode 5 of the finite element model which has displayed elements of rigid body motion. Table 13.13 shows the modal data of the first five modes from measured FRF data.

Table 13.12 Modal data of the first five analysed modes

Mode No.	Natural frequency (Hz)	Mode type
1	7.664	First bending
2	9.096	Second bending
3	10.915	Third bending
4	13.338	Roof
5	17.180	Roof and bending

Table 13.13 Modal data of the first five identified modes

Mode No.	Natural frequency (Hz)	Damping loss factor
1	8.14	0.043
2	9.35	0.030
3	11.03	0.017
4	13.44	0.043
5	17.16	0.017

The natural frequencies from the FE modelling and modal testing reconcile with a satisfactory error range. The difference between the experimentally and analytically determined natural frequencies is lower than 10% for the first ten modes. The structure exhibits relatively low damping. This needs to be taken into consideration when estimating the energy absorption of the structure during rollover or in case of side impact. The data indicate that damping for bending modes is generally less than 5% while damping for roof modes is less than half of that value. This should have a significant effect on the difference in deformation of these components during rollover. In addition, the determined bending modes in the lateral direction of the bus frame indicate an inherent sensitivity of the design to excitation in the lateral direction which has a negative effect on vehicle stability and rollover propensity in general.

These findings will be addressed through structural modification of the bus frame. A strategy could be to conduct initial sensitivity studies. For example, the bus frame can be modified to enable the spreading of the impact deformation over a large distance thus increasing the amount of plastic deformation of the vehicle. The amount of unrecoverable work used to crush the vehicle will be increased, reducing the amount of energy which can contribute to the rolling motion of the vehicle. Also, the structure of the bus can be modified to reduce its vulnerability in the lateral direction and increase the level of damping.

Literature

1. Accorsi, M., Leonard, J. and Benney, R. 2000: Structural modeling of parachute dynamics. *AIAA Journal*, **38**(1), 139–146.
2. Ahn, T.K. and Mote Jr, C.D. 1998: Mode identification of a rotating disk. *Experimental Mechanics*, **38**(4), 250–254.
3. Alampalli, S. 2000: Modal Analysis of a Fiber-reinforced Polymer Composite Highway Bridge. *Proceedings of the International Modal Analysis Conference*, San Antonio, 21–25.
4. Almeida, P.A.O. and Rodrigues, J.F.S. 1999: Modal Analysis of the Structure of a Soccer Stadium. *Proceedings of the 17th International Modal Analysis Conference*, Kissimmee, FL, 1417–1422.
5. Catbas, F.N. *et al.* 1999: Modal Analysis as a Bridge Monitoring Tool. *Proceedings of the 17th International Modal Analysis Conference*, Kissimmee, FL, 1230–1236.
6. Cawley, P. 1985: Non-destructive testing of mass produced components by natural frequency measurements. *Proceedings of Institution of Mechanical Engineers*, 199(B3).
7. Duncan, A.E. 1982: Application of modal modelling and mount system optimisation to light duty truck ride analysis. *SAE Paper* 811313.
8. Fu, Z. 1990: Vibration modal analysis and parameters identification (in Chinese).
9. Gaul, L., Willner, K. and Hurlebaus, S. 1999: Determination of Material Properties of Plates from Modal ESPI Measurements. *Proceedings of the 17th International Modal Analysis Conference*, Kissimmee, FL, 1756–1762.
10. Ginenez, J.G. and Carrascosa, L.I. 1984: Application of modal analysis techniques to the resolution of some dynamic problems in railways. *Proceedings of the 3rd International Modal Analysis Conference*, Orlando, FL, 1075–1084.
11. Idichandy, V.G. and Ganapathy, C. 1990: Modal parameters for structural integrity monitoring of fixed offshore platforms. *Experimental Mechanics,* 382–391.
12. Knight, C.E. Wicks, A.L. and Braunwart, P. 2000: Modeled Dynamic Response of a Golf Club During the Swing. *Proceedings of the International Modal Analysis Conference*, San Antonio, 550–554.
13. Kimmel, E., Peleg, K. and Hinga, S. 1992: Vibration modes of spheroidal fruits. *Journal of Agricultural Engineering Research*, **52**, 201–213.
14. Kronast, M. and Hildebrandt, M. 2000: Vibro-acoustic modal analysis of automobile body cavity noise. *Sound and Vibration*, **34**(6), 20–23.

15. Lomnitz, C. 1997: Frequency response of a strainmeter. *Bulletin of the Seismological Society of America*, **87**, 1978–1980.

16. Larsson, D. 1997: Using modal analysis for estimation of anisotropic material constants. *Journal of Engineering Mechanics*, **123**, 222–229.

17. Messac, A. 1996: Flexible-body dynamics modeling of a vehicle moving on the rails of a structure. *Journal of Guidance, Control and Dynamics*, **19**, 540–548.

18. Proulx, J. and Paultre, P. 1997: Experimental and numerical investigation of dam–reservoir–foundation interaction for a large gravity dam. *Canadian Journal of Civil Engineering*, **24**, 90–105.

19. Saczalski, K.J. and Loverich, E.B. 1991: Vibration analysis methods applied to forensic engineering problems. *DE-Vol. 34, Structural Vibration and Acoustics, SAME*, 197–206.

20. Sung, M.H. and Lee, J.M. 1989: A study on the interior noise of a passenger car. *Proceedings of the 7th International Modal Analysis Conference*, Las Vegas, Nevada, 1482–1488.

21. Talapatra, D.C. and Haughton, J.M. 1984: Modal Testing of a Large Test Facility for Space Shuttle Payloads. *Proceedings of the 2nd International Modal Analysis Conference*, Orlando, Florida, 1–7.

22. Todd, R.V. 1999: Modal Analysis of an Upper Nevada Penstock at Hoover Dam. *Proceedings of the 17th International Modal Analysis Conference*, Kissimmee, FL, 1794–1798.

23. Ulm, S.C. 1984: Application of modal analysis to the design of a large fan-foundation system. *Proceedings of the 3rd International Modal Analysis Conference*, Orlando, FL, 304–310.

24. Volvo GM Heavy Truck Corp. 1997: Dynamic analysis guides truck fairing design, 31.

25. Woensel, G.V., Verdonck, E. and De Baerdemaeker, J. 1988: Measuring the mechanical properties of apple tissue using modal analysis. *Journal of Food Process Engineering*, **10**, 151–163.

26. Wu, T. X. and Thompson, D.J. 2000: The vibration behavior of railway track at high frequencies under multiple preloads and wheel interactions. *The Journal of the Acoustical Society of America*, **108**(3), 1046–1053.

27. Zhang, P.Q. *et al.* 1991: Modal analysis of mini-small object – dental drill. *DE-Vol. 38, Modal Analysis, Modelling, Diagnostics, and Control, SAME*, 17–23.

Index

Printed and bound by CPI Group (UK) Ltd, Croydon, CR0 4YY

08/05/2025

01864808-0001